LONDON MATHEMATICAL SOCIETY LECTURE NOTE SERIE

Managing Editor: Professor J.W.S. Cassels, Dep
and Mathematical Statistics, 16 Mill Lane, Cam.

The books in the series listed below are available from booksellers, or, in case of difficulty, from Cambridge University Press.

4 Algebraic topology, J.F.ADAMS
5 Commutative algebra, J.T.KNIGHT
8 Integration and harmonic analysis on compact groups, R.E.EDWARDS
11 New developments in topology, G.SEGAL (ed)
12 Symposium on complex analysis, J.CLUNIE & W.K.HAYMAN (eds)
13 Combinatorics, T.P.McDONOUGH & V.C.MAVRON (eds)
16 Topics in finite groups, T.M.GAGEN
17 Differential germs and catastrophes, Th.BROCKER & L.LANDER
18 A geometric approach to homology theory, S.BUONCRISTIANO, C.P.ROURKE & B.J.SANDERSON
20 Sheaf theory, B.R.TENNISON
21 Automatic continuity of linear operators, A.M.SINCLAIR
23 Parallelisms of complete designs, P.J.CAMERON
24 The topology of Stiefel manifolds, I.M.JAMES
25 Lie groups and compact groups, J.F.PRICE
26 Transformation groups, C.KOSNIOWSKI (ed)
27 Skew field constructions, P.M.COHN
29 Pontryagin duality and the structure of LCA groups, S.A.MORRIS
30 Interaction models, N.L.BIGGS
31 Continuous crossed products and type III von Neumann algebras,A.VAN DAELE
32 Uniform algebras and Jensen measures, T.W.GAMELIN
34 Representation theory of Lie groups, M.F. ATIYAH et al.
35 Trace ideals and their applications, B.SIMON
36 Homological group theory, C.T.C.WALL (ed)
37 Partially ordered rings and semi-algebraic geometry, G.W.BRUMFIEL
38 Surveys in combinatorics, B.BOLLOBAS (ed)
39 Affine sets and affine groups, D.G.NORTHCOTT
40 Introduction to Hp spaces, P.J.KOOSIS
41 Theory and applications of Hopf bifurcation, B.D.HASSARD, N.D.KAZARINOFF & Y-H.WAN
42 Topics in the theory of group presentations, D.L.JOHNSON
43 Graphs, codes and designs, P.J.CAMERON & J.H.VAN LINT
44 Z/2-homotopy theory, M.C.CRABB
45 Recursion theory: its generalisations and applications, F.R.DRAKE & S.S.WAINER (eds)
46 p-adic analysis: a short course on recent work, N.KOBLITZ
47 Coding the Universe, A.BELLER, R.JENSEN & P.WELCH
48 Low-dimensional topology, R.BROWN & T.L.THICKSTUN (eds)
49 Finite geometries and designs,P.CAMERON, J.W.P.HIRSCHFELD & D.R.HUGHES (eds)
50 Commutator calculus and groups of homotopy classes, H.J.BAUES
51 Synthetic differential geometry, A.KOCK
52 Combinatorics, H.N.V.TEMPERLEY (ed)
54 Markov process and related problems of analysis, E.B.DYNKIN
55 Ordered permutation groups, A.M.W.GLASS
56 Journêes arithmêtiques, J.V.ARMITAGE (ed)
57 Techniques of geometric topology, R.A.FENN
58 Singularities of smooth functions and maps, J.A.MARTINET
59 Applicable differential geometry, M.CRAMPIN & F.A.E.PIRANI
60 Integrable systems, S.P.NOVIKOV et al
61 The core model, A.DODD
62 Economics for mathematicians, J.W.S.CASSELS
63 Continuous semigroups in Banach algebras, A.M.SINCLAIR

64 Basic concepts of enriched category theory, G.M.KELLY
65 Several complex variables and complex manifolds I, M.J.FIELD
66 Several complex variables and complex manifolds II, M.J.FIELD
67 Classification problems in ergodic theory, W.PARRY & S.TUNCEL
68 Complex algebraic surfaces, A.BEAUVILLE
69 Representation theory, I.M.GELFAND et al.
70 Stochastic differential equations on manifolds, K.D.ELWORTHY
71 Groups - St Andrews 1981, C.M.CAMPBELL & E.F.ROBERTSON (eds)
72 Commutative algebra: Durham 1981, R.Y.SHARP (ed)
73 Riemann surfaces: a view towards several complex variables,A.T.HUCKLEBERRY
74 Symmetric designs: an algebraic approach, E.S.LANDER
75 New geometric splittings of classical knots, L.SIEBENMANN & F.BONAHON
76 Spectral theory of linear differential operators and comparison algebras, H.O.CORDES
77 Isolated singular points on complete intersections, E.J.N.LOOIJENGA
78 A primer on Riemann surfaces, A.F.BEARDON
79 Probability, statistics and analysis, J.F.C.KINGMAN & G.E.H.REUTER (eds)
80 Introduction to the representation theory of compact and locally compact groups, A.ROBERT
81 Skew fields, P.K.DRAXL
82 Surveys in combinatorics, E.K.LLOYD (ed)
83 Homogeneous structures on Riemannian manifolds, F.TRICERRI & L.VANHECKE
84 Finite group algebras and their modules, P.LANDROCK
85 Solitons, P.G.DRAZIN
86 Topological topics, I.M.JAMES (ed)
87 Surveys in set theory, A.R.D.MATHIAS (ed)
88 FPF ring theory, C.FAITH & S.PAGE
89 An F-space sampler, N.J.KALTON, N.T.PECK & J.W.ROBERTS
90 Polytopes and symmetry, S.A.ROBERTSON
91 Classgroups of group rings, M.J.TAYLOR
92 Representation of rings over skew fields, A.H.SCHOFIELD
93 Aspects of topology, I.M.JAMES & E.H.KRONHEIMER (eds)
94 Representations of general linear groups, G.D.JAMES
95 Low-dimensional topology 1982, R.A.FENN (ed)
96 Diophantine equations over function fields, R.C.MASON
97 Varieties of constructive mathematics, D.S.BRIDGES & F.RICHMAN
98 Localization in Noetherian rings, A.V.JATEGAONKAR
99 Methods of differential geometry in algebraic topology, M.KAROUBI & C.LERUSTE
100 Stopping time techniques for analysts and probabilists, L.EGGHE
101 Groups and geometry, ROGER C.LYNDON
102 Topology of the automorphism group of a free group, S.M.GERSTEN
103 Surveys in combinatorics 1985, I.ANDERSEN (ed)
104 Elliptic structures on 3-manifolds, C.B.THOMAS
105 A local spectral theory for closed operators, I.ERDELYI & WANG SHENGWANG
106 Syzygies, E.G.EVANS & P.GRIFFITH
107 Compactification of Siegel moduli schemes, C-L.CHAI
108 Some topics in graph theory, H.P.YAP
109 Diophantine analysis, J.LOXTON & A.VAN DER POORTEN (eds)
110 An introduction to surreal numbers, H.GONSHOR
111 Analytical and geometric aspects of hyperbolic space, D.B.A.EPSTEIN (ed)
112 Low dimensional topology and Kleinian groups, D.B.A.EPSTEIN (ed)
113 Lectures on the asymptotic theory of ideals, D.REES
114 Lectures on Bochner-Riesz means, K.M.DAVIS & Y-C.CHANG
115 An introduction to independence for analysts, H.G.DALES & W.H.WOODIN
116 Representations of algebras, P.J.WEBB (ed)
117 Homotopy theory, E.REES & J.D.S.JONES (eds)
118 Skew linear groups, M.SHIRVANI & B.WEHRFRITZ

London Mathematical Society Lecture Note Series. 76

Spectral Theory of Linear Differential Operators and Comparison Algebras

H. O. CORDES

University of California, Berkeley

The right of the University of Cambridge to print and sell all manner of books was granted by Henry VIII in 1534. The University has printed and published continuously since 1584.

CAMBRIDGE UNIVERSITY PRESS

Cambridge

London New York New Rochelle

Melbourne Sydney

CAMBRIDGE UNIVERSITY PRESS
Cambridge, New York, Melbourne, Madrid, Cape Town, Singapore, São Paulo

Cambridge University Press
The Edinburgh Building, Cambridge CB2 8RU, UK

Published in the United States of America by Cambridge University Press, New York

www.cambridge.org
Information on this title: www.cambridge.org/9780521284431

First published 1987
Re-issued in this digitally printed version 2007

A catalogue record for this publication is available from the British Library

Library of Congress Cataloguing in Publication data

Cordes, H. O. (Heinz Otto), 1925–
Spectral theory of linear differential operators and
comparison algebras.

(London Mathematical Society lecture note series ; 76)
1. Differential operators. 2. Linear operators.
I. Title.
QA329.4.C67 1986 515.7'242 85–47935

ISBN 978-0-521-28443-1 paperback

P R E F A C E

The main purpose of this volume is to introduce the reader to the concept of comparison algebra, defined as a type of C^*-algebra of singular integral operators, generally on a noncompact manifold, generated by an elliptic second order differential expression, and certain classes of multipliers and 'Riesz-operators' As for singular integral operators on $\mathbb{R}^n$ or on a compact manifold the Fredholm properties of operators in such an algebra are governed by a symbol homomorphism. However, for noncompact manifolds the symbol is of special interest at infinity. In particular the structure of the symbol space over infinity is of interest, and the fact, that the symbol no longer needs to be complex-valued there.

The first attempts of the author to make a systematic presentation of this material happened at Berkeley (1966) and at Lund (1970/71). Especially the second lecture exists in form of (somewhat ragged) notes [CS] . The cases of the Laplace comparison algebra of R^n and the half-space were presented in [C_1] .

In the course of laying out theory of comparison algebras we had to develop in details spectral theory of differential operators, as well as many of the basic properties of elliptic second order differential operators. This was done in the first four chapters. Comparison algebras (in L^2-spaces and L^2-Sobolev spaces) are discussed in chapters V to IX . Finally, in chapter X we recall the basic facts of theory of Fredholm operators, partly without proofs.

The material has been with the author for more than 20 years and has been subject of innumerable discussions with students and associates. Accordingly it is almost impossible to recall in detail the origin of the various concepts introduced. Especially we are indebted to E. Herman, M. Breuer, E. Luft, M. Taylor, R. McOwen, A. Erkip, D.Williams, H. Sohrab, in chronological, not alphabetical order.

We are indebted to S.H.Doong, S.Melo, R.Rainsberger, M.Arsenovic for help with proof reading. This volume originally was planned under the title 'techniques of pseudodifferential operators', but then split into two parts, with the second yet to appear. We are grateful to the publisher, Cambridge University Press, for cooperation and patience in waiting for the manuscript.

Berkeley, September 1986 Heinz O. Cordes

TABLE OF CONTENTS

TO HILLGIA

CHAPTER 1. ABSTRACT SPECTRAL THEORY IN HILBERT SPACES.

In this chapter we give a short introduction into spectral theory of abstract unbounded operators of a Hilbert space. In sec. 1 we give a discussion of general facts on unbounded operators. In sec.2 we discuss the v.Neumann-Riesz theory of self-adjoint extension of hermitian operators. Sec.3 gives a general discussion of the abstract spectral theorem for unbounded self-adjoint operators. We discuss a proof of the spectral theorem in sec.4. Also, in sec.5 we discuss an extension of a result by Heinz and Loewner useful in the following. Finally an abstract result on Fredholm operators in a certain type of Frechet algebra related to a chain of Hilbert spaces generated by powers of a self-adjoint positive operator is discussed in sec.6. The typical 'HS-chain' is a chain of L^2-Sobolev spaces.

The chapter is self-contained and elementary, and only requires some familiarity with general concepts of analysis and functional analysis of bounded linear operators.

1. Unbounded linear operators on Banach and Hilbert spaces.

The term "(unbounded) linear operator" (between Banach spaces X and Y) is commonly used to denote any linear map $A:\mathrm{dom}\, A \to Y$ from a dense linear subspace dom A of X to Y. The space $\mathrm{dom}\, A \subset X$ then is called the domain of A. Here we distinguish between a linear map $X \to Y$,and a linear operator: A linear map $X \to Y$ by definition has its domain equal to X .

The term "unbounded linear operator" will be used with the meaning "not necessarily bounded linear operator", so that the bounded linear operators are special unbounded operators. A bounded linear operator, satisfying

$$(1.1) \qquad \sup \{\|Au\|/\|u\| : 0 \neq u \in \mathrm{dom}\, A\} < \infty ,$$

is necessasily continuous, hence admits a unique extension to X, in which dom A is dense. This is why we usually assume that a bounded linear operator also is a linear map.

The class of all such unbounded linear operators between two given spaces X and Y will be denoted by $P(X,Y)$. In particular the class $L(X,Y)$ of all continuous linear maps $X \to Y$ then is a subset of $P(X,Y)$.

Since unbounded linear operators are not linear maps from X to Y, (but only from their individual domain to Y) their sum and product needs the following special interpretation: For $A,B \in P(X,Y)$ and $C \in P(W,X)$ we define the sum $A+B \in P(X,Y)$ and the product $AC \in P(W,Y)$ by setting

(1.2) $\text{dom}\,(A+B) = \text{dom}\,A \cap \text{dom}\,B\,,\ (A+B)u = Au+Bu \text{ for } u \in \text{dom}(A+B),$

and

(1.3) $\text{dom}\,AC = \{u \in \text{dom}\,C : Cu \in \text{dom}\,A\}\,,\ (AC)u = A(Cu) \text{ for } u \in \text{dom}\,AC\,,$

where it is assumed that dom(A+B) and dom AC are dense in X (or else, we will say that dom(A+B) or dom AC is not defined). Also we define $cA = (c\cdot 1)A$, with the identity operator $1 \in L(X,X)$.

A linear operator $A \in P(X,Y)$ is uniquely characterized by its <u>graph</u>, defined as the linear subspace

(1.4) $$\text{graph}\,A = \{(u;Au) : u \in \text{dom}\,A\}$$

of the cartesian product $X \times Y = \{(u;v) : u \in X\,,\ v \in Y\}$, where $(u;v)$ denotes the ordered pair. Vice versa, if for any linear subspace $T \subset X \times Y$ the set of all first components is dense in X, and if T does not contain elements of the form $(0;u)$ other than $(0;0)$, then a unique unbounded operator $A \in P(X,Y)$ is defined by setting

(1.5) $\text{dom}\,A = \{u \in X : (u;v) \in T \text{ for some } v \in Y\}\,,\ Au = v\,,\ \text{for } u \in \text{dom}\,A\,,$

and then we have graph $A = T$.

For two linear operators $A,B \in P(X,Y)$ we shall say that A extends B (or that B is a restriction of A) if graph $A \supset$ graph B. We then will write $A \supset B$ (or $B \subset A$).

Notice that the cartesian product $X \times Y$ of two Banach spaces is a Banach space again, for example under the norm $\|(u;v)\| = \|u\| + \|v\|$. Therefore it is meaningful to speak of a closed subspace of $X \times Y$. An unbounded operator A is defined to be <u>closed</u> if its graph is a closed subspace of $X \times Y$. The class of all closed operators in $P(X,Y)$ is denoted by $Q(X,Y)$. It is clear that a continuous linear map $A \in L(X,Y)$ is closed, since $(u_k;Au_k) \to (u;v)$ implies $v = \lim Au_k = Au$, hence $(u;v) = (u;Au) \in$ graph A.

An operator $A \in P(X,Y)$ is called <u>preclosed</u> if the closure of graph A is a graph again. Then A^c with graph $A^c = (\text{graph } A)^{\text{closure}}$ is called the <u>closure</u> of A .

In the following we will be mainly interested in unbounded linear operators $A \in P(H) = P(H,H)$, where H is an infinite dimensional separable Hilbert space with inner product (u,v) and norm $\|u\| = \{(u,u)\}^{1/2}$. In that case the graph space $H \times H$ becomes a Hilbert space again, under the inner product and norm

$$(1.6) \quad ((u;w),(v;z)) = (u,v) + (w,z) \ , \ \|(u;w)\| = (\|u\|^2 + \|w\|^2)^{1/2} \ .$$

The following facts, regarding adjoint and closure all work for general Banach spaces X , Y , with proper amendments. However we will get restricted to the case $X = Y = H$, in all of the following. For an operator $A \in P(H)$ we will say that the (Hilbert space) <u>adjoint</u> $A^* \in P(H)$ exists if the space $T_A = J(\text{graph } A)^{\perp}$ is a graph again. Here $J: H \times H \to H \times H$ denotes the map $(u;w) \to (w;-u)$, and "$\perp$" denotes the orthogonal complement in the graph space (with respect to the inner product (1.6) . Then we define the adjoint $A^* \in P(H)$ of A by setting

$$(1.7) \qquad \text{graph } A^* = T_A \ .$$

Notice that A^*, if it exists, is closed, since all orthogonal complements are necessarily closed and since J is inverted by $-J$. It is clear that $A \subset B$ implies $B^* \subset A^*$, assuming that A^* and B^* exist. The proposition, below, translates the definition of the adjoint into a more transparent form. (The proof is left to the reader.)

<u>Proposition 1.1</u>. Assume that $A^* \in Q(H)$ exists, for some $A \in P(H)$. Then dom A^* consists precisely of all $u \in H$ for which there exists an element $v \in H$ such that

$$(1.8) \qquad (u,Aw) = (v,w) \ , \ \text{for all } w \in \text{dom } A \ .$$

Moreover the element v thus defined for each $u \in \text{dom } A^*$ is uniquely determined, and we have $A^*u = v$.

<u>Proposition 1.2</u>. An operator $A \in P(H)$ admits an adjoint A^* if and

only if it is preclosed. Moreover, then A^* also admits an adjoint A^{**}, and the closure A^c of A equals A^{**}.

Proof. Let first $A \in P(H)$ have an adjoint $A^* \in Q(H)$. Then let $T =$ (graph A)$^{clos.}$ contain the element $(0;z)$. It follows that there exists a sequence $w_k \in \text{dom } A$ with $(w_k;Aw_k) \to (0;z)$. Substitute $w = w_k$ in (1.8), and conclude that $(u,z) = 0$ for all $u \in \text{dom } A^*$, since the inner products in (1.8) allow a passing to the limit. Since dom A^* is dense, it follows that $z=0$, so that no element of the form $(0;z)$ is in T, except $(0;0)$. Also, $T \supset$ graph A, which implies that the set of first components of elements in T is dense. Therefore T indeed is a graph of some $A^c \in P(H)$, and A is preclosed.

Vice versa, let A be preclosed, and let again T be the closure of graph A. If the set V^* of all first components of elements in $T_A = J(\text{graph } A)^{\perp}$ (i.e., of all second components of $(\text{graph } A)^{\perp}$) is not dense in H then there exists $0 \neq z \in H$ with $((0;z),(u;v)) = (z,v) = 0$ for all $(u;v) \in (\text{graph } A)^{\perp}$. But this implies that $(0;z) \in (\text{graph } A)^{\perp\perp} = (\text{graph } A)^{clos} = T$. However, for preclosedness of A it is required that T does not contain such elements. Thus the set V^* must be dense in H. On the other hand if $(0;v) \in T_A = \text{graph } A^*$, then (1.8) yields $(v,w)=0$ for all $w \in \text{dom } A$, so that $v=0$, since dom A is dense. This shows that then indeed A^* is a well defined operator in $P(H)$, hence in $Q(H)$, q.e.d.

All continuous linear maps in $L(H)$ are closed, hence have an adjoint. Moreover $L(H)$ is adjoint invariant, and, for an $A \in L(H)$, the above adjoint coincides with the well known Hilbert space adjoint A^* of the bounded operator A.

Also, if $A \in P(H)$ is (pre-)closed, then $\gamma A+B$ is (pre-)closed, for every $B \in L(H)$, $0 \neq \gamma \in \mathbb{C}$. Then $(\gamma A+B)^c = \gamma A^c + B$, $(\gamma A+B)^* = \bar{\gamma} A^* + B^*$ in the sense of (1.2).

Our main interest, in the following, will focus on self-adjoint unbounded operators. Here an operator A is called self-adjoint if (A^* exists, and) $A^* = A$. First of all a self-adjoint operator allows a spectral decomposition, as a direct generalization of the principal axis transformation of a symmetric matrix. Second we will learn about important classes of differential operators which are unbounded self-adjoint operators, and therefore allow such a spectral decomposition. Third, we will show how results on unbounded non-selfadjoint differential operators can be achieved by "comparing" them with certain standard self-adjoint differential operators.

Note that a bounded operator (i.e., a continuous linear map) A is self-adjoint if and only if it satisfies the relation

(1.9) $(u,Av) = (Au,v)$ for all $u,v \in \text{dom } A$,

(where dom A = H).A general unbounded operator A satisfying (1.9) needs not to be self-adjoint, because (1.9) just implies that $A^* \supset A$, not that $A^* = A$. Such an operator is called <u>hermitian</u> . If the closure of a hermitian operator A is self-adjoint then we speak of an <u>essentially self-adjoint operator</u>(i.e., $A^*=A^{**}$). Note that a hermitian operator A indeed has an adjoint:If $u_k \in \text{dom } A$ $u_k \to 0$, $Au_k \to w$, then we may substitute $u=u_k$ into (1.9), for fixed v, and pass to the limit, resulting in $0 = (w,v)$, for all $v \in \text{dom } A$. It follows that $w=0$, so that A is preclosed, and A^* exists, by prop. 1.2. Comparing (1.8) and (1.9), it then follows at once that $A \in P(H)$ is hermitian if and only if $A \subset A^*$.

If A is (essentially) self-adjoint, and B bounded hermitian then $\gamma A+B$ is (essentially) self-adjoint for all $0 \neq \gamma \in \mathbb{R}$.

More generally, two operators $A, B \in P(H)$ will be said to be <u>in adjoint relation</u> if

(1.10) $(u,Av) = (Bu,v)$, for all $u \in \text{dom } B$, $v \in \text{dom } A$.

The above conclusion, showing that hermitian operators have adjoints, can be repeated to prove that operators A, B in adjoint relation must have adjoints (both) ,and that,moreover, A and B are in adjoint relation if and only if $A \subset B^*$ (or if and only if $B \subset A^*$).

One of the first major problems occurring in our discussion of differential operators, in later sections, will be the construction of all self-adjoint extensions of a given hermitian operator. It turns out that not all hermitian operators possess self-adjoint extensions. On the other hand, the problem of characterizing all self-adjoint extensions was solved by v. Neumann [vN_1] and F. Riesz [Ri_1]. We will discuss the v. Neumann-Riesz theory in section 2, together with some other constructions of self-adjoint extensions.

2. Self-adjoint extensions of Hermitian operators.

In this section we discuss the v. Neumann-Riesz theory of self-adjoint extensions of hermitian operators.

It is at once clear that a hermitian operator A has a hermitian closure $A^c = A^{**}$, because $A \subset A^*$ implies $A^{**} \subset A^*$. Since a self-adjoint extension $B = B^*$ of A is necessarily closed, it also must be an extension of the closure A^{**}. Therefore, in looking for self-adjoint extensions of a given hermitian operator A we may look for such extensions of the closure, and thus assume that A is closed, without loss of generality.
Also if $A \subset B = B^*$, then $B^* = B \subset A^*$, so that every self-adjoint extension B of A is a restriction of A^*, as well: We have

$$(2.1) \qquad A \subset B = B^* \subset A^* .$$

Proposition 2.1. A hermitian operator A satisfies the identity

$$(2.2) \quad \|(A-\lambda)u\|^2 = \|(A-\text{Re }\lambda)u\|^2 + (\text{Im }\lambda)^2\|u\|^2 , \text{ for all } u\in\text{dom }A , \lambda\in\mathbb{C} .$$

Proof. We have $\|(A-\lambda)u\|^2=((A-\mu-i\nu)u, (A-\mu-i\nu)u)=((A-\mu)u,(A-\mu)u) + \nu^2(u,u) - 2\text{Re }((A-\mu)u,i\nu u)$, where we have written $\mu = \text{Re }\lambda$, $\nu = \text{Im }\lambda$. Here the last term vanishes, due to $(Au,iu) + (iu,Au) = i((Au,u)-(u,Au)) = 0$, using that A is hermitian, q.e.d.

For a closed hermitian operator A it is implied by prop.2.1 that im $(A-\lambda)$ is closed for every nonreal $\lambda \in \mathbb{C}$. Indeed, consider a sequence $u_k \in \text{dom }A$ such that $(A-\lambda)u_k \to v$. It follows that $(A-\lambda)(u_k-u_l) \to 0$, as $k, l \to \infty$. But (2.2) implies the inequality

$$(2.3) \qquad \|(A-\lambda)u\| \geq |\text{Im }\lambda|\|u\| , \ u \in \text{dom }A .$$

Substituting $u=u_k-u_l$ into (2.3) yields $\|u_k-u_l\| \to 0$, since Im $\lambda \neq 0$, by assumption. Hence $u_k \to u$ for some $u \in H$, and $(u_k;Au_k) \to (u;v)$ in $H\times H$. But graph A is closed ,since A is closed. Thus it follows that $(u;v) \in$ graph A, or, $u \in \text{dom }A$, $v=Au$.

In the following we first consider the special case $\lambda = \pm i$. Since we have im $(A\pm i)$ closed, by the above, we obtain a pair of orthogonal direct decompositions

$$(2.4) \qquad H = \text{im}(A\pm i) \oplus \mathcal{D}_\pm , \ \mathcal{D}_\pm = (\text{im}(A\pm i))^\perp .$$

The two spaces $\mathcal{D}_\pm=\mathcal{D}_\pm(A)$ are called the defect spaces of the closed hermitian operator A , and their dimensions are called the defect indices of A . We write

$$(2.5)\qquad \text{def } A = (\dim(\text{im}(A+i))^{\perp}, \dim(\text{im}(A-i))^{\perp}) = (\nu_A^+, \nu_A^-) .$$

Note that the defect spaces $\mathcal{D}_\pm$ are just the eigenspaces of the adjoint operator A^* to the eigenvalues $\pm i$; We have

$$(2.6)\qquad \mathcal{D}_\pm = (\text{im}(A\pm i))^{\perp} = \ker(A^* \mp i) .$$

Indeed, $f \in \mathcal{D}_+$,for example,amounts to $0 = (f,(A+i)u)$ for all $u \in \text{dom } A$. We write this as $(f,Au) = (if,u)$, $u \in \text{dom } A$, and compare with (1.8) , concluding that $f \in \text{dom } A^*$, $A^*f = if$.

As another consequence of prop.2.1 we note that (2.2) ,for $\lambda = \pm i$, implies

$$(2.7)\qquad \|(A\pm i)u\| = (\|Au\|^2+\|u\|^2)^{1/2} = \|(u;Au)\| , \quad u \in \text{dom } A .$$

Note that graph A, as a closed subspace of $H\times H$ is a Hilbert space under the norm and inner product of $H\times H$. Moreover graph A is in linear 1-1-correspondence with dom A ,which may be used to transfer that Hilbert space structure of graph A to dom A . In other words, dom A is a Hilbert space under the (stronger) norm

$$(2.8)\qquad \|u\|_A = (\|u\|^2+\|Au\|^2)^{1/2} ,$$

with inner product

$$(2.9)\qquad (u,v)_A = (u,v) + (Au,Av) , \quad u,v \in \text{dom } A .$$

The latter is true for every closed operator $B \in \mathcal{Q}(H)$, not only for hermitian operators. In particular we may apply it to the adjoint $B=A^*$ of our closed hermitian operator A , obtaining a corresponding norm $\|u\|_{A^*}$ and inner product $(u,v)_{A^*}$, $u,v \in \text{dom } A^*$. In fact, we then get $\|u\|_{A^*} = \|u\|_A$,for $u \in \text{dom } A \subset \text{dom } A^*$, and dom A appears as a closed subspace of dom A^* , under graph norm.

Note that (2.7) may be interpreted as follows: The two operators $(A\pm i)$ are isometries $\text{dom } A \to \text{im}(A\pm i) = \mathcal{D}_\pm^{\perp}$. In fact these isometries are 'onto'. Therefore $V = (A+i)(A-i)^{-1} : \text{im}(A-i)\to\text{im}(A+i)$ defines an isometry between the two closed subspaces of H .

Proposition 2.2. For a closed hermitian operator A we have

$$(2.10)\qquad \text{dom } A^* = \text{dom } A \oplus \mathcal{D}_+(A) \oplus \mathcal{D}_-(A) ,$$

as an orthogonal direct decomposition of the Hilbert space dom A^* under its norm and inner product.

Proof. We already noticed that dom A is a closed subspace of dom A^*, under graph norm. The eigenspaces $\mathcal{D}_\pm$ of A^* are closed subspaces of H, as nulspaces of the closed operators $A^* \mp i$. Moreover, on $\mathcal{D}_\pm$ we have $\|u\|_{A^*} = \sqrt{2}\|u\|$, so that $\mathcal{D}_\pm$ are also closed under graph norm. For $f_\pm \in \mathcal{D}_\pm$ one confirms that $(f_+,f_-)_{A^*} = 0$, using that $A^* f_\pm = \pm i f_\pm$. Also, for $u \in$ dom A, $(u,f_\pm)_{A^*} = (Au, \pm i f_\pm) + (u,f_\pm)$ $= \pm i((A \pm i)u, f_\pm) = 0$, so that the three spaces in the decomposition (2.10) are orthogonal. Suppose $f \in$ dom A^* satisfies $0 = (f,u)_{A^*}$ $= (A^*f, A^*u) + (f,u)$ for all $u \in$ dom A. Comparing this with (1.8) it is found that $A^*f \in$ dom A^*, hence $f \in$ dom $(A^*)^2$, and $(A^*)^2 f + f = 0$. One also may write this as $(A^*+i)(A^*-i)f = (A^*-i)(A^*+i)f = 0$. In particular, $f \in$ dom $(A^*+i)(A^*-i)$ = dom $(A^*-i)(A^*+i)$, in the sense of (1.3). Now we write $f = (A^*+i)f/2i - (A^*-i)f/2i = f_+ + f_-$, noting that $(A^* \mp i)f_\pm = 0$. This proves that every f orthogonal to dom A, under graph inner product of A^*, is in $\mathcal{D}_+ \oplus \mathcal{D}_-$, and thus completes the proof.

Corollary 2.3. A closed hermitian operator A is self-adjoint if and only if its defect indices vanish (i.e. def A = 0). Or, equivalently, if and only if im $(A \pm i) = H$, for both, "+" and "-".

Indeed, $A = A^*$ implies dom A = dom A^*, so that (2.10) gives $\mathcal{D}_\pm = \{0\}$, hence def A = 0. Vice versa, if def A = 0, (2.10) gives dom A = dom A^*, hence $A^* = A$, since $A^* \supset A$, q.e.d.

We now state the v. Neumann-Riesz extension theorem.

Theorem 2.4. The closed hermitian extensions B of a given closed hermitian operator A are in 1-1-correspondence with the extensions $W : \mathcal{W}_- \to \mathcal{W}_+$ of the isometry $V = (A+i)(A-i)^{-1}$: im(A-i) → im(A+i) as an isometry between the closed subspaces $\mathcal{W}_\pm \supset$ im$(A \pm i)$. This correspondence is established by assigning to $B \supset A$ the operator $W = V_B = (B+i)(B-i)^{-1}$ (which is an isometry extending V between the spaces $\mathcal{W}_\pm$ = im$(B \pm i) \supset$ im$(A \pm i)$). Vice versa, given an isometry $W : \mathcal{W}_- \to \mathcal{W}_+$ extending the isometry V, one must observe that W, as an extension of V, is determined by its restriction $W_0 = W|\mathcal{W}^0_-$, where $\mathcal{W}^0_- = \mathcal{W}_- \cap (\text{im}(A-i))^\perp = \mathcal{W}_- \cap \mathcal{D}_-$ is a subspace of the defect space $\mathcal{D}_-$, and where W_0 is just any isometry $\mathcal{W}^0_- \to \mathcal{W}^0_+ = \mathcal{W}_+ \cap \mathcal{D}_+ \subset \mathcal{D}_+$. Then we have the closed hermitian extension B given by

(2.11) $\quad \operatorname{dom} B = \operatorname{dom} A \oplus \{W_0\phi - \phi : \phi \in \mathcal{W}^0_-\}$, $B(W_0\phi-\phi)=i(W_0\phi+\phi)$,

where the direct sum again is orthogonal in $(.,.)_{A^*}$.

The proof is almost self-explanatory. It is clear from the above that $W = (B+i)(B-i)^{-1}$ is an isometric extension of V , for every closed hermitian extension B of A . Vice versa, that W_0 determines W , as described, follows from the well known fact that isometries preserve orthogonality. Then, of course, the operator B , if it exists, should satisfy

(2.12) $\quad W(B-i)u = (B+i)u$, $u \in \operatorname{dom} B = \operatorname{dom} A \oplus Z_0$,

with a certain subspace $Z_0 \subset \mathcal{D}_+ \oplus \mathcal{D}_-$, because $\operatorname{dom} A \subset \operatorname{dom} B \subset \operatorname{dom} A^*$ and due to (2.10) . For $u \in Z_0$ let $\phi = Bu-iu$, $\chi = Bu + iu = W_0\phi$. It follows that $u = (W_0\phi-\phi)/2i$, $Bu = (W_0\phi+\phi)/2$, in agreement with (2.11). Now one simply must verify that the operator B of (2.11) is closed and hermitian. The closedness follows if we show that $Z_0 = \{W_0\phi-\phi : \phi \in \mathcal{W}^0_-\}$ is a closed space, under graph norm of A^* . But we have

(2.13) $\quad \|W_0\phi-\phi\|_{A^*}^2 = \|W_0\phi-\phi\|^2 + \|W_0\phi+\phi\|^2 = 2\|W_0\phi\|^2 + 2\|\phi\|^2 = 4\|\phi\|^2$,

which shows that the map $\phi \to W_0\phi-\phi$, taking $\mathcal{W}^0_-$ onto Z_0 , is, in fact an isometry (up to the factor 4). Since $\mathcal{W}^0_-$ is closed, Z_0 also is closed. To verify that B of (2.11) is hermitian is only a calculation; since we know that $B|\operatorname{dom} A = A$ is hermitian one must show that $(A^*u,v)=(u,A^*v)$ for all $u,v \in Z_0$, and for $u \in \operatorname{dom} A$, $v \in Z_0$. Both follow trivially, q.e.d.

Theorem 2.4 has the following important consequence.

<u>Corollary 2.5</u>. A closed hermitian operator A admits a self-adjoint extension if and only if def A = (ν,ν), with $\nu=0,1,2,\ldots,\infty$ arbitrarily given. In other words, we must have

(2.14) $\quad \operatorname{codim} \operatorname{im}(A+i) = \operatorname{codim} \operatorname{im}(A-i)$.

Then every self-adjoint extension B of A is obtained by picking an arbitrary isometry $W_0 : \mathcal{D}_- \to \mathcal{D}_+$,between the two defect spaces (2.6), and then defining B with (2.11) , and W_0 , with $W^0_\pm=\mathcal{D}_\pm$.

The proof is evident.

Although the v.Neumann-Riesz theory completely clarifies

the problem of self-adjoint extensions, other criteria are useful, of course, because the construction of isometries between defect spaces is not always practical. In particular not every closed hermitian operator A satisfies the condition (2.14), so that a self-adjoint extension need not always to exist. There are two well known general criteria giving existence of self-adjoint extensions. Shortly, 'real' hermitian operators as well as 'semi-bounded' hermitian operators always have self-adjoint extensions.

The concept of real operator refers to a given involution $u \to \bar{u}$ of the Hilbert space H. In most applications we will have $H = L^2(X, d\mu)$ with some measure space X and measure $d\mu$, and then refer to the complex conjugation $u(x) \to \bar{u}(x)$ of the complex-valued function $u(x) \in H = L^2$. However, one may think of an abstract space H and an involution map $u \to u^-$, with the properties

$$(2.15)\qquad (u^-)^- = u\ ,\ (c_1u+c_2v)^- = \bar{c}_1u^- + \bar{c}_2v^-\ ,$$
$$(u^-,v^-) = (v,u)\ ,\ \text{for } u,v \in H\ ,\ c_j \in \mathbb{C}\ ,\ j=1,2.$$

Then a real operator is defined as an operator $A \in P(H)$ satisfying

$$(2.16)\qquad (\text{dom } A)^- = \text{dom } A\ ,\text{and}\quad (Au)^- = Au^-\ ,\ \text{for all } u \in \text{dom } A\ .$$

Now, if a closed hermitian operator A is real with respect to any such involution of H, then one confirms at once that

$$(2.17)\qquad D_+^- = \{u^- : u \in D_+\} = D_-\ .$$

Indeed if $f \in D_+$, i.e., $(f,(A+i)u)=0$ for all $u \in \text{dom } A$, then we get $0 = ((A+i)u,f) = (f^-,((A+i)u)^-) = (f^-,(A-i)u^-)$, hence $f^- \in (\text{im}(A-i))^\perp$, using (2.16). This conclusion may be reversed, so that (2.17) follows. Also it is clear that D_+ and D_+^- have the same dimension.

We have proven:

Proposition 2.6. A closed hermitian operator A which is real with respect to some involution of H has equal defect indices and hence admits a self-adjoint extension.

A hermitian operator $A \in P(H)$ is called semi-bounded below, if there exists a real constant c such that

(2.18) $(Au,u) \geq c(u,u)$ for all $u \in \text{dom } A$.

Similarly one speaks of semi-boundedness above if (2.18) holds with "$\geq$" replaced by "$\leq$". In such case c is called an (upper or lower) (semi-) bound of A .

Theorem 2.7. A semi-bounded hermitian operator (regardless whether above or below) admits a self-adjoint extension.

We will prove thm.2.7 by constructing a distinguished self-adjoint extension B of a semi-bounded operator A , where B has the same semi-bounds as A. This operator B will be referred to as the Friedrichs extension of A (cf. K.O.Friedrichs [Fr_2], and M.H.Stone [St_1]). In the construction of B we assume that A is semi-bounded below, and that c=1 is a lower bound. The general case may be reduced to this by considering the operator $\pm A + \gamma$ with suitable real γ and sign $\pm$.

In the case of c=1 we conclude from (2.18) that $(u,v)_\sim=(Au,v)=(u,Av)$ defines a positive definite inner product in the space $\text{dom } A \subset H$. One may complete dom A under the corresponding norm $\|u\|_\sim = ((u,u)_\sim)^{1/2}$, and obtain a new Hilbert space $H^\sim$, containing dom A as a dense subspace. It is easily confirmed that every class of equivalent Cauchy sequences under $\|.\|_\sim$ is contained in a unique class of equivalent Cauchy sequences under $\|.\|$. This provides an imbedding of $H^\sim$ into H , so that we get $\text{dom } A \subset H^\sim \subset H$.

Now the Friedrichs extension B of A is defined as the restriction of A^* to the space $(\text{dom } A^*)\cap H^\sim$. In other words,

(2.19) $\text{dom } B = (\text{dom } A^*)\cap H^\sim \ , \ Bu = A^*u \ , u \in \text{dom } B$.

First of all, B still is hermitian: for $u,v\in\text{dom } B = H^\sim\cap\text{dom } A^*$ there exist sequences u_k , $v_k \in \text{dom } A$ with $u_k\to u$, $v_k\to v$, in $H^\sim$, by construction of $H^\sim$. Then $(u,Bv) = (u,A^*v) = \lim (u_k,A^*v) = \lim (Au_k,v) = \lim (u_k,v)_\sim = (u,v)_\sim = \lim (u,v_k)_\sim = \ldots = (Bu,v)$. Also it follows that

(2.20) $(u,Bu) = (u,u)_\sim \geq (u,u)$, for all $u \in \text{dom } B$,

so that B has the same lower bound 1 as A .

In order to show that B is self-adjoint refer to (1.8) and let $(f,Bu)=(g,u)$, for all $u\in\text{dom } B$, and a given $f,g\in H$. Now consider the linear functional $l(u) = (g,u)$, for g as above ,and all $u \in \text{dom } A$. We get $|l(u)| \leq \|g\|\|u\| \leq \|g\|\|u\|_\sim$, using (2.18). Hence l(u) is a bounded linear functional over a dense subspace of $H^{\sim}$, and we may write $l(u) = (g,u) = (h,u)_{\sim}$, with some $h \in H^{\sim}$, by the Frechet-Riesz theorem (or Hahn-Banach-theorem). As above, we find that $(g,u) = (h,u)_{\sim} = (h,Au)$, for all $u \in \text{dom } A$. Therefore it also follows that $h \in \text{dom } A^*$, and that $A^*h=g$. Therefore, $h \in H^{\sim} \cap \text{dom } A^*$ $= \text{dom } B$, and $Bh = g$. We get $(f,Bu) = (g,u) = (Bh,u) = (h,Bu)$, or $(f-h,Bu) = 0$, for all $u \in \text{dom } B$. Note that we also have shown, with the above conclusion, that $\text{im } B = H$, because the construction of h works for every $g \in H$. Thus it follows that $f-h=0$. Or $f = h$ $\in \text{dom } B$, and $Bf = Bh = g$. This proves that the Friedrichs extension B is self-adjoint, and completes the proof of thm.2.7 .

Note that, in the special case of the lower bound 1 considered we find that the operator B^{-1} is bounded and hermitian, since we proved that $\text{im } B = H$, while (2.20) implies injectiveness of B, so that B^{-1} exists. Also $(Bu,v)=(u,Bv)$, $u,v \in \text{dom } B$ implies $(f,B^{-1}g) = (B^{-1}f,g)$, $f,g \in H$. Furthermore,

$$(2.21) \quad \|Bu\|^2 = \|(B-1)u+u\|^2 = \|(B-1)u\|^2+\|u\|^2+2((Bu,u)-(u,u)) \geq \|u\|^2 ,$$

or, $\|Bu\| \geq \|u\|$, $u \in \text{dom } B$, which yields $\|B^{-1}f\| \leq \|f\|$, $f \in H$, or $B^{-1} \in L(H)$, $\|B^{-1}\| \leq 1$. The proposition, below, will be useful.

<u>Proposition 2.8</u>. Let $C = B^{-1/2} = \sum_{j=0}^{\infty} \binom{1/2}{j} (B^{-1}-1)^j$ be the unique bounded positive self-adjoint square root of B^{-1}, satisfying $C^2 = B^{-1}$. Then $C: H \to H^{\sim}$ is an isometry between H and $H^{\sim}$:

$$(2.22) \quad H^{\sim} = \text{im } C , \text{ and } \|u\| = \|Cu\|^{\sim}, \text{ for all } u \in H .$$

We have C^{-1}, with $\text{dom } C^{-1} = H^{\sim} \subset H$ a self-adjoint operator in $P(H)$, with lower bound 1. Moreover, the restriction $C^{-1}|\text{dom } A$ still is essentially self-adjoint.

The proof is left to the reader (cf.also sec.4).

<u>3. On the spectral theorem for self-adjoint operators.</u>

The <u>resolvent</u> of a closed operator $A \in Q(H)$ is commonly defined as the inverse $R(\lambda) = (A-\lambda)^{-1}$, similarly as for linear maps. More precisely, the resolvent set Rs(A) is defined as the set of all $\lambda \in \mathbb{C}$ such that

$$(3.1) \quad \text{im}(A-\lambda) = H , \text{ and } \|(A-\lambda)u\| \geq c\|u\| , u \in \text{dom } A ,$$

with a positive constant c . It is clear that (3.1) holds if and only if the linear map $A-\lambda$ between the spaces dom A and H has an inverse $(A-\lambda)^{-1} : H \to \text{dom } A \subset H$ which constitutes a bounded operator of H, with $\|(A-\lambda)^{-1}\| \le c^{-1}$. Therefore $R(\lambda) = (A-\lambda)^{-1}$ is well defined in this sense, for all $\lambda \in Rs(A)$. The spectrum Sp(A) of A is defined as the complement of Rs(A) : $Sp(A) = \mathbb{C} \setminus Rs(A)$

It is easily seen that Rs(A) is an open set, so that Sp(A) must be closed: Note that (3.1), for $\lambda=\lambda_0$, and $|\lambda-\lambda_0| \le c/2$ implies

$$(3.2) \qquad \|(A-\lambda)u\| \ge \|(A-\lambda_0)u\| - |\lambda-\lambda_0| \|u\| \ge (c/2)\|u\| \ , \ u \in \text{dom } A \ ,$$

so that the second condition holds for a neighbourhood of λ_0 . Regarding the first condition we observe that

$$(3.3) \qquad (A-\lambda)u = (1+(\lambda_0-\lambda)R(\lambda_0))(A-\lambda_0)u \ , \ u \in \text{dom } A \ ,$$

by a simple calculation. We know that $\text{im}(A-\lambda_0) = H$, and boundedness of $R(\lambda_0)$. For small $|\lambda-\lambda_0|$ the first factor at right of (3.3) is of the form $1+ E$, $\|E\| < 1$, hence is invertible in $L(H)$, and takes H onto H . Thus (3.3) shows that $\text{im}(A-\lambda)=H$ for all small $|\lambda-\lambda_0|$, and Rs(A) is open.

Now we conclude from (3.3) that

$$(3.4) \qquad R(\lambda) = R(\lambda_0)(1+(\lambda_0-\lambda)R(\lambda_0))^{-1} \ , \ |\lambda-\lambda_0| < c \ ,$$

where only bounded operators in $L(H)$ occur. It is evident that the right hand side of (3.4) provides a norm convergent power series expansion of the operator $R(\lambda)$ in powers of $(\lambda-\lambda_0)$, for λ close to λ_0 . Therefore it follows that $R(\lambda)$ is an analytic function from the resolvent set Rs(A) to $L(H)$.

Let us now return to self-adjoint operators.

Theorem 3.1. For a self-adjoint $A \in Q(H)$ we have $Rs(A) \supset \mathbb{C} \setminus \mathbb{R}$, i.e., $Sp(A) \subset \mathbb{R}$, and

$$(3.5) \qquad \|R(\lambda)\| \le |\text{Im } \lambda|^{-1}$$

If in addition $A \ge c$ (or $A \le c$) then $Rs(A) \supset (-\infty, c)$ (or $Rs(A) \supset (c,\infty)$), and

$$(3.6) \qquad \|R(\lambda)\| \le (\text{dist}(\lambda,(c,\infty)))^{-1} \ , \ (\text{or } \|R(\lambda)\| \le (\text{dist}(\lambda,-\infty,c)))^{-1})$$

A hermitian $A \in P(H)$ is essentially self-adjoint if(f) for some λ, $\text{Im } \lambda \ne 0$, we have both $\text{im}(A-\lambda)$, $\text{im}(A-\bar{\lambda})$ dense in H. If in addition $A \ge c$ (or $A \le c$) then A is essentially self-adjoint if(f) for some

real ($\gamma<c$) (or $\gamma>c$) we have $\mathrm{im}(A-\gamma)$ dense in H .
Proof. The estimate (3.1), for a non-real λ is an immediate consequence of prop.2.1, which also gives a value of the constant c in agreement with (3.5). On the other hand, for $\lambda = \pm i$ the first condition (3.1) is a consequence of cor.2.3. In fact, it is noticed that the conclusions leading to cor.2.3 can be repeated for every pair of non-real complex conjugate numbers. Therefore we indeed get the first statement and (3.5). On the other hand, for a semi-bounded operator we must repeat the conclusions leading to prop.2.8., particularly (2.21). Details are left to the reader.

The spectral theorem for unbounded self-adjoint operators provides a Fourier type integral representation of every self-adjoint operator by a so-called spectral family $\{E(\lambda) : \lambda \in \mathbb{R}\}$. In this section we only will give a general discussion of the spectral theorem, while the proof will be discussed in sec.4.

A bounded hermitian operator A semi-bounded below by 0 will be called positive, and for two bounded hermitian operators A , B we will write $A \leq B$ (or $B \geq A$) if B-A is positive.

An orthogonal projection (here shortly 'projection') of the Hilbert space H is a bounded hermitian operator P satisfying $P^2 = P$ (i.e. an idempotent). One easily verifies :

Proposition 3.2. The orthogonal projections P of H correspond to the orthogonal direct decompositions of H :

$$H = \mathrm{im}\, P \oplus \ker P \quad . \tag{3.7}$$

For $u \in H$ we get $u = v + w$, corresponding to (3.7) , where $v = Pu$ and $Pw = 0$. For any direct decomposition $H = M \oplus N$ with M , N orthogonal the assignment $u \to v$ defines a bounded hermitian idempotent operator P such that $M = \mathrm{im}\, P$, $N = \ker P$, Pu=u in M .

Moreover, for two orthogonal projections P,Q we have $P \leq Q$ if and only if $\mathrm{im}\, P \subset \mathrm{im}\, Q$, and then Q-P is the orthogonal projection onto the complement $(\mathrm{im}\, Q) \cap (\mathrm{im}\, P)^{\perp}$. In particular this holds if and only if

$$PQ = QP = P \quad . \tag{3.8}$$

Proof. For $u \in \mathrm{im}\, P$ get $u=Pv$, $Pu=P^2v=Pv=u$. Hence for $w \in \ker P$ get $(u,w) = (Pv,w) = (v,Pw) = 0$, so that im P and ker P are orthogonal. Also, for $z \in H$ write $z=Pz+(1-P)z$, where clearly $Pz \in \mathrm{im}\, P$ $P(1-P)z = (P-P^2)z=0$, hence $(1-P)z \in \ker P$, so that we get the desired orthogonal direct decomposition. Vice versa we confirm easily that for an orthogonal decomposition $H=M\oplus N$ the map given is a hermitian idempotent.

If P,Q are projections, and im P⊂im Q , then we get $H=\text{im } P \oplus (\text{im } Q)\cap(\text{im } P)^{\perp} \oplus \ker Q$, an orthogonal direct decomposition, where P = 1,0,0 , and Q = 1,1,0 , in the three spaces. Hence Q-P= 0,1,0 , which shows that Q-P projects onto the second space. Also we then clearly get (3.8).
Vice versa, for projections P,Q, let $P\leq Q$, which implies $1-Q\leq 1-P$, so that Pu=u implies $\|(1-Q)u\|^2 = ((1-Q)u,(1-Q)u)=(u,(1-Q)^2u) = (u,(1-Q)u) \leq (u,(1-P)u) = 0$, hence Qu=u , and im P ⊂ im Q . Then, of course Q-P projects as described, and we have QP=PQ=P.

Finally, if PQ=QP=P , one computes $(Q-P)^2=Q^2+P^2-QP-PQ=Q-P$, so that Q-P is a projection. If (Q-P)u=u then find Qu=Q(Q-P)u=u, Pu=0 , and vice versa, so that Q-P projects onto (im Q)∩(ker P), as stated, q.e.d.

A <u>spectral family</u> is defined to be an increasing one parameter family $\{E(\lambda) : \lambda\in \mathbb{R}\}$ of orthogonal projections which is left continuous, in strong operator convergence of $L(H)$,and such that

$$\text{(3.9)} \qquad \lim_{\lambda\to-\infty}E(\lambda) = 0 \ , \ \lim_{\lambda\to\infty}E(\lambda) = 1 \ ,$$

also in strong convergence. In particular,

$$\text{(3.10)} \qquad E(\lambda) \leq E(\mu) \ , \text{ as } \lambda \leq \mu \ ,$$

and

$$\text{(3.11)} \qquad \lim_{\lambda\to\mu-0}E(\lambda) = E(\mu) \ .$$

It is clear from prop.3.2 that the projection operators $E(\lambda)$ of a spectral family all commute, and that $H_\lambda = \text{im } E(\lambda)$ forms an increasing family of closed subspaces : We have

$$\text{(3.12)} \qquad H_\lambda \subset H_\mu \ , \text{ as } \lambda < \mu \ .$$

For similar reason it is found that, for a spectral family $E(\lambda)$, the two limits

$$\text{(3.13)} \qquad E(\lambda+0) = \lim_{\mu\to\lambda,\mu>\lambda}E(\mu) \ , \ E(\lambda-0) = \lim_{\mu\to\lambda,\mu<\lambda}E(\mu)$$

exist at each $\lambda \in \mathbb{R}$ in strong operator convergence of H similarly as for a real-valued function. In fact for $\mu > \lambda$ we have

(3.14) $\|(E(\mu)-E(\lambda))u\|^2 = (u,E(\mu)u) - (u,E(\lambda)u)$, $u \in H$,

hence the upper and lower limits will exist, just as they do for the monotone real-valued functions $(u,E(\lambda)u)$, $u \in H$. Using (3.14) for an orthonormal basis of (the separable space) H one easily concludes that $E(\lambda)$ is strongly continuous over all of $\mathbb{R}$, except perhaps at certain 'jumps', and that there are at most countably many jumps. Then the existence of the limits $E(\pm\infty)=\lim_{\lambda\to\pm\infty} E(\lambda)$ in strong convergence may be derived from the monotony of $E(\lambda)$ only, without the explicit requirements (3.9) .

For an interval $\Delta = (\lambda',\lambda'']$ of the real axis, left open and right closed, let $E_\Delta=E(\lambda'')-E(\lambda')$. If we have a partition

(3.15) $\lambda' = \lambda_0 < \lambda_1 < \dots < \lambda_{N-1} < \lambda_N = \lambda''$,

of Δ into finitely many sub-intervals $\Delta_j = (\lambda_{j-1},\lambda_j]$, $j=1,\dots,N$, then we get a corresponding orthogonal decomposition

(3.16) $\operatorname{im} E_\Delta = \oplus_{j=1}^N (\operatorname{im} E_{\Delta_j})$.

For arbitrary $u \in H$ we get

(3.17) $\|E_\Delta u\|^2 = \sum_{j=1}^N \|E_{\Delta_j} u\|^2$,

in view of the orthogonality. One may use this fact to define the Riemann-Stieltjes integral $\int_\Delta \phi(\lambda)dE(\lambda)$, for a continuous function $\phi(\lambda)$, as the limit, in norm convergence, $\lim_{\delta\to 0} S$, with Riemann sum S and maximum interval length δ (with points $\lambda_j^* \in [\lambda_{j-1},\lambda_j]$), where

(3.18) $S = \sum_{j=1}^N \phi(\lambda_j^*)E_{\Delta_j}$, $\delta = \operatorname{Max}\{(\lambda_j-\lambda_{j-1}) : j=1,\dots,N\}$.

Here the existence of the limit for continuous functions ϕ is easily derived from the observation that, for the Riemann sum S' of any refinement of the partition (3.15), we have

(3.19) $\|(S-S')u\| \le \sqrt{\delta}\|E_\Delta u\|$,

by the same orthogonality. (Since any pair of partitions, with sums S,S" , and maximal lengthes δ,δ'' has a common refinement sum S', we get $\|S-S''\| \le \sqrt{\delta}+\sqrt{\delta}$,insuring convergence of the integral.)

Also,

(3.20) $\|\int_\Delta \phi(\lambda)dE(\lambda)u\|^2 = \int_\Delta |\phi(\lambda)|^2 d\|E(\lambda)u\|^2 = \int_\Delta |\phi(\lambda)|^2 d(E(\lambda)u,u)$, with real-valued Riemann-Stieltjes integrals at right. For $u \in H$ and a function $\phi \in C(\mathbb{R})$ one then may look for existence of the improper Riemann-Stieltjes integral $Iu = \int_{-\infty}^{\infty} \phi(\lambda)dE(\lambda)u$. It is found that Iu exists if and only if

$$(3.21) \qquad \|Iu\|^2 = \int_{-\infty}^{\infty} |\phi(\lambda)|^2 d(E(\lambda)u,u) < \infty .$$

(This follows by applying (3.20) to the difference $I_\Delta u - I_{\Delta'} u$, with two large intervals Δ , Δ' , and $I_\Delta u = \int_\Delta \phi(\lambda)dE(\lambda)u$.)

In the theorem, below, we will say that an unbounded operator A and a bounded operator B commute , if we have $BA \subset AB$, with products defined by (1.3) . One confirms easily that this holds true if and only if B commutes with the resolvent $R(\lambda)=(A-\lambda)^{-1}$ for every $\lambda \in Rs(A)$.

We now can state the spectral theorem.

Theorem 3.3. For every self-adjoint operator $A \in \mathcal{Q}(H)$ there exists a unique spectral family $\{E(\lambda)\}$, with $E(\lambda)$ commuting with every bounded hermitian operator B , commuting with A , such that

$$(3.22) \qquad A = \int_{-\infty}^{\infty} \lambda dE(\lambda) ,$$

in the sense that

$$(3.23) \qquad \text{dom } A = \{u \in H : \int_{-\infty}^{\infty} \lambda^2 d(E(\lambda)u,u) < \infty \} ,$$

and

$$(3.24) \qquad Au = \int_{-\infty}^{\infty} \lambda dE(\lambda)u ,$$

with an improper Riemann-Stieltjes integral at right.

Vice versa, for every spectral family $E(\lambda)$, formula (3.22) defines an unbounded operator A , and $E(\lambda)$ commutes with every bounded hermitian operator, commuting with A .

4. Proof of the spectral theorem.

In this section we discuss a proof of thm.3.3. Note that there is a selection of proofs, mainly differing by the type of construction used for the family $E(\lambda)$. Here we define $E(\lambda)$ as the orthogonal projection onto the null space of the bounded operator

$$(4.1) \qquad S_\lambda = T_\lambda + (T_\lambda^2)^{1/2} \ , \ T_\lambda = (R(\lambda+i)+R(\lambda-i))/2 \ ,$$

with the resolvent $R(\mu) = (A-\mu)^{-1}$ of the self-adjoint operator A. More precisely, since the above would give a right (not left) continuous spectral family, the above family of projections is called P_λ , and then $E(\lambda) = \lim_{\mu\to\lambda,\mu<\lambda} P_\mu = P_{\lambda-0}$.
For a similar proof cf. Nagy [NSz_1], [RN].

For a self-adjoint operator A the resolvent $R(\mu)$ is defined (and holomorphic) in all of $\mathbb{C}\setminus\mathbb{R}$ (thm.3.1), and we have $(A-\mu)^* = A-\bar{\mu}$, hence $R(\mu)^* = R(\bar{\mu})$. Therefore the operator T_λ of (4.1) is bounded and self-adjoint, since we may write $T_\lambda = (R(\mu)+R(\mu)^*)/2$, with $\mu=\lambda+i$. In fact we get $\|T_\lambda\| \le 1$, using (3.5) . Accordingly we have T_λ^2 bounded self-adjoint and $0 \le T_\lambda^2 \le 1$, and the binomial series of prop.2.8 , i.e.,

$$(4.2) \qquad U_\lambda = (T_\lambda^2)^{1/2} = \sum_{j=0}^{\infty} \binom{1/2}{j} (T_\lambda^2 - 1)^j$$

defines a bounded self-adjoint operator U_λ satisfying $0 \le U_\lambda \le 1$, and $U_\lambda^2 = T_\lambda^2$. Indeed, the binomial series $\sum \binom{1/2}{j} x^j$ converges absolutely for x=1, as well known. We get $-1 \le (T_\lambda^2 - 1) \le 0 < 1$, hence $\|T_\lambda^2 - 1\| \le 1$ (using prop 4.3), so that the series (4.2) converges absolutely. Thus the Cauchy-type reordering of the product $U_\lambda^2 = U_\lambda \cdot U_\lambda$ is correct, just as for a series of numbers, and we get $U_\lambda^2 = (1+(T_\lambda^2 - 1)) = T_\lambda^2$. The coefficients of the series are real , hence all terms are self-adjoint. Moreover, all terms but the 0-th power are ≤ 0 , and one estimates $-\sum_1^\infty \le 1$, so that $U_\lambda = 1 + \sum_1^\infty \ge 0$.

It is clear that T_λ and U_λ commute, by definition of U_λ (We have $T_\lambda U_\lambda = U_\lambda T_\lambda$). Accordingly, $0 = (T_\lambda^2 - U_\lambda^2) = (T_\lambda + U_\lambda)(T_\lambda - U_\lambda)$. For a $u \in \operatorname{im} T_\lambda$ we get $u = T_\lambda v = (T_\lambda + U_\lambda)v/2 + (T_\lambda - U_\lambda)v/2 = w+z$, where $w = (T_\lambda + U_\lambda)v/2 \in \ker(T_\lambda - U_\lambda)$, $z = (T_\lambda - U_\lambda)v/2 \in \ker(T_\lambda + U_\lambda)$. Also, $(w,z) = 4^{-1}(v,(T_\lambda^2 - U_\lambda^2)v) = 0$, using that T_λ and U_λ are self-adjoint. This means that w and z are orthogonal.

Since T_λ is self-adjoint, we have

$$(4.3) \qquad \mathcal{H} = \ker T_\lambda \oplus (\operatorname{im} T_\lambda)^{\text{clos.}} \ ,$$

which is an orthogonal direct decomosition. Also, $\ker T_\lambda = \ker U_\lambda$: If $T_\lambda u=0$, then $U_\lambda{}^2 u = T_\lambda{}^2 u = 0$, hence $\|U_\lambda u\|^2 = (u,U_\lambda{}^2 u) = 0$, so that $\ker T_\lambda \subset \ker U_\lambda$, and by symmetry we get equality. It follows that $\ker(T_\lambda+U_\lambda) \cap \ker(T_\lambda-U_\lambda) = \ker T_\lambda = \ker U_\lambda$. Define

$$(4.4)\qquad H_\lambda = \ker(T_\lambda+U_\lambda)\ ,\ N_\lambda = \ker T_\lambda\ ,\ K_\lambda = \ker(T_\lambda-U_\lambda) \cap (N_\lambda^\perp),$$

and conclude from the above that

$$(4.5)\qquad H = H_\lambda \oplus K_\lambda\ ,$$

with an orthogonal direct decomposition. Also,

$$(4.6)\qquad \begin{aligned} (u,T_\lambda u) &= -(u,U_\lambda u) \le 0 \text{ on } H_\lambda = \ker(T_\lambda+U_\lambda),\\ (u,T_\lambda u) &= (u,U_\lambda u) \ge 0 \text{ on } K_\lambda \subset \ker(T_\lambda-U_\lambda)\ . \end{aligned}$$

Finally, in this chain of arguments, note that

$$\begin{aligned} 2T_\lambda u &= R(\lambda+i)u + R(\lambda-i)u = R(\lambda+i)((A-\lambda+i)+(A-\lambda-i))R(\lambda-i)u\\ &= 2R(\lambda+i)(A-\lambda)R(\lambda-i)u\ ,\ u\in H\ , \end{aligned}$$

implying that

$$(4.7)\qquad T_\lambda = R(\lambda-i)^*(A-\lambda)R(\lambda+i)\ .$$

Combining this with (4.6) we get

$$(4.8)\qquad (R(\lambda-i)u,(A-\lambda)R(\lambda-i)u) \le 0\ ,\ \text{as } u\in H_\lambda\ ,\ \ge 0\ ,\ \text{as } u\in K_\lambda\ .$$

Furthermore we observe that $R(\nu)R(\mu) = R(\mu)R(\nu)$, for every $\mu,\nu\in\mathbb{C}$, which also implies that $R(\mu)$ commutes with T_λ and U_λ . We conclude from there that, for all $\mu \in \mathbb{C}\setminus\mathbb{R}$, we have

$$(4.9)\qquad R(\mu)H_\lambda = \text{dom } A \cap H_\lambda\ ,\ R(\mu)K_\lambda = \text{dom } A \cap K_\lambda\ .$$

Indeed, for $u\in H_\lambda$ we get $R(\mu)u\in$ dom A , and $(T_\lambda+U_\lambda)R(\mu)u = R(\mu)(T_\lambda+U_\lambda)u = 0$, so that $R(\mu)u \in \text{dom } A \cap H_\lambda$. For $u\in \text{dom } A \cap H_\lambda$ one may write $u=R(\mu)v$, $v=(A-\mu)u$, and then $0=(T_\lambda+U_\lambda)u=R(\mu)(T_\lambda+U_\lambda)v$. Since $R(\mu)$ is 1-1, it follows that $(T_\lambda+U_\lambda)v=0$, hence $v\in H_\lambda$, confirming the first equation (4.9). Similarly one finds that $R(\mu)(\ker(T_\lambda-U_\lambda)) = \text{dom } A \cap \ker(T_\lambda-U_\lambda)$. Also it is clear that $R(\mu)(\ker T_\lambda) = \ker T_\lambda = \ker(A-\lambda) \subset \text{dom } A$, using (4.7) and (4.8), so that also the second (4.9) follows.

One then finds that

$$\text{dom } A = (\text{dom } A \cap H_\lambda) \oplus (\text{dom } A \cap K_\lambda)$$

in the sense that every $u \in \text{dom } A$ allows a unique sum decomposition $u=v+w$ with v,w in the spaces at right. Such decomposition arises if $(A-\mu)u$ is subject to the decomposition (4.5), and then $R(\mu)$ is applied. Let $R(\mu)^\lambda = R(\mu)|H_\lambda$, then it follows that $R(\mu)^\lambda : H_\lambda \to H_\lambda$, and im $R(\mu)^\lambda$ is dense in H_λ, while $R(\mu)^\lambda$ is 1-1, hence has an inverse in $Q(H_\lambda)$. One gets $(R(\mu)^\lambda)^* = R(\bar{\mu})^\lambda$, and thus finds that

$$A^\lambda = (R(\mu)^\lambda)^{-1} + \mu$$

defines an unbounded self-adjoint operator in $Q(H_\lambda)$ with domain $\text{dom } A \cap H_\lambda$. Moreover, A_λ is the restriction of A to that space, and is independent of μ. Similarly for the restriction of A to K_λ.

We summarize the above:

Theorem 4.1. We have

(4.10) $\quad \ker(A-\lambda) \subset H_\lambda$, $\ker(A-\lambda) \cap K_\lambda = \{0\}$,

and

(4.11) $(u,Au) \geq \lambda(u,u)$, $u \in \text{dom } A \cap H_\lambda$, and $(u,Au) \leq \lambda(u,u)$, $u \in \text{dom } A \cap K_\lambda$.

Moreover, all the spaces H_λ, K_λ, for $\lambda \in \mathbb{R}$, are left invariant by the resolvent $R(\mu)$, and by A, where the latter means that

(4.12) $\quad A:(\text{dom } A \cap H_\lambda) \to H_\lambda$, $A:(\text{dom } A \cap K_\lambda) \to K_\lambda$,

while $\text{dom } A \cap H_\lambda$, and $\text{dom } A \cap K_\lambda$ are dense in H_λ and K_λ, respectively, and the restrictions of A to these spaces are unbounded self-adjoint operators of the Hilbert spaces H_λ, and K_λ, resp.

Thm.4.1 already holds the principal information of the spectral theorem, at least as far as existence of the spectral resolution is concerned. We complete this as follows.

Corollary 4.2. For every self-adjoint operator $B \in L(H)$, commuting with every $R(\mu)$, $\mu \in \mathbb{C} \setminus \mathbb{R}$, the restrictions $B^\lambda = B|H_\lambda$, and $B_\lambda = B|K_\lambda$ are maps $H_\lambda \to H_\lambda$ and $K_\lambda \to K_\lambda$, respectively.

In other words, every bounded self-adjoint operator B, commuting with the resolvent $R(\mu)$, is reduced by the direct decomposition (4.5).

Indeed, such an operator B commutes with T_λ, hence with U_λ, given as a power series in $T_\lambda - 1$. As above we thus conclude that

$(T_\lambda+U_\lambda)Bu=B(T_\lambda+U_\lambda)u=0$, as $u \in H_\lambda$, so that $B: H_\lambda \to H_\lambda$. Similarly $B: K_\lambda \to K_\lambda$.

It is convenient now to introduce the orthogonal projections P_λ by setting

(4.13) $$P_\lambda u=u \text{ on } H_\lambda \text{ , } P_\lambda u=0 \text{ on } K_\lambda \text{ .}$$

Then, as a consequence of cor.4.2 we find that

(4.14) $$P_\lambda P_{\lambda'} = P_{\lambda'} P_\lambda \text{ , for all } \lambda, \lambda' \in \mathbb{R} \text{ .}$$

Indeed, the operator P_λ is bounded and self-adjoint, and commutes with $R(\mu)$. The latter is an immediate consequence of thm.4.1: For $u \in H_\lambda$ we get $R(\mu)u \in H_\lambda$, and $u \in K_\lambda$ implies $R(\mu)u \in K_\lambda$, hence $P_\lambda R(\mu)u=R(\mu)u=R(\mu)Pu$, $u \in H_\lambda$, and $(1-P_\lambda)R(\mu)u=R(\mu)u=R(\mu)(1-P_\lambda)u$, $u \in K_\lambda$, or also $P_\lambda R(\mu)u=R(\mu)P_\lambda u$, for $u \in K_\lambda$. Since $H=H_\lambda \oplus K_\lambda$ we get $P_\lambda R(\mu) = R(\mu)P_\lambda$. Self-adjointness and boundedness of P_λ is a consequence of the orthogonality of H_λ and K_λ .

Using cor.4.2 it thus follows that B is reduced by any of the other direct decompositions (4.5) . Thus again, we get

(4.15) $$P_{\lambda'} P_\lambda u=P_\lambda u=P_\lambda P_{\lambda'} u,\ u \in H_{\lambda'},\ (1-P_{\lambda'})P_\lambda u=P_\lambda(1-P_{\lambda'})u,\ u \in K_{\lambda'},$$

implying (4.14).

Next let $\lambda', \lambda \in \mathbb{R}$, $\lambda' < \lambda$. For $u \in H_{\lambda'}$ write

(4.16) $$u = P_{\lambda'} u = P_{\lambda'} P_\lambda u + P_{\lambda'}(1-P_\lambda)u = v+w \text{ .}$$

The first term, at right, is $\in H_\lambda$, and the second term $w \in H_{\lambda'} \cap K_\lambda$, using the commutativity (4.14) . Thus the second term satisfies

(4.17) $$\lambda(w,w) \leq (w,Aw) \leq \lambda'(w,w) \text{ .}$$

The latter is only possible if $w=0$, since $\lambda' < \lambda$, using (4.11). Hence it follows that $u \in H'$, and we have proven that

(4.18) $$H_{\lambda'} \subset H_\lambda \text{ , as } \lambda' < \lambda \text{ .}$$

Now we invoke prop.3.2, and conclude that $\{P_\lambda : \lambda \in \mathbb{R}\}$, is a nondecreasing family; we have (3.10). As discussed in sec.3, this implies existence of the monotone limits (3.9) and (3.13) and these limits must be orthogonal projections, since the functional equations $P^2=P$ and $P^* = P$ are continuous. Thus we may set $E(\lambda) = P_{\lambda-0}$, as postulated.

In particular $E(-\infty)$ and $E(\infty)$ are the projections onto $H_{-\infty} = \cap H_\lambda$, and $H_\infty = (\cup H_\lambda)^c$, respectivey, with union and intersection over $\mathbb{R}$. For $u \in H_{-\infty}$ we must have $(u,Au) \leq \lambda(u,u)$ for all $\lambda \in \mathbb{R}$, which is only possible, if $u=0$. For $u \in K_\infty = (H_\infty)^\perp$ we must have $(u,Au) \geq \lambda(u,u)$ for all $\lambda \in \mathbb{R}$, implying $u=0$ again. This yields $H_{-\infty} = (K_\infty)^\perp = 0$, or, $E(-\infty) = 1-E(\infty) = 0$, confirming (3.9). Similarly we conclude (3.11), by construction. Therefore $E(\lambda)$ indeed is a (left continuous) spectral family, while $\{P_\lambda\}$ is right continuous.

Next consider an interval $\Delta=(\lambda',\lambda''] \subset \mathbb{R}$, with finite λ',λ'', and a partition (3.15). With E_Δ, E_{Δ_j} as in sec.3 observe that $E_{\Delta_j} = E_{\lambda_j}(1-E_{\lambda_{j-1}})$. Thus (4.10) implies

$$(4.19) \qquad \lambda_{j-1}\|E_{\Delta_j}u\|^2 \leq (E_{\Delta_j}u, AE_{\Delta_j}u) \leq \lambda_j\|E_{\Delta_j}u\|^2 \ , \ u \in H .$$

In particular one gets (4.19) first for $u \in \text{dom } A \cap H_{\lambda_j} \cap K_{\lambda_{j-1}}$, a space dense in $\text{im } E_{\Delta_j} = H_{\lambda_j} \cap H_{\lambda_{j-1}}$, by a direct sum argument, as in the proof of thm.4.1. This implies boundedness of (the restriction to $\text{im } E_{\Delta_j}$ of) A in $\text{im } E_{\Delta_j}$, using prop.4.3, below.

Since that restriction is a closed operator, it follows $\text{dom } A \cap \text{im } E_{\Delta_j} = \text{im } E_{\Delta_j}$, hence $\text{im } E_{\Delta_j} \subset \text{dom } A$.

From (4.19) we conclude that

$$(4.20) \qquad 0 \leq (E_{\Delta_j}u, (A-\lambda_{j-1})E_{\Delta_j}u) \leq (\lambda_j-\lambda_{j-1})(u, E_{\Delta_j}u) \ , \ u \in H .$$

Summing this over j one has

$$(4.21) \qquad 0 \leq (E_\Delta u, (A-S)E_\Delta u) \leq \delta \ (u, E_\Delta u) \ ,$$

with S and δ of (3.18), for $\lambda_j^* = \lambda_{j-1}$, and $\phi(\lambda) = \lambda$. Here we used that $(E_{\Delta_j}u, E_{\Delta_l}u) = (E_{\Delta_j}u, AE_{\Delta_l}u) = 0$ for $j \neq l$.

Now we use (4.21) and prop.4.3, below, and conclude that

$$(4.22) \qquad AE_\Delta = \int_\Delta \lambda \ dE(\lambda)$$

with an operator norm convergent Riemann-Stieltjes integral. Then, for $u \in \text{dom } A$, we get $Au = \lim_{\lambda' \to -\infty, \lambda'' \to \infty} AE_\Delta u$, which equals the improper Riemann-Stieltjes integral of (3.22). Vice versa, if that integral exists, or, equivalently, if $\int_{-\infty}^{\infty} \lambda^2 d\|E(\lambda)u\|^2 < \infty$, then

$E_\Delta u \to u$, $AE_\Delta u \to \int_{\mathbb{R}} \lambda dE(\lambda)u$, so that $u \in H$, and (3.24) holds, by closedness of the operator A.

This completes the proof of thm.3.3, except for uniqueness of $E(\lambda)$.

Now, if Q_λ is another spectral family such that Q_λ commutes with every self-adjoint $B \in L(H)$ commuting with A (i.e., with every $R(\mu)$), then we get

$$Q_\lambda P_\mu = P_\mu Q_\lambda \ , \tag{4.23}$$

using that P_μ commutes with $R(\nu)$, by construction. Then the argument leading to (4.14) may be repeated, showing that also

$$\text{im } Q_{\lambda'} \subset H_\lambda \subset \text{im } Q_{\lambda''} \ , \text{ as } \lambda' < \lambda < \lambda'' \ . \tag{4.24}$$

Then, by monotony of Q_λ we get

$$Q_{\lambda-0} \le P_\lambda \le Q_{\lambda+0} \ , \tag{4.25}$$

resulting in equality $Q_\lambda = P_\lambda = E(\lambda)$ at all points of continuity of Q_λ. Since both $E(\lambda)$ and Q_λ are left continuous we get $E(\lambda) = Q_\lambda$ also at the 'jumps' of Q_λ , using the fact that $\text{im}(Q_{\lambda+0} - Q_{\lambda-0}) = \ker(A-\lambda)$. This completes the proof of thm.3.3.

Proposition 4.3. Assume that a hermitian operator T satisfies

$$|(u,Tu)| \le c(u,u) \ , \text{ for all } u \in \text{dom } T \ , \tag{4.26}$$

with some nonnegative constant c. Then T is bounded, and we have $\|T\| \le c$.

Proof. Apply (4.26) to u+v and u-v both, and subtract the two inequalities, for

$$-2c(\|u\|^2 + \|v\|^2) \le 4\text{Re}(u,Tv) \le 2c(\|u\|^2 + \|v\|^2), \ u,v \in \text{dom } T \ . \tag{4.27}$$

Here we introduce $u = Tv/c$ (if necessary, let u run through a sequence converging to Tv/c). It follows that

$$(2/c)\|Tv\|^2 \le (1/c)\|Tv\|^2 + c\|v\|^2 \ , \ v \in \text{dom } T \ , \tag{4.28}$$

which implies the statement.

5. A result on powers of positive operators.

In the lemma,below, let H be a separable Hilbert space.

Lemma 5.1. Let $B \in L(H)$ be self-adjoint, and let A be an unbounded self-adjoint operator with domain dom A . Assume A positive, and B positive definite, so that B^{-1} exists as an unbounded self-adjoint operator, and the positive powers A^s , B^s , $s>0$, are well defined positive self-adjoint operators.
If AB is bounded, then also $A^s B^s$ is bounded, for every $0 \le s \le 1$. Moreover, we have

$$(5.1) \qquad \|A^s B^s\| \le \|AB\|^s .$$

If AB is compact, then also $A^s B^s$ is compact, for every $0 < s \le 1$.

Remark (a): The boundedness of $A^s B^s$,and (5.1) is a well known result, due to E.Heinz [Hz_1],(cf. also [CQ], Dixmier [Dx_2], Kato [K_3], Loewner [Lw_1]). The compactness of $A^s B^s$ seems to be new.

Remark (b): Note that the condition of boundedness of AB is equivalent to the condition dom A $\supset$ im B , and again equivalent to the condition that the unbounded operator product AB has dense domain and extends continuously to H . Indeed, if dom A $\supset$ im B , then the product AB has domain H , while we get $(AB)^* \supset BA$, due to self-adjointness. Hence $\text{dom}(AB)^*$ is dense, which implies boundedness of AB , by the closed graph theorem. If AB has dense domain and a continuous extension to H , then for all $u \in H$ there exists $u_k \in$ dom (AB) with $u_k \to u$, $ABu_k \to v = (AB)^{**}u$. Since B is bounded we also get $Bu_k \to Bu$, $A(Bu_k) \to v$. The selfadjoint operator is closed. Thus we conclude that $Bu \in$ dom A , hence dom A $\supset$ im B . Thus indeed all 3 above conditions are equivalent.

Remark (c): Clearly boundedness of $A^s B^s$ in lemma 5.1 again may be interpreted in each of the ways of rem.(b). Also, clearly, since compactness of AB implies boundedness of AB , all of remark (b) applies in the case of $AB \in K(H)$, where $K(H)$ denotes the (closed 2-sided) ideal of compact operators in $L(H)$.

The proof of the lemma will require some preparations.
Let $B \in L(H)$ be self-adjoint, positive definite. Let H_0 be the completion of H under the norm $\|u\|_0 = \{(u,u)_0\}^{1/2}$, with inner product $(u,v)_0 = (u,Bv) = (Bu,v)$, $u,v \in H$. From simple arguments it

follows that H may be regarded as a dense subspace of H_0. Now let us assume that A with domain dom A is an unbounded closed operator of H which is hermitian with respect to the inner product $(.,.)_0$. That is, we have $(u,Av)_0 = (Au,v)_0$, for all $u,v \in$ dom A. Also let the resolvent set Rs(A) of A contain at least one real point λ_0.

Proposition 5.2. The linear map $A:\text{dom } A \to H_0$ defines an essentially selfadjoint operator of H_0. If A_0 denotes the closure of A in H_0, then the resolvent set $\text{Rs}(A_0)$ contains Rs(A). Moreover,

$$(5.2)\qquad \|(A_0-\lambda)^{-1}\|_0 \le \|(A-\lambda)^{-1}\| \text{ for all } \lambda \in \text{Rs}(A) .$$

Proof. Let $R = R(\lambda_0) = (A-\lambda_0)^{-1} \in L(H)$. We get $((A-\lambda_0)u,v)_0 = (u,(A-\lambda_0)v)_0$ for all $u,v \in$ dom A, which implies $(f,Rg)_0 = (Rf,g)_0$ for all $f,g \in H$. In other words, the operator $R \in L(H)$ is hermitian in H_0.

Proposition 5.3. We have R bounded in H_0, and, $\|R\|_0 \le \|R\|$.
Proof. With a spectral family $E(\lambda)$ of the operator B let $P_\varepsilon = 1 - E(\varepsilon)$, for $\varepsilon > 0$, and let $S_\varepsilon = \text{im } P_\varepsilon$. In the closed subspace S_ε of H the norms $\|.\|$ and $\|.\|_0$ are equivalent. We have $\varepsilon\|u\|^2 \le \|u\|_0^2 \le \|B\|\|u\|^2$. The projection $P_\varepsilon: H \to S_\varepsilon$ is orthogonal with respect to both inner products $(.,.)$ and $(.,.)_0$. Define $R_\varepsilon = P_\varepsilon A|S_\varepsilon$. For $u,v \in S_\varepsilon$ we get $(u,Rv) = (P_\varepsilon u,Rv) = (u,(P_\varepsilon R|S_\varepsilon)v) = (u,R_\varepsilon v)$, $(u,R_\varepsilon v)_0 = (P_\varepsilon u,Rv)_0 = (u,Rv)_0 = (Ru,v)_0 = (R_\varepsilon u,v)_0$. Now it is easy to see that we must have $\|R_\varepsilon\|_0 \le \|R_\varepsilon\|$. Indeed, the spectrum, and hence the spectral radius ρ of R_ε is independent of the norm under which S_ε is considered. For a bounded hermitian operator we get $\|R_\varepsilon\|_0 = \rho$, while generally $\|R_\varepsilon\| \ge \rho$. Thus we indeed have the above inequality.

It follows that

$$(5.3)\qquad \sup\{|(u,Ru)|/(u,u) : 0\neq u\in S_\varepsilon\} \le \|P_\varepsilon R\| \le \|R\| .$$

But the space of all u with $u = P_\varepsilon u$, for some $\varepsilon > 0$, is dense in H, since B was assumed positive definite. The proof is complete.

Continuing in the proof of prop.5.2 it now follows that the linear map $A:\text{dom } A \to H_0$ is an essentially self-adjoint operator, since $(A-\lambda_0)^{-1}$ is bounded and densely defined (cf. thm.3.1). The closure of this operator (in H_0) is self-adjoint. Its resolvent exists for all non-real points λ, and is a normal operator wherever it exists. Also, at each point λ of the resolvent set

$Rs(A_0)$ we find that $A-\lambda$ must be invertible as a linear map, at least, and the inverse is the restriction of $(A_0-\lambda)^{-1}$ to H. We trivially have $Rs(A)\setminus\mathbb{R} \subset Rs(A_0)$, since $Rs(A_0)$ contains $\mathbb{C}\setminus\mathbb{R}$. For every real $\lambda \in Rs(A)$ prop.5.2 applies as well as to λ_0. Hence we also get $Rs(A)\cap\mathbb{R} \subset Rs(A_0)$, thus $Rs(A) \subset Rs(A_0)$, as stated. Finally, for real $\lambda \in Rs(A)$ prop.5.3 implies the inequality (5.2). For non-real $\lambda \in Rs(A)$ we know that $\lambda \in Rs(A_0)$, and that $(A_0-\lambda)^{-1}$ is normal. If ρ and ρ_0 denote the spectral radii of $(A-\lambda)^{-1}$ and $(A_0-\lambda)^{-1}$, then again $\rho_0 = \|(A_0-\lambda)^{-1}\|$, $\rho \leq \|(A-\lambda)^{-1}\|$, and we must have $\rho_0 \leq \rho$, since ρ and ρ_0 are just the reciprocal minimum distances from λ to $Sp(A)$ and $Sp(A_0)$, and since $Sp(A) = (Rs(A))^{comp} \supset (Rs(A_0))^{comp} = Sp(A_0)$. This completes the proof of prop.5.2 as well.

Proposition 5.4. Under the assumptions of prop.5.2, if we have $A \in K(H)$ then also $A_0 \in K(H_0)$.

Proof. The resolvent set $Rs(A)$ of an operator $A \in K(H)$ is known to be the complement in $\mathbb{C}^* = \mathbb{C}\setminus\{0\}$ of an at most countable set of eigenvalues, each of finite algebraic multiplicity, and with only possible cluster point at zero. In particular, for each eigenvalue $\lambda \neq 0$, the spectral projection

$$(5.4) \qquad P_\lambda = i/(2\pi)\int_{\Gamma_\lambda} R(\mu)d\mu \ , \ R(\mu) = (A-\mu)^{-1} \ ,$$

with a sufficiently small positively oriented circle Γ_λ of center λ, not containing any other eigenvalue, is a bounded operator of finite rank. The resolvent $R(\lambda)$ is a holomorphic map $Rs(A) \to L(H)$, and the integral (5.4) exists in norm convergence of $L(H)$.

Now we apply prop. 5.2 to conclude that $Sp(A_0) \subset Sp(A)$, so that $Sp(A_0)\setminus\{0\}$ is a collection of some of the eigenvalues λ of A. For each such λ we get $R_0(\mu) = (A_0-\mu)^{-1}$ as the continuous extension of $R(\mu)$ onto H_0, in view of (5.2), and uniformly so over Γ_λ. It follows that

$$(5.5) \qquad P^0_\lambda = i/(2\pi)\int_{\Gamma_\mu} R_0(\mu)d\mu$$

is the continuous extension of P_λ to H_0. Since P_λ is of finite rank, this is true for P^0_λ as well. It follows that all non-vanishing points of $Sp(A_0)$ of the self-adjoint operator A_0 are isolated eigenvalues of finite multiplicity. This proves that A_0 is compact and our proof is complete.

Proof of lemma 5.1. First we introduce the inner product $(.,.)_0$

and space H_0 with B of lemma 5.1 . Then if $AB \in L(H)$, it may be observed that $(u,ABv)_0 = (u,BABv) = (ABu,Bv) = (ABu,v)_0$, for all $u,v \in H$, so that AB satisfies the conditions of prop.5.3. It follows that $\|A^{1/2}B^{1/2}\|^2 = \|B^{1/2}AB^{1/2}\| = \|AB\|_0 \leq \|AB\|$, where we have used the C^*-algebra property of $L(H)$, (i.e. $\|X\|^2 = \|X^*X\|$, for $X \in L(H)$), and the fact that $B^{1/2}$ defines an isometry $H \to H_0$, so that $\|X\|_0 = \|B^{1/2}XB^{-1/2}\|$, for $X \in L(H_0)$. Accordingly, we have proven (5.1) for s = 1/2 . Similarly, if $AB \in K(H)$, we may apply prop.5.4 instead of prop.5.3 to conclude that $B^{1/2}AB^{1/2} \in K(H)$ .Since $B^{1/2}AB^{1/2} = (A^{1/2}B^{1/2})^*(A^{1/2}B^{1/2})$ it then follows that $A^sB^s \in K(H)$ for s=1/2 .

In a similar way one now proves the lemma for all rationals of the form $s = k/2^l$, $0 < k \leq 2^l$, as follows. (We only indicate how to proceed for s=1/4 and s=3/4 to make the method evident) . For s=1/4 we may just repeat the above, now using $(u,v)_0 = (u,B^{1/2}v)$,and $A^{1/2}B^{1/2}$ as the operator A of prop.5.2 . For s = 3/4 , if $AB \in L(H)$,we already know that $A^{1/2}B^{1/2} \in L(H)$, so that $C = B^{1/2}A^{3/2}B = (A^{1/2}B^{1/2})^*AB \in L(H)$. The operator C is hermitian with respect to $(u,v)_0 = (B^{1/2}u,v)$, as easily checked. Hence it also is bounded in the corresponding norm. With the isometry $B^{1/4}:H_0 \to H$ we get that $\|A^{3/4}B^{3/4}\|^2 = \|B^{3/4}A^{3/2}B^{3/4}\| = \|C\|_0 \leq \|C\| \leq \|A^{1/2}B^{1/2}\|\|AB\| \leq \|AB\|^{3/2}$. Taking square roots, we thus get (5.1) for s=3/4 . If $AB \in K(H)$ we find that $C \in K(H)$, so that prop.5.4 yields $A^sB^s \in K(H)$ for s=3/4 .

Note finally that boundedness of $A^sB^s = E_s$ is equivalent to the inequality

(5.6) $\|A^su\| \leq \|E_s\|\|B^{-s}u\|$ for all $u \in \text{dom } B^{-s}$.

If (5.6) with $\|E_s\| \leq \|E_1\|^s$ is correct for all rationals r=p/q, $q=2^l$, $0<s\leq 1$, then it must be true for all $0\leq s\leq 1$. Indeed, since B is bounded selfadjoint we have $\text{dom } B^{-s} = \text{im } B^s$ decreasing, as s increases. Let s_j be a sequence of rationals with denominators 2^{l_j} with $s_j \to s$, $s_j<s$. Then $u \in \text{dom } B^{-s} \subset \text{dom } A^s$ implies $u \in \text{dom } B^{-s_j} \subset \text{dom } A^{s_j}$, and $B^{-s_j}u \to B^{-s}u$, $A^{s_j}u \to A^su$, as $j \to \infty$. Using this in (5.6) we conclude that this inequality is correct for all $0 \leq s \leq 1$ if it is true for s=1 .

Finally, regarding compactness of A^sB^s for general s we will have to discuss the function $F(z)=A^zB^z$ of the complex argument z .

For u , v $\in \mathcal{H}$ define the function $\phi(z) = (u, A^z B^z v)$, where it may be noted that the general powers A^z and B^z are well defined also for complex z . The expression $\phi(z)$ is meaningful for $0 \leq \operatorname{Re} z \leq 1$, assuming that AB is bounded, and using our knowledge on boundedness of $A^s B^s$,for $0 \leq s \leq 1$. In fact, for $z=s+it$, $0 \leq s \leq 1$, we get $A^z B^z v = A^{it}(A^s B^s)(B^{it} v)$ well defined, since the operators A^{it} and B^{it} are unitary, of course. We get

$$(5.7) \qquad |\phi(z)| \leq \|AB\|^s \|u\| \|v\| \ , \ 0 \leq s \leq 1 \ , \ -\infty < t < \infty \ .$$

Note that $\phi(z)$ also is analytic in the open strip $0<s<1$, if we assume that, for some $0<\varepsilon<1$, we have $u \in \operatorname{dom} A^{\varepsilon}$. With a partition $1 = \chi(\lambda) + \omega(\lambda)$, $0 \leq \chi, \omega \in C^{\infty}(\mathbb{R})$, $\chi=1$ for $\lambda \leq 1$, $\omega=1$, $\lambda \geq 2$, we write, with some $0<\delta<\operatorname{Min}\{\varepsilon, \operatorname{Re} z\}$,

$$(5.8) \qquad \begin{aligned} f'(z) = {} & (((\log A)A^{\delta}\chi(A))u, (A^{z-\delta}B^{z-\delta})(B^{\delta}v)) + \\ & ((\log A)\omega(A))u, (A^z B^z)v) + ((A^{\delta}u), (A^{z-\delta}B^{z-\delta})(B^{\delta}\log B)v) \ . \end{aligned}$$

All expressions at right are meaningful, in view of the selfadjointness of A and B , using that B is bounded and positive definite, while A is possibly unbounded, but still positive. In particular, the function $\chi(\lambda)\lambda^{\delta} \log \lambda$ extends bounded and continuously to $[0,\infty)$, while $\omega(\lambda)\log \lambda = O(\lambda^{\varepsilon})$ in $[0,\infty)$,and $\lambda^{\delta}\log \lambda = O(1)$ on the bounded set $\operatorname{Sp} B \subset [0,\infty)$.

Note that the restriction $u \in \operatorname{dom} A^{\varepsilon}$ can be removed. For a general $u \in \mathcal{H}$ let the sequence $u_k \in \operatorname{dom} A^{\varepsilon}$ converge to u in $\mathcal{H}$. We get $\phi_k(z)$, with u_k instead of u, analytic in $0<s<1$, and

$$(5.9) \qquad |\phi(z)-\phi_k(z)| = |(u-u_k, A^z B^z v)| \leq \|AB\|^s \|v\| \|u-u_k\| \ ,$$

which implies uniform convergence $\phi_k(z) \to \phi(z)$ in the entire strip Accordingly $\phi(z)=\phi_{u,v}(z)$ is bounded and analytic in $0< \operatorname{Re} z <1$, for any arbitrary $u,v \in \mathcal{H}$.

For a given $0<s<1$ we may apply the Cauchy estimates on the first derivative $\phi'(s)$. With $\eta = \operatorname{Min}\{s,1-s\}$ we get

$$(5.10) \quad |\phi'(s)| \leq \eta^{-1} \operatorname{Max}\{|\phi(\zeta)| : |\zeta-s|=\eta\} \leq \|u\| \|v\| \operatorname{Max}\{1, \|AB\|\}/\eta .$$

From this estimate we conclude that

$$(5.11) \qquad (u, (A^s B^s - A^t B^t)v) = \int_t^s d\tau \phi'_{u,v}(\tau) = O(|s-t| \|u\| \|v\|) \ ,$$

with $O(|s-t| \|u\| \|v\|) \leq c(s,t)|s-t| \|u\| \|v\|$, the constant $c(s,t)$

independent of u,v ,and bounded in compact subintervals of (0,1).

Note that (5.11) implies the proposition,below.

Proposition 5.5. Under the assumptions of lemma 5.1, if $AB \in L(H)$, then the operator family $F(s) = A^sB^s$, $0 < s < 1$, is norm continuous (even Lipschitz continuous) . We have

$$(5.12) \qquad \|A^sB^s - A^tB^t\| \leq c(s,t)|s-t| \quad .$$

Moreover, the corresponding is true for the more general family $F(z)=A^zB^z$, $0 < \mathrm{Re}\, z < 1$. We then get (5.12) with s,t replaced by z,z', with c(z,z') bounded in conpact subsets of the strip.

We will use only (5.12), not its complex extension.

Prop.5.5 also provides the final link in the proof of lemma 5.1 . If $AB \in K(H)$ then certainly $AB \in L(H)$, so that prop.5.5 applies and A^sB^s is norm continuous in the open interval (0,1) . Since A^sB^s is compact in the dense subset of all rationals of the form $s=k/2^l$ it must be compact for every $s \in (0,1)$, since the compact ideal is closed in norm topology.

Notice that $A^0B^0=I$, of course, is not in $K(H)$. Accordingly it is clear that $F(z) = A^zB^z$ cannot be norm continuous at $z=0$.

6. On HS-chains.

In this section we shall discuss an abstract result on certain families of Hilbert spaces similar to the so-called Sobolev spaces. For a concrete example cf. [C_1],ch.III and IV, where corresponding facts are derived for the Sobolev spaces over $\mathbb{R}^n$ (we shall apply the abstract result in VI,3 and ch.IX below).

Let H be a (separable) Hilbert space and let H be a self-adjoint operator of H with domain dom H . We assume H semi-bounded below and that 1 be a lower semi-bound, i.e., $H \geq 1$. Also we assume H properly unbounded, i.e., that its spectrum extend to $+\infty$. Note that the complex powers H^s, $s \in \mathbb{C}$ are well defined closed operators. For example we may use the spectral family $E(\lambda)$ of H and define (with $\lambda^s = e^{s \log \lambda}$, $\log \lambda$ real-valued)

$$(6.1) \qquad H^s = \int_1^\infty \lambda^s dE(\lambda), \ \mathrm{dom}\, H^s = \{u \in H: \int_1^\infty \lambda^{2\mathrm{Re}(s)} d\|E(\lambda)u\|^2 < \infty\} \ ,$$

For a formal convenience, useful later on, we introduce the inverse positive square root $\Lambda = H^{-1/2}$, and henceforth will also write $H^s = \Lambda^{-s/2}$.

For every $s\in\mathbb{R}$ let H_s denote the Hilbert space obtained by completing dom $H^{s/2}$ = dom $\Lambda^{-s} \subset H$ under the norm

$$(6.2) \qquad \|u\|_s = \|\Lambda^{-s}u\| .$$

Evidently we get H_s = dom $\Lambda^{-s} \subset H$ for $s>0$, and the norm $\|.\|_s$ is equivalent to the graph norm of the unbounded operator Λ^{-s}. For $s<0$ the norm $\|.\|_s$ is weaker that the norm of H , whence $H_s \supset H$. For example one may consider H_s , $s<0$, represented by the collection of all 1-parameter families $U=\{u_\lambda : \lambda\in[1,\infty)\}$ such that $E(\mu)u_\lambda = u_\mu$, as $\mu\le\lambda$, and that $\|u\|_s^2 = \int \lambda^{2\mathrm{Re}(s)} d\|u_\lambda\|^2 < \infty$. The same representation of $u\in H_s$ as a function of λ also holds true for general $s\in\mathbb{R}$.

In this way one finds that H_s is naturally imbedded in H_t , as $s>t$, so that $\{H_s : s\in\mathbb{R}\}$ is a decreasing family of spaces. We still introduce the locally convex spaces

$$(6.3) \qquad H_\infty = \cap\{H_s : s\in\mathbb{R}\} \ , \ H_{-\infty} = \cup\{H_s : s\in\mathbb{R}\} \ ,$$

where the union carries the inductive limit topology, while H_∞ carries the Frechet topology induced by all the norms $\|.\|_s$: $s\in\mathbb{R}$.

A collection of spaces H_s, $-\infty\le s\le\infty$, as constructed above will be referred to as an <u>HS-chain</u>.

It is natural to extend a projection $E(\lambda)$ of the spectral family to an operator $E(\lambda)_{-\infty}: H_{-\infty} \to H_\infty$ by defining $E(\lambda)_{-\infty}u=u_\lambda$ for $u=\{u_\mu \mu\ge 1\}$, as $\lambda\ge 1$, and $E(\lambda)_{-\infty}u=0$ for $\lambda<1$.

<u>Proposition 6.1</u>. The space H_∞ is dense in every H_s, $|s|\le\infty$, under its topology.

<u>Proof</u>. For $u \in H_s$, $s<\infty$, we clearly have $u_\lambda = E(\lambda)_{-\infty}u \in H_\infty$,and

$$(6.4) \qquad \|u-u_\lambda\|_s^2 = \int_\lambda^\infty \mu^{2\mathrm{Re}\ s} d\|E(\mu)_{-\infty}u\|^2 \to 0 \ , \text{ as } \lambda\to\infty \ .$$

For general $u\in H_{-\infty}$ we must have $u\in H_s$ for some $s\in\mathbb{R}$. Then $u_\lambda\to u$ in H_s as above, implying $u_\lambda\to u$ in $H_{-\infty}$, q.e.d.

We now define the <u>order classes</u> $O(m)$.

<u>Definition 6.2</u>. For $m\in\mathbb{R}$ we denote by $O(m)$ the class of all linear operators $A \in L(H_\infty)$ such that, for every $s\in\mathbb{R}$ the induced unbounded operator $A\in P(H_s,H_{s-m})$ with dense domain $H_\infty \subset H_s$ extends to a (continuous) operator $A\in L(H_s,H_{s-m})$ (from H_s to H_{s-m}) .

It is clear that an operator $A\in O(m)$ also admits a continuous extension to $H_{-\infty}$. In particular we must have $A_s u=A_t u$ for $u \in$

$H_s \cap H_t = H_r$, $r = \mathrm{Max}\{s,t\}$. Clearly every $O(m)$ is a linear space, and

(6.5) $$AB \in O(m+m') \text{ for all } A \in O(m) \text{ , } B \in O(m') \text{ .}$$

Accordingly $O(0)$ and $O(\infty) = \cup O(m)$ are algebras and $O(-\infty) = \cap O(m)$ is an ideal of both algebras.

In particular we have $\Lambda^{-m} \in O(m)$ (i.e., more precisely, the restrictions $\Lambda^{-m}|H_\infty$ are contained in $O(m)$, for every $m \in \mathbb{R}$). Moreover the operators $\Lambda^{-\zeta}$ are in $O(\mathrm{Re}\ \zeta)$ for every $\zeta \in \mathbb{C}$. Also it is found that $(\Lambda^{-\zeta})_s$ constitutes an isometry $H_s \to H_{s-\xi}$, $\xi = \mathrm{Re}\ \zeta$, for every real s and complex ζ . Also $E(\lambda)$ (or rather $E(\lambda)|H_\infty$) is an operator of $O(-\infty)$, and $E(\lambda)_{-\infty}$ above is its canonical extension.

One may introduce a locally convex topology on $O(m)$, $m \in \mathbb{R}$, using the collection of all operator norms

(6.6) $$\|A\|_{s,s-m} = \|A_s\|_{s,s-m} = \sup\{\|A_s u\|_{s-m} : u \in H_s, \|u\|_s \le 1\} \text{ , } s \in \mathbb{R} \text{ .}$$

<u>Theorem 6.3</u>. The topology of (6.6) is a Frechet topology, and $O(m)$ is a Frechet space. In particular we have

(6.7) $$\|A\|_{r,r-m} \le \|A\|_{s,s-m}^{(t-r)/(t-s)} \|A\|_{t,t-m}^{(r-s)/(t-s)} \text{ , } s < r < t \text{ .}$$

<u>Proof</u>. It is sufficient to discuss the last statement. Moreover, we may consider the case m=0 only, since $\|A\|_{p,p-m} = \|\Lambda^m A\|_{p,p} = \|\Lambda^m A\|_p$. This is a matter of Calderon interpolation (a straight application of the Phragmen Lindeloef principle, cf.[Se_1]): For fixed s<t and and $u,v \in H_\infty$ define $s(z) = (1-z)s + zt$, $z \in \mathbb{C}$, and the function

(6.8) $$f(z) = (u, \Lambda^{-s(z)} A \Lambda^{s(z)} v) \text{ , } z \in \mathbb{C} \text{ .}$$

It was seen before that Λ^s is well defined for real and complex s. We have $2d/dz(\Lambda^{s(z)}w) = -(t-s)\Lambda^{s(z)}\log H\, w$, with complex differention, for every $w \in H_\infty$. Thus f(z) is analytic in $\mathbb{C}$. Write $f = (p, \Lambda^{m-s(z)}, A\Lambda^{m+s(z)} q)$, $p = \Lambda^{-m}u$, $q = \Lambda^{-m}v$, $p,q \in H_\infty$, $m = \max\{|s|,|t|\}$, to see that f is bounded. For Re z=0, z=iy, get $s(z) = s + iy(t-s)$, hence,

(6.9) $$|f(z)| = |(\Lambda^{iy(t-s)}u, \Lambda^{-s} A \Lambda^s \Lambda^{iy(t-s)} v)| \le \|A\|_s \|u\|_0 \|v\|_0 \text{ ,}$$

using that $\Lambda^{i\rho}$ is a unitary operator of $H_0 = H$, for all real ρ , and that $\|A\|_s = \|\Lambda^{-s} A \Lambda^s\|_0$. Similarly, for Re z = 1,

(6.10) $$|f(z)| \le \|A\|_t \|u\|_0 \|v\|_0 \text{ .}$$

From the Phragmen-Lindeloef principle we thus conclude that

$$(6.11)\qquad |f((r-s)/(t-s))| \le \|u\|_0 \|v\|_0 \|A\|_s^{(t-r)/(t-s)} \|A\|_t^{(r-s)/(t-s)} .$$

Also note that $f((r-s)/(t-s)) = (u, \Lambda^{-r} A \Lambda^{r} v)$. Setting $u = \Lambda^{-r} A \Lambda^{r} v$ and using that

$$(6.12)\qquad \|A\|_r = \|\Lambda^{-r} A \Lambda^{r}\| = \sup\{\|\Lambda^{-r} A \Lambda^{r} u\| : u \in H_\infty\ \|u\|_0 \le 1\} ,$$

we conclude (6.7), q.e.d.

An operator $A \in O(m)$ is said to have order m . Since A is determined by any one of its extensions A_s we often will refer to an operator $A: H_r \to H_{r-m}$ and say that $A \in O(m)$ (or that A is of order m) whenever $A|H_\infty \in O(m)$.

Next let us discuss adjoints of operators in $L(H_s)$. For $A \in L(H_s)$ the standard interpretation of adjoint would be that of Hilbert space adjoint, an operator of $L(H_s)$ again. However, for the present purposes it is more practical to regard H_{-s} as the adjoint space of H_s , using the 'pairing'

$$(6.13)\qquad \{u,v\} \to (u,v) = (\Lambda_s u, \Lambda_{-s} v) , \text{ where } u \in H_s ,\ v \in H_{-s} .$$

Here we write $\Lambda_s = (\Lambda^{-s})_s$ for the isometry $H_s \to H_0 = H$ mentionned above and (.,.) denotes the inner product of $H = H_0$. A continuous linear functional $l: H_s \to \mathbb{C}$ may be uniquely written in the form $l(u) = (f,u) = (\Lambda_{-s} f, \Lambda_s u)$, with some $f \in H_{-s}$, as an immediate consequence of the Frechet-Riesz theorem. One then may regard the adjoint Banach space H_s^* identified with the space H_{-s}.

Accordingly for $A \in L(H_s)$ we define the adjoint operator A^* as an operator in $L(H_{-s})$, defined by the relation

$$(6.14)\qquad (u,Av) = (A^* u, v) ,\ u \in H_{-s} ,\ v \in H_s ,$$

with the pairing (.,.) of (6.13). Similarly, for $A \in L(H_s, H_t)$ there exists a unique operator $A^* \in L(H_{-t}, H_{-s})$ such that (6.14) holds for $u \in H_{-t}$, $v \in H_s$.

<u>Proposition 6.4</u>. For an operator $A \in O(m)$ there exists a unique operator $A^* \in O(m)$ such that

$$(6.15)\qquad (A_s)^* = (A^*)_{m-s} .$$

<u>Proof</u>: For $A \in O(m)$ all the adjoints $(A_s)^*$ exists and define continuous maps $H_{m-s} \to H_{-s}$. Moreover, these maps coincide for s and t,

on $u \in H_{m-s} \cap H_{m-t}$, due to

$$(6.16) \qquad (A_s^* u - A_t^* u, v) = (u, (A_s - A_t)v) = 0 \ , \ v \in H_\infty \ ,$$

since H_∞ is dense in every H_r . It follows that $(A_s)^*|H_\infty = A^*$ takes H_∞ to itself and is independent of s . Moreover, A^* admits continuous extensions A^*_s defined by (6.15), and hence has all norms $\|A\|_{s,m}$ finite. This implies that $A^* \in L(H_\infty)$, q.e.d.

Now we come to the main result of this section, a result on Fredholm operators in $O(m)$.

Recall that an operator $A \in L(H_s, H_t)$ is called Fredholm if we have $\alpha(A) = \dim \ker A < \infty$, $\beta(A) = \dim(H_t/\mathrm{im}\ A) < \infty$. A Fredholm operator always has closed image, and it follows that $\beta(A) = \dim \ker A^*$. Here we may interpret A^* as the above adjoint operator $A^* : H_{-t} \to H_{-s}$.

The Fredholm index of A is defined as $\mathrm{ind}\ A = \alpha(A) - \beta(A)$. It is known that Fredholm property and index are invariant under small bounded and arbitrary compact perturbations. For a more detailed discussion cf.X,1. For fully detailed proofs cf.[C_1],AI.

Definition 6.5. An operator $A \in O(0)$ is said to be K-invariant under H-conjugation if we have

$$(6.17) \qquad \Lambda_s A_s \Lambda_s^{-1} - A_0 = K(s) \in K(H) \ , \quad s \in \mathbb{R} \ ,$$

with the above isometry $\Lambda_s : H_s \to H_0$, defined as $\Lambda_s = (\Lambda^{-s})_s$.

Theorem 6.6. Let $A \in O(0)$ be K-invariant under H-conjugation, and let A_0 be Fredholm. Then all operators A_s, $s \in \mathbb{R}$, are Fredholm, and we have $\mathrm{ind}\ A_s$ = const. ,independent of s . Moreover, we even have $\ker A_s$ and $\ker(A^*)_s$ independent of s , and both spaces are subspaces of H_∞ .

Proof. Since Λ_s and its inverse are isometries it is trivial that A_s is Fredholm if and only if $\Lambda_s A \Lambda_s^{-1} = A_0 + K(s)$ is Fredholm. The latter is correct, since A_0 is Fredholm, and $K(s)$ is compact. Similarly we get $\mathrm{ind}\ A_s = \mathrm{ind}\ \Lambda_s A_s \Lambda_s^{-1} = \mathrm{ind}(A_0 + K(s)) = \mathrm{ind}\ A_0$. In particular $\mathrm{ind}\ A_s$ is independent of s .

On the other hand

$$(6.18) \qquad \mathrm{ind}\ A_s = \dim \ker(A_s) = \dim \ker A_s - \dim \ker(A^*)_{-s} \ ,$$

using (6.15). Here both terms at right cannot increase, as s increases: Indeed we must have $\ker A_s \supset \ker A_t$ as $s \le t$, so that $\alpha(A_s) = \dim \ker A_s$ is nonincreasing. Similarly, $\ker(A^*)_{-s} \supset \ker(A^*)_{-t}$, as $s \leq t$, so that $\beta(A_s) = \dim \ker(A^*)_{-s}$ is nondecreasing.

It follows that both dimensions at right of (6.18) also must be constant. Theorefore the two null spaces do not depend on s and must be contained in H_∞ . This completes the proof.

Remark 6.7. Let P and Q denote the orthogonal projections in $H=H_0$ onto ker A_0 and ker A^*_0, respectively. Recalling that

$$(6.19) \qquad Pu=\sum \phi_j(\phi_j,u) \ , \ Qu = \sum \psi_j(\psi_j,u) \ , \ u \in H \ ,$$

with orthonormal bases $\{\phi_j\}$ and $\{\psi_j\}$ of the finite dimensional spaces ker A_0 and ker A^*_0 , respectively, it is clear at once that $P,Q \in O(-\infty)$, using that $\phi_j, \psi_j \in H_\infty$, and using the pairing (6.13).

Later on we shall construct a distinguished Fredholm inverse B_0 of A_0, such that $B_0A_0 = 1-P$, $A_0B_0 = 1-Q$. It follows thus that such operator B_0 is an inverse even mod operators $H_{-\infty} \to H_\infty$ of finite rank (called a Green inverse, later on).

Problem 6.8. Let F_∞ denote the class of all operators $T \in O(-\infty)$ of the form $T = \sum f_j)(g_j$, $f_j, g_j \in H_\infty$, i.e.,

$$(6.20) \qquad Tu = \sum f_j(g_j,u) \ , \ u \in H_{-\infty} \ , \text{ with } (.,.) \text{ of } (6.13) \ .$$

with given f_j , g_j , and a finite sum. Show that the closure of F_∞ in the Frechet topology of $O(0)$ coincides with the class of all $A \in O(0)$ such that $A_s \in K(H_s)$, for all $s \in \mathbb{R}$.
Hint: Repeat the proof of $[C_1]$,p.144, lemma 3.2 in abstract form.

CHAPTER 2. SPECTRAL THEORY OF DIFFERENTIAL OPERATORS.

In the present chapter we will apply some of the abstract principles developed in ch.1 to the case of a differential operator. Here a differential operator appears as a realization of a differential expression

$$L = \sum_{|\alpha|\leq N} a_\alpha(x) D^\alpha \ , \ x\in\Omega \ , \tag{0.1}$$

where Ω is an open subset of $\mathbb{R}^n$, and multi-index notation is used (cf.sec.1). The coefficients a_α will be complex-valued functions over Ω . Later on we also will consider matrix-valued coefficients, or differential expressions on manifolds, acting on crosssections of vector bundles (cf. X,3). The term realization is discussed in sec.1, below.

All of our discussion will be restricted to hypo-elliptic expressions only. In fact, our main effort will be devoted to elliptic expresions, and even only ordinary differential expressions. In sec.1 we discuss the v.Neumann-Riesz extension theory for the minimal operator of a hypo-elliptic self-adjoint expression. In sec.2 we discuss generalized boundary conditions, in the regular and singular case. In sec's 3 and 4 we focus on the spectral theorem of an essentially self-adjoint realization of an ordinary differential expression. Specifically a detailed proof is discussed in the case of a second order expression.

1. Linear differential operators on a subdomain of $\mathbb{R}^n$.

In the present section we will be concerned with differential operators, as a special class of unbounded linear operators, regarding their spectral theory. It is convenient to use multi-index notation of $\mathbb{R}^n$, as follows .

$$\partial_j = \partial_{x_j} = \partial/\partial_{x_j} \ , \ D_j = -i\partial_j \ , \ D = (D_1,\dots,D_n) \ , \ \alpha = (\alpha_1,\dots,\alpha_n)$$

$\alpha_j = 1,2,\ldots$, $D^\alpha = D_1^{\alpha_1} D_2^{\alpha_2} \ldots D_n^{\alpha_n}$, $\partial^\alpha = \partial_1^{\alpha_1} \ldots \partial_n^{\alpha_n}$, $x^\alpha = x_1^{\alpha_1} \cdots x_n^{\alpha_n}$, for $x = (x_1,\ldots,x_n)$, etc.

For multi-indices α,β it also will be convenient to abbreviate

$$|\alpha| = \alpha_1 + \ldots + \alpha_n \ , \ \alpha! = \alpha_1! \cdots \alpha_n! \ , \ \binom{\alpha}{\beta} = \binom{\alpha^1}{\beta^1}\binom{\alpha^2}{\beta^2}\ldots \ .$$

Suppose we are given a <u>differential expression</u>

$$(1.1) \qquad L = \sum_{|\alpha| \le N} a_\alpha(x) D^\alpha \quad , \ a_\alpha \in C^\infty(\Omega)$$

over some open set $\Omega \subset \mathbb{R}^n$. For a differential expression a <u>formal adjoint</u> L^* is defined, also a differential expression,by

$$(1.2) \qquad (Lu,v) = (u,L^*v) \quad , \ u,v \in C_0^\infty(\Omega) \quad ,$$

with the inner product and norm of the Hilbert space $H = L^2(\Omega)$

$$(1.3) \qquad (u,v) = \int_\Omega \bar{u}(x)v(x)dx \ , \quad \|u\| = \{\int_\Omega |u|^2 dx\}^{1/2}$$

By a partial integration one finds that, for all $u,v \in C_0^\infty$,

$$(1.4) \qquad (Lu,v) = \int_\Omega dx \sum_{|a| \le N} \bar{a}_\alpha(x)\overline{D^\alpha u}\, v = \int_\Omega dx \sum_{|\alpha| \le N} (-1)^{|\alpha|} \bar{a}_\alpha D^\alpha \bar{u} v$$

$$= \int_\Omega dx \sum_{|\alpha| \le N} \bar{u} D^\alpha(\bar{a}_\alpha v) = (u,w) \ , \ w = \sum_{|\alpha| \le N} D^\alpha(\bar{a}_\alpha v) = Mv$$

since all boundary terms cancel, for such functions. A differential expression is determined by the values Mu, for all $u \in C_0^\infty$, since they determine all a_α.Thus M is determined by all values of (u,Mv), for $u,v \in C_0^\infty$, since C_0^∞ is dense in H.Accordingly we have

$$(1.5) \qquad L^* = \sum_{|\alpha| \le N} D^\alpha \bar{a}_\alpha \quad .$$

Using Leibnitz'formula

$$(1.6) \qquad P(D)(au) = \sum_\beta (D^\beta a)(P^{(\beta)}(D)u)/\beta! \ , \ P^{(\gamma)}(\xi) = (\partial^\gamma P)(\xi)$$

one may write L^* in the form (1.1) again, with new coefficients $a^*_\alpha \in C_0^\infty$. Note that $a^*_\alpha = \bar{a}_\alpha$ for $|a|=N$, so that the <u>principal part</u> L^*_N (i.e,the sum of highest order terms) of L^* is obtained by conjugating the coefficients of L_N .

Generally, a <u>differential operator</u> will be a (linear) map $A:X \to Y$ between spaces of functions or distributions X and Y , such that Au=Lu , for all $u \in X$, with some differential expression L. Normally, unless stated otherwise, we shall assume the expression L to have C^∞-coefficients, and that $C_0^\infty(\Omega) \subset X \subset \mathcal{D}'(\Omega)$, with the space $\mathcal{D}'(\Omega)$ of all distributions over Ω.

Conventionally spectral theory is considered only for operators defined by hypo-elliptic differential expressions. An expression L (of the form (1.1)) is called hypo-elliptic if the differential equation Lu=f admits only infinitely differentiable solutions, whenever f is C^∞ , for all open subsets $\Omega' \subset \Omega$. (Speaking precisely, every distribution $u \in \mathcal{D}'(\Omega')$ solving Lu = f in Ω' necessarily is $C^\infty(\Omega')$, and a classical solution of Lu = f.)

In [C_3] we will discuss criteria for hypo-ellipticity of a differential expression (cf. also [Ho_1] [Ho_2]) . Here we note that an elliptic differential expression necessarily is hypo-elliptic. A differential expression L of the form (1.1) is called elliptic if, for all $x \in \Omega$, and all $0 \neq \xi \in \mathbb{R}^n$,

$$(1.7) \qquad \sum_{|\alpha|=N} a_\alpha(x)\xi^\alpha \neq 0 \quad .$$

Particularly an ordinary differential expression

$$(1.8) \qquad L = \sum_{0 \leq j \leq N} a_j(x)D^j \quad , \alpha < x < \beta \ ,$$

with $x=(x_1) \in \mathbb{R}^1$, and $\Omega = (\alpha,\beta) \subset \mathbb{R}$, an open interval, is elliptic if and only if $a_N(x) \neq 0$, for all $x \in (\alpha,\beta)$.

We shall require hypo-ellipticity only in the slightly weakened form of lemma 1.2, below. Detailed proofs of such weaker regularity property, in the cases of an ordinary expression L, or a second order strongly elliptic expression are discussed in II,2, and III,1.

Presently we shall study unbounded operators of the Hilbert space H above, in the sense of I,1, which are differential operators. For a given differential expression L one first introduces the minimal operator L_0 and the maximal operator L_1 , defined by setting $L_j u = Lu$ for $u \in \text{dom } L_j$, j=0,1 , where

$$(1.10) \qquad \text{dom } L_0 = C_0^\infty(\Omega) \subset H \ , \ \text{dom } L_1 = \{u \in C^\infty(\Omega) : u, Lu \in H\}$$

Clearly dom L_0 and dom L_1 are dense in H , so that L_0 and L_1 are unbounded linear operators in the sense of I,1. We have $L_0 \subset L_1$. An operator $A \in P(H)$ with $L_0 \subset A \subset L_1$ will be called a realization of the expression L. (In later chapters we will use the term 'realization' somewhat more generally for an operator $A \in P(H)$ with $L_0^{**} \subset A \subset L_1^{**}$. Actually this notation is in general use for unbounded linear operators A of more general Hilbert or Banach spaces in a similar relation to the (closures of) the minimal and maximal operator. For the present chapter we appoint that $L_0 \subset A \subset L_1$ is required for a realization.)

For a realization A we must have $L_0^* \supset A^* \supset L_1^* \supset (L^*)_0$, implying that dom $A^* \supset$ dom $(L^*)_0$ is dense in H. Thus A is preclosed. Note that realizations normally are not closed. Apart from realizations also the (restrictions of) their closures and (Hilbert pace) adjoints are differential operators in the distribution sense, as easily verified.

A differential expression L is called self-adjoint if $L^*=L$. For a self-adjoint expression L the minimal operator clearly is hermitian, but the maximal operator in general is not hermitian. For spectral theory of differential operators one will be interested in the self-adjoint extensions of L_0 . As a restriction of L_1 a realization cannot be self-adjoint, but at most will be essentially self-adjoint. In view of the Riesz-v.Neumann extension theory of I,2 the question arises whether all self-adjoint extensions of L_0 are closures of essentially self-adjoint realizations of the given expression L.

As a <u>general assumption</u> let us assume that all expressions $L-\lambda$, for $\lambda \in \mathbb{C}$,are hypo-elliptic. Then L will be called <u>strongly hypo-elliptic</u>. If L is not of order 0 and is either elliptic or of constant coefficients and hypo-elliptic, then it is strongly hypo-ellitic. Similarly for the general formally hypo-elliptic expressions fitting into the definitions of [C_3],III, (having a local parametrix of negative order). All this is discussed in detail in [C_3] .

<u>Lemma 1.1.</u> Let the expression L be self-adjoint and strongly hypo-elliptic. Then the <u>defect spaces</u> $\mathcal{D}_\pm = (\mathrm{im}(L_0 \pm i))^\perp$ are contained in dom L_1 and are just the eigenspaces of L_1 to the eigenvalues $\pm i$, respectively. Moreover, every eigenvector of L_0^* also is an eigenvector of L_1.

<u>Proof.</u> It suffices to prove the last statement.Let $L_0^* u = \lambda u$, then

$$(1.11) \qquad (u,(L-\lambda)v) = 0 \ , \text{ for all } v \in C_0^\infty(\Omega) = \text{dom } L_0 \ .$$

Since L is self-adjoint and $u \in H \subset \mathcal{D}'(\Omega)$, (1.11) implies that

$(L-\lambda)u=0$, $x\in\Omega$, in the distribution sense. Thus $u \in C^\infty(\Omega)$, since by assumption $L-\lambda$ is hypo-elliptic ,and the function 0 is $C^\infty(\Omega)$. Then $Lu=\lambda u$ in the sense of standard partial derivatives. Since $u\in H$ we get $Lu=\lambda u \in H$, hence $u \in \text{dom } L_1$ which completes the proof.

The lemma, below, is a straight extension of lemma 1.1.

Lemma 1.2. Let L be hypo-elliptic, not necessarily self-adjoint. Suppose for some $f \in H$ and $g \in C^\infty(\Omega)\cap H$ we have

$$(1.12)\qquad (f,L^*u) = (g,u) \text{ , for all } u \in C_0^\infty(\Omega) \text{ .}$$

Then we have $f \in C^\infty(\Omega)$, and, moreover, $f \in \text{dom } L_1$, $Lf=g$, $x \in \Omega$, with standard partial derivatives, possibly after changing f on a set of measure zero.

Again the proof is immediate if we realize that (1.12) for an L^2-function f implies that f solves $Lf=g$ in the distribution sense. A function $f \in H$ satisfying (1.12) (for some $g\in H$) is called a weak L^2-solution of the differential equation $Lf=g$. Evidently such f is just a distribution solution which also belongs to H .

Lemma 1.2, for the special case of a second order elliptic expression, is generally known as Weyls Lemma. It was proven by H.Weyl for the potential equation [We$_2$] . We will refer to the property of an expression to have at least every weak L^2-solution in C^∞ (for $g\in C^\infty$) as (the property) of Weyl's lemma. This property is weaker than hypo-ellipticity, but it will be sufficient for all of our purposes. It will be more convenient to use since we do not have to refer to distribution calculus.

For the remainder of sec.1 let us assume that he expression L is self-adjoint. For the closure L_0^{**} of the minimal operator (a closed hermitian operator) we now have the well known direct decom position (1.13), below. Moreover, it is orthogonal with respect to the "graph inner product" $(u,v)_{L_0^*} = (u,v)+(L_0^*u,L_0^*v)$ (cf.I,(2.9)).

$$(1.13)\qquad \text{dom } L_0^* = \text{dom } L_0^{**} \oplus \mathcal{D}_+ \oplus \mathcal{D}_- \quad .$$

Let us introduce the amended (slightly larger) minimal operator $\tilde{L}_0$ as the restriction of L_0^{**} to $\text{dom}\,\tilde{L}_0 = C^\infty \cap \text{dom } L_0^{**}$. As a restriction of the closure L_0^{**} of L_0 , $\tilde{L}_0$ can be reached from L_0 by closing . That is, each $u \in \text{dom } \tilde{L}_0$ is the limit of a sequence $u_n \in C_0^\infty$ such that also $Lu_n \to \tilde{L}_0u$. We get $(\tilde{L}_0u,v)=(L_0^{**}u,v)=(u,L_0v)$ $=(u,Lv)=(Lu,v)$ for $u \in \text{dom } \tilde{L}_0$, $v\in C_0^\infty(\Omega)$, using (1.4). Since $C_0^\infty(\Omega)$

is dense in H, we get $\tilde{L}_0 u = Lu$, for all $u \in \text{dom } \tilde{L}_0$. In other words, $\tilde{L}_0$ is a realization of the expression L.

Notice that also L_1 is the restriction of L_0^* to $C^\infty(\Omega) \cap \text{dom } L_0^*$, by the same conclusion: We get $\text{dom } L_1 \subset C^\infty \cap \text{dom } L_0^*$, using (1.4). Also, $L_0^* f = g$, for some $f \in C^\infty$, amounts to (1.14), below, using (1.4) again.

(1.14) $$(g,u) = (f,L_0 u) = (Lf,u) \quad , \text{ for all } u \in C_0^\infty .$$

This implies $L_0^* f = Lf \in H$, so that $f \in \text{dom } L^1$, hence $\text{dom } L_1 \supset C^\infty \cap \text{dom } L_0^*$. Thus L_1 indeed is the stated restriction.

Theorem 1.3. For a self-adjoint, strongly hypo-elliptic expression L we have the algebraic decomposition

(1.15) $$\text{dom } L_1 = \text{dom } L^\sim{}_0 + \mathcal{D}_+ + \mathcal{D}_- \quad ,$$

where every pair of terms at right intersects at 0 only.

Every self-adjoint extension A of L_0 (or $L^\sim{}_0$) is the closure of a unique essentially self-adjoint realization B determined as the restriction of A to $\text{dom } A \cap C^\infty(\Omega) = \text{dom } B$.

Moreover, B is determined as a restriction of L_1 to the space

(1.16) $$\text{dom } B = \text{dom } \tilde{L}_0 + \{ (u-Wu) : u \in \mathcal{D}_- \} ,$$

where $W: \mathcal{D}_- \to \mathcal{D}_+$ is an isometry onto $\mathcal{D}_+$, with $\mathcal{D}_\pm$ Hilbert spaces, as closed subspaces of H. Also B is uniquely determined by W, and

(1.17) $$Bu = Lu \text{ on dom } \tilde{L}_0 \ , \ B(u-Wu) = -i(u+Wu) \ , \ u \in \mathcal{D}_- .$$

The decomposition (1.16) of dom B is orthogonal, with respect to the inner product $(u,v)_L=(u,v)+(Lu,Lv)$, defined for $u,v \in \text{dom } L_1$.

The proof of (1.15) is a consequence of Lemma 1.1, and the facts on $\tilde{L}_0$ and L_1 derived. The remainder of the theorem then follows from the v.Neumann-Riesz theorem (cf. I, thm.2.4).

Theorem 1.3 amounts to a strong simplification of spectral theory, for hypo-elliptic expressions, because all discussions may be restricted to C^∞-functions, and no generalized derivatives, Sobolev spaces, etc., have to be considered. This is under the assumption that simple criteria for hypo-ellipticity can be derived, which in turn do not use generalized derivatives. Note, that the concept of distribution may be avoided, above, because all conclusions already follow, if the statement of Weyl's Lemma is known to be true.

Let us still mention the Carleman alternatives. Carleman [Ca_1], in essence, arrived at the conclusions of thm.1.3 in the case of a Schroedinger operator $L = H = -\Delta+q$, for $\Omega=\mathbb{R}^n$, with a 'potential' $q(x)$. He distinguishes between the definite case , where def $L_0= 0$, and L_0 possesses precisely one self-adjoint extension, and the indefinite case, where def $L_0 = (\nu,\nu)$, $\nu>0$. Similarly, in the case of an ordinary differential expression, H.Weyl speaks of the limit point case, and the limit circle case, respectively, with a notation referring to his special construction.

2.Generalized boundary problems;ordinary differential expressions.

In section 1 we derived a direct decomposition (1.15) of the domain of the maximal operator, assuming that the expression (1.1) is strongly hypo-elliptic and self-adjoint. In particular we concluded that all self-adjoint extensions of the (hermitian) minimal operator L_0 are given as closures of certain restrictions of the maximal operator L_1, called e.s.a. realizations of L. In turn the e.s.a. realizations are characterized by an isometry $W:\mathcal{D}_- \to \mathcal{D}_+$ between the defect spaces $\mathcal{D}_\pm$, similar as in the v. Neumann-Riesz theorem (I, thm.2.4).

For a given e.s.a.-realization A of the expression L the condition ' $u\in\text{dom } A$ ', imposed on a function $u\in\text{dom } L_1$, amounts to a generalized boundary condition, because if $u \in \text{dom } L_1$ satisfies $u \in \text{dom } A$, then so does $u + v = w$, for all $v \in C_0^\infty$, due to $C_0^\infty \subset \text{dom } A$. In other words, the condition $u \in \text{dom } A$ is not influenced by the behaviour of u away from the boundary, it depends only on the properties of u in some (arbitrarily small) neighbourhood of the boundary of the domain Ω. In this sense we associate to every e.s.a.-realization of L a (generalized) boundary problem .

One will be tempted to ask for classes of e.s.a.-realizations with boundary conditions of the conventional type. For an investigation of this kind involving partial differential operators, and more generally, dissipative and accretive boundary conditions cf. [CFO] .

In this and the following section we will get restricted to n=1, i.e. , to the case of an ordinary differential operator. Spectral theory of "ODE's" was completed to a high degree of perfection, starting with the work of H.Weyl [We_1]. From the large number of contributions to this subject we mention the work of Hilb [Hl_1], K.Kodaira [Ko_1], M.G.Krein [Kr_1], N.Levinson [Le_1], E.C.Titchmarsh [Ti_1], M.A.Naimark [Ne_1], E.A.Coddington [Cd_1], and others (cf. also the monographs [Ne_2], [Gl_1], [Bz_1], [RS], [CdL_1])

We write

$$(2.1)\qquad L = \sum_{j=0}^{N} a_j(x)\partial^j \quad , \quad \partial = d/dx \quad , \alpha < x < \beta \ ,$$

with extended real numbers $\alpha<\beta$. For simplicity assume that $a_j \in C^\infty((\alpha,\beta))$ are real-valued. The adjoint of L is given by

$$(2.2)\qquad L^* = \sum_{j=0}^{N}(-1)^j\partial^j a_j(x) = (-1)^N a_N\partial^N + \text{terms of lower order} \ .$$

Assume $L^*=L$, and L elliptic (i.e., $a_N\neq 0$). Particularly $a_N=(-1)^N a_N$ follows from (2.2), so that N must be even. Also assume $N\neq 0$, so that $N=2,4,6,\ldots$.

We give a proof of Weyl's lemma (i.e., Lemma 1.2), for this case.

<u>Lemma 2.1.</u> If $(L_0^*-\lambda)f = g$, for some $g \in H\cap C^\infty(\Omega)$, then $f\in \mathrm{dom}\, L_1$.
<u>Proof.</u> We require a <u>fundamental solution</u> of L, that is, in distribution terms, a solution e(x,t) of the differential equation

$$(2.3)\qquad (L_x-\lambda)e(x,t) = \delta(x-t) \quad , \quad \alpha < x,t < \beta \quad ,$$

with the 'Delta-function' δ. In simple terms this is a C^{N-2}-function in the square of (2.3) being C^N in each of the two triangular regions $\alpha < x \le t < \beta$, and $\alpha<t\le x<\beta$ and solving $(L_x-\lambda)e=0$ there, where writing L_x expresses that L is to be applied onto x (not t). Moreover, at x=t we require the 'jump condition'

$$(2.4)\qquad \partial^{N-1}e(x+0,x) - \partial^{N-1}e(x-0,x) = 1/a_N(x) \quad , \quad \alpha < x < \beta \quad .$$

It then is easily verified that (2.3) holds.

Note that the general existence and uniqueness theorem for ODE-s, and the known results on differential dependence of solutions on initial data at once supply the existence of (an infinite family of) fundamental solutions: One may prescribe arbitrary initial data $\partial_x^j e(t\pm 0,t)$, subject to (2.4) and $\partial_x^j e(t+0,t)=\partial_x^j e(t-0,t)$, $0\le j\le N-2$. For any such choice (in $C^N(\Omega)$, of course) one obtains a unique function e(x,t) satisfying the above requirements. If, in addition, we require the data in C^∞ then also e will be C^∞ in the two triangles. The continuity of the derivatives then also gives

$$\text{(2.5)}\qquad \partial_x^{N-1}e(x,x-0) - \partial_x^{N-1}e(x,x+0) = 1/a_N(x)\ ,\ \alpha < x < \beta\ .$$

In the proof of Lemma 2.1 we may assume that $\lambda=0$, due to $N\geq 2$. Suppose we have $L_0^*f = g$, $g \in C^\infty$, or,

$$\text{(2.6)}\qquad (f,Lu) = (g,u)\ ,\ \text{for all } u \in C_0^\infty\ .$$

Let us focus on some point $x^0 \in \Omega = (\alpha,\beta)$. We can find $\chi \in C_0^\infty(\Omega)$ with $\chi=1$ near x^0, and $\phi\in C_0^\infty(\mathbb{R})$, $\phi\geq 0$ in $\mathbb{R}$, $\phi = 0$, $|x|\geq 1$, $\int\phi dx=1$, and then let $\phi_\varepsilon(x) = \phi(x/\varepsilon)/\varepsilon$. The function

$$\text{(2.7)}\qquad u(x) = u_\varepsilon(x) = \chi(x)\int_\alpha^\beta \phi_\varepsilon(x^0-y)e(x,y)dy\quad ,\ 0 < \varepsilon\ ,$$

then will be in $C_0^\infty(\Omega)$,and may be substituted into (2.6). Indeed, we have $v(x) = \int_\alpha^\beta \phi_\varepsilon(x^0-y)e(x,y)dy = \int_\alpha^x + \int_x^\beta$, with both integrands being C^∞, so that $v \in C^\infty$, hence $u \in C^\infty$. Moreover (2.3) suggests that

$$\text{(2.8)}\quad Lv(x) = L_x\int_\Omega\phi_\varepsilon(x^0-y)e(x,y)dy = \int_\Omega\phi_\varepsilon(x^0-y)\delta(x-y)dy = \phi_\varepsilon(x^0-x),$$

as also may be easily verified by a calculation, using (2.4). Moreover, for sufficiently small ε ,

$$\text{(2.9)}\quad Lu(x) = \phi_\varepsilon(x^0-x) + \int_\Omega dy\,\gamma(x,y)\phi_\varepsilon(x^0-y),\ \gamma=\{L_x(\chi(x)e(x,y))\}_f$$

with the 'function part' $\{d\}_f$ of the distribution d (i.e., $\{d\}_f$ omits the delta-function term arising from differentiating at the discontinuity x=y). Clearly the integrand vanishes identically for small $|x-x^0|$, using that $\gamma = (L_x e(x,y))_f = (\delta(x-y))_f = 0$ (due to $\xi=1$). Moreover, we have $|x^0-y|$ small in supp $\phi_\varepsilon(x^0-y)$, so that the possible singularity of $\gamma(x-y)$ is outside supp ϕ_ε , unless $|x^0-x|$ is small, in which case $\xi\equiv 0$ anyhow. Thus, for small $\varepsilon>0$ we have the integrand C^∞ for all $x,y\in\Omega$.

Substituting into (2.6) and interchanging integrals we get

$$\text{(2.10)}\qquad \begin{aligned}\int \phi_\varepsilon(x^0-y)dy\int\chi(x)g(x)e(x,y)dx &= \int f(x)\phi_\varepsilon(x^0-x)dx\\ &+\int dy\phi_\varepsilon(x^0-y)\int dxf(x)\gamma(x,y)\end{aligned}$$

Clearly (2.10) remains valid if x^0 is replaced by x^1 close to x^0, where the functions χ and ϕ may be kept the same. In the limit $\varepsilon \to 0$, $\varepsilon > 0$, one obtains the equation

$$(2.11) \qquad f(x^1) = \int_\Omega e(x,x^1)\chi(x)g(x)dx - \int_\Omega dxf(x)\gamma(x,x^1) ,$$

for almost all $x^1 \in \Omega$, with x^1 close enough to x^0 , so that $\chi(x)=1$ still holds near x^1 . Evaluation of these limits requires a common technique known as regularization (or mollification) (cf. [C_1],I).

Note that the right hand side is C^∞, as a function of x^1 , since the first term looks like v of (2.7) , while the kernel γ of the second term is C^∞ , as seen above. Thus by changing f on a null set, keeping it within the equivalence class, representing the given element of H , we find that $f \in C^\infty$, near x_0 . Since the construction is valid for every $x^0 \in \Omega$, the proof is complete.

The above proof is valid without the assumption of self-adjointness for L or reality of coefficients, as long as L is elliptic, as easily checked. Also, along similar arguments, one easily confirms the following.

Corollary 2.2. The space dom L_0^* consists precisely of all $u \in C^{N-1}(\Omega)$ with absolutely continuous N-1-st derivative such that $L^*u \in H$, where the N-th derivative $u^{(N)}$ is defined as derivative almost everywhere of $u^{(N-1)}$. Moreover, for $f \in$ dom L_0^* , $L_0^*f=g$ we still have (2.11) valid, for any $C_0^\infty(\Omega)$-functions $\chi(x)$, equal to 1 near $x^1 \in \Omega$, and the C^∞-function $\gamma(x,y)=L_x(\chi(x)e(x,y))$, (ignoring the delta-function arising from the discontinuity of the (N-1)-st derivative of e(x,y) at x=y).

Indeed, we have proven (2.11) for all $f \in$ dom L_0^* with $L_0^*f=g$, where $g \in C^\infty \cap H$, but notice that exactly the same derivation works for $f \in$ dom L_0^*, $g=L_0^*f$, regardless whether or not $g \in C^\infty$. By differentiating (2.11) it then follows at once that such f must be $C^{N-1}(\Omega)$, and that its (N-1)-st derivative must be still absolutely continuous. Also the differentiation implies that $L^*f=g \in H$. Vice versa, if $f \in C^{N-1}(\Omega)$ has its (N-1)-st derivative absolutely continuous, and if $g=L^*f$, with $f^{(N)}$ existing only almost everywhere, is in H, then (2.6) follows by partial integration, so that $f \in$ dom L_0^* , and $L_0^*f=g$, q.e.d.

Note that, after Lemma 2.1, we now have complete control of sec.1 , without having to rely on the hypo-ellipticity of L (which in general is discussed in [C_3].).

If an endpoint α or β is finite and has the property that all coefficients a_j may be extended up to it, as C^∞-functions, keeping $a_N \neq 0$, also at that endpoint, then we shall speak of a regular endpoint. Otherwise the endpoint is called singular.

Lemma 2.3. If α is a regular endpoint, then dom $L_1 \subset C^{N-1}([\alpha,\beta))$. Similarly for a regular endpoint β . If both endpoints are regular then dom $L_1 \subset C^{N-1}([\alpha,\beta])$.

Proof. If α is a regular endpoint, then all functions of the defect spaces $D\pm$ are $C^\infty([\alpha,\beta))$, as solutions of the differential equation $(L\pm i)u=0$. Indeed, it is well known that a solution of a linear differential equation with C^∞-coefficients stays C^∞ in the entire interval where the coefficients are defined, as long as $a_N \neq 0$, also at the boundary (c.f. [CdL_1] for example). In view of the decomposition (1.13) it thus is sufficient to show that dom $L^\sim_0 \subset C^{N-1}([\alpha,\beta))$.

We again use a fundamental solution $e(x,y)$. Since the solutions of $Lw=0$ extend smoothly into the left endpoint the same is true for the fundamental solution $e(x,y)$ we constructed. In fact, we may assume $e(x,y)$ defined for all $x,y \in [\alpha',\beta)$, with $\alpha'<\alpha$, a slightly larger interval. Let first $f \in C_0^\infty$, $g=Lf$. Then (2.11) follows even for the more general case of a cut-off function $\chi \in C^\infty(\mathbb{R})$, $\chi=1$ near α , $=0$ near β , and is valid for x^1 near α . The reason for this extension is that (2.6) now holds not only for u with compact support, but for general u with $Lu \in H$, hence u of the form (2.7), with our new function χ . Of course we also must define γ with this new χ , and then get $\gamma \in C^\infty([\alpha',\beta) \times [\alpha',\beta))$, $\gamma=0$ near $x=\beta$.

The same formula then holds for general $f \in$ dom $\tilde{L}_0$, because a sequence $f_j \in C_0^\infty$ may be choosen with $f_j \to f$, $Lf_j \to g$, and setting $f=f_j$ in (2.11), and passing to the limit $j \to \infty$ will give that formula. Given the new representation of f , near its left endpoint, one may differentiate N-1 times to confirm that $f \in C^{N-1}$ near α. Similarly for the other endpoint. Q.e.d.

The extended formula (2.11) of the proof of lemma 2.3 also gives the following.

Corollary 2.4. If α is regular then

(2.12) dom $\tilde{L}_0 = \{u \in \text{dom } L_1 :\ u=u'=\ldots=u^{(N-1)}=0 \text{ at } \alpha,\ u \text{ satisfies } \mathcal{B}_\beta\}$,

where $\mathcal{B}_\beta$ denotes a certain generalized boundary condition at β.

A corresponding formula holds, if β is regular. If both, α and β , are regular, then dom $L^{\sim}_0$ consists precisely of all $u \in$ dom L_1 with derivatives up to order N-1 including vanishing at α and β .

<u>Proof.</u> The statement may be expressed as follows: (i) a function $u \in \text{dom } \tilde{L}_0$ must have all derivatives at α equal to zero, up to order N-1 . Moreover (ii) if $v \in$ dom L_1 has that property near α and vanishes near β , then we get $v \in \text{dom } \tilde{L}_0$. To prove this we apply our modified (2.11) (of the proof of lemma 2.3). For $u \in \text{dom } \tilde{L}_0$, $w = \tilde{L}_0 u$ we get

$$(2.13) \quad u(\alpha)=\lim_{x\to\alpha}u(x)=\int_\Omega e(y,\alpha)\chi(y)w(y)dy-\int_\Omega \gamma(y,\alpha)u(y)dy \; .$$

Any such u is the limit of a sequence $u_j \in C_0^\infty$ for which the right hand side of (2.13) vanishes. Hence we also get the left hand side zero. Similarly we may differentiate up to N-1-times, for

$$(2.14) \quad u^{(l)}(x) = \int_\Omega \kappa_l(y,x)w(y)dy - \int_\Omega \lambda_l(y,x)u(y)dy \; , \; l \leq N-1,$$

with kernels κ_l, λ_l given as the l-th x-derivatives of the kernels in (2.13). These kernels still are L^2 in the integration variable y, and as L^2-functions vary continuously with x , as x is near α. Therefore we also get $u^{(l)}(\alpha)=0$, $l=1,\dots,N-1$. This proves (i) .

To verify (ii), let $v \in C^\infty$ be 0 near β , and let $v=v'=\dots=v^{(N-1)}=0$ at α. (That is, $\lim_{x\to\alpha}v(x)=\dots=\lim_{x\to\alpha}v^{(N-1)}(x)=0$.) Then we just show that $v \in$ dom $L^{\sim}_0$, by constructing a sequence $v_j \in C_0^\infty$ with $v_j \to v$, $Lv_j \to Lv$. Let $\omega \in C^\infty(\mathbb{R})$, $\omega = 0$ near α , $= 1$ near β. Then $v_j(x) = v(x)\omega(\alpha+j(x-\alpha))$, $j=1,2,\dots$, has the desired properties, as a calculation shows. Similarly if β is regular or if both α and β are regular. Q.E.D.

<u>Remark 2.5.</u> Note that, for an elliptic ordinary differential expression L a characterization of dom L_0^{**} is possible along the lines of cor.2.4: If the endpoint α is regular, then dom L_0^{**} is given as the set of (2.12), with dom L_1 replaced by dom L_0^* . Similar for a regular endpoint β , or regular α and β .

Let us first assume that both endpoints α and β of a given expression L are regular. We introduce the sesqui-linear form

$$(2.15) \qquad Q(u,v) = (u,Lv) - (Lu,v) \; , \; u,v \in \text{dom } L_1 \; .$$

<u>Proposition 2.6.</u> For $u \in$dom L_1 let us introduce the column vectors

$$(2.16)\qquad u_\alpha^{\sim} = (u(\alpha),\ldots,u^{(N-1)}(\alpha))\ ,\ u_\beta^{\sim} = (u(\beta),\ldots,u^{(N-1)}(\beta))\ ,$$

(well defined in view of lemma 2.3), and the (2N-column-) vector

$$(2.17)\qquad u^{\sim} = (u_\alpha^{\sim}, u_\beta^{\sim})\ .$$

There exists a nonsingular skew-symmetric 2N×2N-matrix

$$(2.18)\qquad Q^{\sim} = \begin{pmatrix} -Q_\alpha^{\sim}, & 0 \\ 0 & , Q_\beta^{\sim} \end{pmatrix}$$

with N×N-matrices $Q_\alpha^{\sim}$, $Q_\alpha^{\sim}$ of real coefficients such that, with the inner product of $\mathbb{C}^{2N}$, or $\mathbb{C}^{N}$, respectively,

$$(2.19)\qquad Q(u,v) = (u^{\sim},Q^{\sim}v^{\sim}) = (u_\beta^{\sim},Q_\beta^{\sim}v_\beta^{\sim}) - (u_\alpha^{\sim},Q_\alpha^{\sim}v_\alpha^{\sim}),\ u,v\in \text{dom } L_1.$$

<u>Proof.</u> Letting $\alpha < \alpha' < \beta' < \beta$, we get

$$(2.20)\qquad Q(u,v) = \lim_{\alpha',\beta'\to\alpha,\beta} \int_{\alpha'}^{\beta'} (\bar{u}Lv - L^*\bar{u}v)dx\ .$$

Using a partial integration as in (1.4) we find the right hand integral equal to a sum of boundary expressions of the form (2.19) with α,β replaced by α',β'. The matrices $Q_\alpha^{\sim}$, $Q_\beta^{\sim}$, are composed from derivatives of the coefficients a_j. Only products $\bar{u}^{(j)}v^{(l)}$ with $j+l\le N-1$ can occur, so that $q_{jl}=0$ for $j+l>N-1$. Moreover, for $j+l=N-1$ the coefficient is either $\pm a_N(\alpha')\neq 0$ or $\pm a_N(\beta')$. In the limit $\alpha'\to\alpha$, $\beta'\to\beta$ we get (2.19). the matrix coefficients remain the same, except that the values of $a^{(j)}$ are taken at α,β now. Hence $Q_\alpha^{\sim}$ and $Q_\beta^{\sim}$ are nonsingular as triangle matrices with non-vanishing diagonal terms, referring to the 'anti-diagonal' $j+l=N-1$. Finally,

$$(2.21)\qquad (Q^{\sim}u^{\sim},v^{\sim}) = \overline{(v^{\sim},Q^{\sim}u^{\sim})} = -((u,Lv)-(Lu,v)) = -(u^{\sim},Q^{\sim}v^{\sim})\ ,$$

confirming that $Q^{\sim}$ is skew-symmetric. (All 2N-vectors $u^{\sim},v^{\sim}$ are assumed, as u,v vary through dom L_1.)

A linear subspace $\mathcal{W}$ of $\mathbb{C}^{2N}$ will be called a <u>maximal null space</u> of the matrix $Q^{\sim}$ if (i) $(u^{\sim},Q^{\sim}u^{\sim})=0$ for all $u^{\sim}\in\mathcal{W}$, and (ii) every subspace $\mathcal{V}$ with $\mathcal{W}\subset\mathcal{V}\subset\mathbb{C}^{2N}$ and with property (i) coincides with $\mathcal{W}$.

<u>Theorem 2.7.</u> If both endpoints of an even order expression L with real coefficients are regular, then the class of all e.s.a.-reali-

zations A of L is in 1-1-correspondence with the class of all maximal null spaces $\mathcal{W}$ of the form $Q^\sim$, as follows. For a maximal null space $\mathcal{W}$ an e.s.a.-realization is defined by

(2.22) $\quad \mathrm{dom}\, A = \{u \in \mathrm{dom}\, L\colon u^\sim \in \mathcal{W}\}\ ,\ Au = Lu \text{ for } u \in \mathrm{dom}\, A\ .$

Vice versa, for an e.s.a-realization A the space $\mathcal{W}$ is defined by

(2.23) $\quad \mathcal{W} = \{u^\sim\colon \text{there exists } u \in \mathrm{dom}\, A \text{ with boundary values } u^\sim\}.$

A proof of Theorem 2.7 will be discussed in section 3. Here let us get some insight into the nature of a maximal null space $\mathcal{W}$. The non-singular skew-symmetric matrix $Q^\sim$ has pure imaginary eigenvalues. Since $Q^\sim$ is real there are precisely N eigenvalues with positive and negative imaginary part each, and the corresponding eigenvectors span spaces $\mathcal{W}_\pm$ with $\dim \mathcal{W}_\pm = N$. We have the orthogonal direct decomposition $\mathbb{C}^{2N} = \mathcal{W}_+ \oplus \mathcal{W}_-$. We get positive definite sesqui-linear forms $F_\pm$ defined by

(2.24) $$F_\pm(u^\sim,v^\sim) = \mp iQ^\sim(u^\sim,v^\sim)\ ,\ u^\sim,v^\sim \in \mathcal{W}_\pm\ .$$

That is, the matrix $Q^\sim$ is reduced by the above direct decomposition, and, up to a factor $\pm i$, $F_\pm$ are the forms induced in $\mathcal{W}_\pm$. Then the maximal null spaces of Q are given by

(2.25) $$\mathcal{W} = \{u^\sim + W^\sim u^\sim\colon u \in \mathcal{W}_-\}\ ,$$

where $W^\sim$ is any isometry $\mathcal{W}_- \to \mathcal{W}_+$, the latter two spaces considered under norm and inner product given by the forms (2.24).

Note that Q also is reduced by the direct decomposition $\mathbb{C}^{2N} = \mathbb{C}^N \oplus \mathbb{C}^N$, as manifested by (2.18). Some (but not all of its) maximal null spaces are of the form

(2.26) $$\mathcal{W} = \{u^\sim = (u_\alpha{}^\sim, u_\beta{}^\sim)\colon u_\alpha{}^\sim \in \mathcal{W}_\alpha\ ,\ u_\beta{}^\sim \in \mathcal{W}_\beta\}\ ,$$

with maximal null spaces $\mathcal{W}_\alpha$, $\mathcal{W}_\beta$ of $Q_\alpha{}^\sim$ and $Q_\beta{}^\sim$, resp. The corresponding boundary condition then will be called <u>separated</u> .We get

(2.27) $$\mathrm{dom}\, A = \{u \in \mathrm{dom}\, L_1\colon u_\alpha{}^\sim \in \mathcal{W}_\alpha,\ u_\beta{}^\sim \in \mathcal{W}_\beta\}\ .$$

Note that (2.25) implies that all maximal null spaces $\mathcal{W}$ of Q have dimension N, and all maximal null spaces of $\tilde{Q}_\alpha$, $\tilde{Q}_\beta$ have dimension N/2 (an integer). As easily seen, any N-dimensional null space of Q is maximal. Similarly for $Q_\alpha{}^\sim$, $Q_\beta{}^\sim$.

Theorem 2.7 then simply is a consequence of the fact that, for regular endpoints α,β, every e.s.a.-realization has defect-index (N,N), and the e.s.a.-realizations are characterized as the hermitian extensions A , $L_0 \subset A \subset L_1$, with dim (dom A/dom L_0)=N. This will be discussed in sec.3.

3. Singular endpoints of a 2r-th order Sturm-Liouville problem.

Let us next discuss the defect index of L_0 , under the assumptions of section 2.

Lemma 3.1. Under the general assumptions of section 2 we have

$$\text{(3.1)} \qquad \text{def } L_0 = \text{def } L_0^{\sim} = (\nu,\nu) \ , \quad 0 \le \nu \le N.$$

Moreover, if one endpoint is regular, we get $r=N/2 \le \nu \le N$. If both endpoints are regular we have $\nu = N$.

Proof. Recall that the functions of $\mathcal{D}_{\pm}$ are C^{∞}-solutions of $Lu=\pm iu$ in Ω . It is well known that there are precisely N linearly independent such solutions of either equation. Clearly $\mathcal{D}_{\pm}$ will consist of precisely the squared integrable solutions, hence must have dimension between 0 and N. Both dimensions are equal since $\mathcal{D}_+ = \overline{\mathcal{D}_-}$, using that L has real coefficients.

If both endpoints are regular then the solutions of $Lu=\pm iu$ are $C^{\infty}([\alpha,\beta])$, as already observed. Hence all solutions are in H, and $\nu = N$ follows.

From (1.13) we get $\dim(\text{dom } L_1/\text{dom } L_0^{\sim})=2\nu$. On the other hand, if one endpoint, for example the point α, is regular, then we have $C_0^{\infty}([\alpha,\beta)) \subset \text{dom } L_1$, while Corollary 2.4 implies that all functions of dom $L_0^{\sim}$ are $C^{N-1}([\alpha,\beta))$ and have their derivatives up to order N-1 zero at α . It follows that any two functions u,v $\in$ dom $L_0^{\sim}$ with different boundary values $u_{\alpha}^{\sim} \neq v_{\alpha}^{\sim}$ are incongruent , modulo dom $L_0^{\sim}$, so that $2\nu = \dim(\text{dom } L_1/\text{dom } L_0^{\sim}) \ge \dim \mathbb{C}^N = N$, q.e.d.

Now we discuss the proof of Theorem 2.7. First of all, for any hermitian extension A , $L_0^{\sim} \subset A \subset L_1$, the space $\mathcal{W}=\{u^{\sim} : u \in \text{dom } A\}$ is a linear subspace of $\mathbb{C}^{2N}$, and a null space of the form Q, by definition of the hermitian operator. From (1.15) and lemma 3.1 it follows that dim (dom A/dom $L_0^{\sim}$) = N . It follows from cor.2.4 that $u \in$dom A is in dom $L_0^{\sim}$ if and only if $u^{\sim} = 0$. Accordingly, for an e.s.a.-realization, we must have $\dim \mathcal{W} = \dim(\text{dom } A/\text{dom } L_0^{\sim})=N$. so that $\mathcal{W}$ is a maximal null space. Therefore (2.23) indeed defines a maximal null space for every e.s.a.-realization.

Vice versa, for a maximal null space $\mathcal{W}$, let a linear operator A be defined as restriction of L_1 with domain (2.22). Then A is hermitian, in view of (2.21), since $\mathcal{W}$ is a null space of $Q^{\sim}$. Also, dim (dom A/dom $L_0^{\sim}$)=dim $\mathcal{W}$ =N, so that A must be essentially self-adjoint, again by lemma 3.1 and (1.15), q.e.d.

Let us consider the boundary form Q of (2.15), in case where endpoint α is regular, while β may be singular. From (2.20) we get

$$(3.2) \qquad Q(u,v) = -(u_\alpha^{\sim},Q_\alpha^{\sim}u_\alpha^{\sim}) + \lim_{\beta'\to\beta-0} (u_{\beta'}^{\sim},Q_{\beta'}^{\sim}u_{\beta'}^{\sim}) ,$$

where the limit at right exists, although it no longer be expressed in the form of (2.19). Define

$$(3.3) \quad Q_\beta(u,v) = \lim_{\beta'\to\beta}(u_{\beta'}^{\sim},Q_{\beta'}^{\sim}v_{\beta'}^{\sim}) = Q(u,v) + (u_\alpha^{\sim},Q_\alpha^{\sim}v_\alpha^{\sim}) .$$

Note that $Q_\beta(u,v) = 0$ if either u or v are in $\mathcal{L}_\beta$ = dom $\tilde{L}_0$ $\oplus$ {u $\in$ dom L_1: u=0 near β} . Similar as in the proof of Lemma 3.1 we get dim (dom $L_1/\mathcal{L}_\beta$) = $2\nu-N = 2(\nu-r) = 2\nu_\beta$, $r = N/2$, $0 \leq \nu_\beta \leq r$. One may refer to $\mathcal{M}_\beta$ = dom $L_1/\mathcal{L}_\beta$ as to the (abstract) boundary space at the right boundary point β . Clearly Q_β induces a form over $\mathcal{M}_\beta$; we get $Q_\beta(u,v)=(u_\beta^{\sim},Q_\beta^{\sim}v_\beta^{\sim})$, with a certain skew-hermitian operator $Q_\beta^{\sim}$ of $\mathcal{M}_\beta$, and the cosets $u_\beta^{\sim}$, $v_\beta^{\sim} \in \mathcal{M}_\beta$ of u,v.

With this new meaning of $u_\beta^{\sim},v_\beta^{\sim},Q_\beta^{\sim}$, (3.2) assumes the form (2.10), and the construction of e.s.a.-realizations may be carried out similar as in section 2, with $\mathbb{C}^{2N}$ replaced by $\mathbb{C}^N\oplus \mathcal{M}_\beta= \mathcal{M}$, and $Q^{\sim}=Q_\alpha^{\sim}\oplus Q_\beta^{\sim}$. One obtains the same 1-1-correspondence between e.s.a. realizations and maximal null spaces of $Q^{\sim}$ as stated for the regular case in Theorem 2.6, with the only difference, that a maximal null space now has dimension $r+\nu_\beta \leq N$. Also the construction of maximal null spaces from isometries remains unchanged from (2.25). We may distinguish separated boundary conditions, etc. It is easy to construct explicit e.s.a.-boundary conditions by first choosing a basis in $\mathcal{M}_\beta$, a matrix for $Q^{\sim}$ and $Q_\beta^{\sim}$, and then constructing an explicit maximal null space.

In case of two singular endpoints it is convenient to first consider the expression L in the two subintervals $(\alpha,\gamma]$ and $[\gamma,\beta)$, with some point γ , $\alpha < \gamma < \beta$. Let the restricted expressions be denoted by L^α and L^β , respectively.

<u>Proposition 3.2.</u> We have, with $\mathcal{L}_\alpha$, $\mathcal{L}_\beta$ as above, for L^α and L^β , respectively,

(3.4) $\mathcal{M} = \text{dom } L_1/\text{dom } L_0^{\sim} = \text{dom } L_1^{\alpha}/\mathcal{L}_{\alpha} \oplus \text{dom } L_1^{\beta}/\mathcal{L}_{\beta} = \mathcal{M}_{\alpha} \oplus \mathcal{M}_{\beta}$.

Moreover, the defect index $\nu = \nu_+ = \nu_-$ of L_0 now is given by $\nu = \nu_{\alpha} + \nu_{\beta}$, with ν_{α} and ν_{β} of L^{α} and L^{β} , the expressions with one regular endpoint.

The simple proof is left to the reader.

Clearly the projections $u \to u_{\alpha}^{\sim}$ and $u \to u_{\beta}^{\sim}$ may be defined, just as above. Then, with two abstract boundary spaces $\mathcal{M}_{\alpha}$ and $\mathcal{M}_{\beta}$ the construction of e.s.a.-realizations from maximal null spaces assumes the old form.

We now will state the spectral theorem for singular (even order) Sturm-Liouville problems. (Note that, conventionally, the eigenvalue problem $Au = \lambda u$, for an e.s.a.-realization A of a self-adjoint elliptic ordinary differential expression of even order, and with real coefficients is called a Sturm-Liouville problem.)

The special point will be that, for an ordinary expression L , the spectral family of A^{**} can be linked to the finite dimensional space of solutions of the differential equation $Lu=\lambda u$.

For the remainder of this section we fix a fundamental set

(3.5) $$u_1(x;\lambda) , \dots , u_N(x;\lambda)$$

(i.e.,a basis of the solution space) for the differential equation $Lu = \lambda u$ by imposing the condition

(3.6) $$\partial^j u_l(\gamma;\lambda) = \delta_{j+1,l} \ , \ j+1,l = 1,\dots,N \ ,$$

where γ is any point, $\alpha < \gamma < \beta$, kept fixed for all the discussion From standard results on ODE's we conclude that $u_j \in C^{\infty}(\alpha,\beta)$, for fixed $\lambda \in \mathbb{C}$, and that $\partial^j u_l(x;\lambda)$ is an entire function of λ , for every fixed $x \in (\alpha,\beta)$ and every j,l. For the following the conditions (3.6) are not essential, but it is important that the u_j form a fundamental set with the above smoothness and analyticity.

Let $\sigma(\lambda)=((\sigma_{jl}))_{j,l=1,\dots,N}$, be a matrix-valued function, for $\lambda \in \mathbb{R}$. The coefficients are assumed real -valued, and $\sigma(\lambda)$ is symmetric, for every λ. We shall call σ non-decreasing if $\sigma(\lambda) \le \sigma(\mu)$, as $\lambda \le \mu$, in the matrix sense (i.e., $\sigma_{\Delta} = \sigma(\mu) - \sigma(\lambda)$ satisfies $\sum \sigma_{\Delta jl}\xi_j\xi_l \ge 0$, for all $\xi \in \mathbb{C}^N$). It is clear that the diagonal elements σ_{jj} are real-valued, non-decreasing functions. Moreover,

$$(3.7) \qquad |\sigma_{jl\Delta}| \leq (\sigma_{\Delta jj}\cdot\sigma_{\Delta ll})^{1/2} \leq 1/2(\sigma_{\Delta jj} + \sigma_{\Delta ll})$$

follows, so that all elements of σ are of bounded variation, in any compact subinterval of $\mathbb{R}$. An inner product and a norm on $\mathbb{C}^N$-valued C^∞-functions may be defined by setting

$$(3.8) \quad (u,v)_\sigma = \int_{-\infty}^{+\infty} \sum_{j,l=1}^{N} \bar{u}_j(\lambda)v_l(\lambda)d\sigma_{jl}(\lambda) \ , \ \|u\| = (u,u)_\sigma^{1/2},$$

with an improper Riemann-Stieltjes integral, at right. The linear space H'_σ of all C^∞-function with finite norm may be completed to a Hilbert space H_σ. It is known that a Riesz-Fischer theorem holds, so that $H = L^2_\sigma(\mathbb{R},d\sigma)$ with a corresponding Lebesgue-Stieltjes integral. We shall not use this fact, however.

In all of the following we assume σ continuous from the left.

Theorem 3.3. (Spectral theorem for an e.s.a.-realization A)
For every e.s.a.-realization A of L there exists a non-decreasing matrix-valued function $\rho(\lambda)$, $-\infty < \lambda < \infty$, called the spectral density matrix, such that $f \to \phi = (\phi_1,\ldots,\phi_N)$, defined by

$$(3.9) \qquad \phi_j(\lambda) = \int_\alpha^\beta u_j(x;\lambda)f(x)dx \ , \ j=1,\ldots,N, \ f \in C_0^\infty(\alpha,\beta) \ ,$$

and $\phi = (\phi_1,\ldots,\phi_N) \to f$, defined by

$$(3.10) \quad f(x) = \int_{-\infty}^{+\infty} \sum_{j,l=1}^{N} u_l(x;\lambda)\phi_j(\lambda)d\rho_{jl}(\lambda) \ , \ \phi = (\phi_j) \in H'_\rho \ ,$$

give a pair of mutually inverse unitary operators $H \leftrightarrow H_\rho$, defined by continuous extension of the maps (3.9) and (3.10). Moreover, if $U: H \to H_\rho$ denotes the operator (3.9), then UAU^* coincides with the operator of multiplication by λ in H_ρ. That is,

$$(3.11) \qquad UAU^*\phi\ (\lambda) = \lambda\phi(\lambda) \ , \ \phi \in U(\text{dom } A) \ , \ \lambda \in \mathbb{R} \ .$$

After choosing the fundamental set $\{u_j\}$ the spectral density matrix is uniquely determined as a left continuous function, up to an additive constant matrix, by the realization A .

A proof of theorem 3.3 will be discussed in section 4, (but only for the simpler case of N=2 , in the interval $[0,\infty)$, with 0 a regular endpoint, and for the Dirichlet boundary condition at x=0). Our proof is that of M.G.Krein, c.f. the book of Neumark [Ne_2]. An extension of the proof to general N would meet only minor technical obstacles.

4. The spectral theorem for a second order expression.

Let us consider a second order self-adjoint expression L in the interval $[0,\infty)$, (i.e., with a regular endpoint at 0). From the self-adjointness we conclude easily that

$$L = -d/dx\, p(x)\, d/dx + q(x) \;,\; x \in [0,\infty) \;, \tag{4.1}$$

with real-valued p, q $\in C_0^\infty([0,\infty))$, and $p(x) \neq 0$. Usually one assumes $p>0$. Expressions of this form are known as (second order) Sturm-Liouville expressions.

From section 3 it follows that all realizations with separated boundary conditions have domains of the form

$$\{u \in \text{dom}\, L_1 : \tilde{u}_0 \in W_0 \;,\; \tilde{u}_\infty \in W_\infty\} \;, \tag{4.2}$$

with certain maximal nulspaces W_0 , W_∞ . We get $\tilde{u}_0 = (u(0),u'(0))$, while $W_\infty \subset M_\infty = \text{dom}\, L_1/L_\infty$. We have dim $W_0 = 1 = N/2$, so that the condition '$\tilde{u}_0 \in B_0$ ' may be expressed as a conventional boundary condition (with certain constants a,b , $|a|^2 + |b|^2 = 1$),

$$au(0) + bu'(0) = 0 \qquad . \tag{4.3}$$

A calculation shows that the real boundary conditions, i.e., the conditions (4.3) with real a,b , precisely characterize all maximal nul spaces at 0. On the other hand, at the singular endpoint ∞ , the space W_∞ has dimension ν_∞ , where $\nu_\infty + 1$ is the number of linearly independent squared integrable solutions of the differential equation Lu = iu (or Lu = -iu). By lemma 3.1 and proposition 3.2 the defect index of L_0^{**} must be $\nu_\infty+1 \geq N/2 = 1$, so that there are always either one or two linearly independent squared integrable solutions of Lu=±iu , each. In the first case dim W_∞= dim M_∞ =0, so that $\tilde{u}_\infty \in W_\infty$ holds for all u $\in$ dom L_1. That is, we need no boundary condition at ∞ in this case.(H.Weyl calls this the limit point case.) In the other case (the limit circle case,after Weyl), we get dim $W_\infty = \nu_\infty = 1$, dim $M_\infty = 2\nu_\infty = 2$, so that '$u_\infty \in W_\infty$ ' amounts to one boundary condition at ∞ . To express this condition explicitly requires a basis in the two-dimensional space M_∞ . We refer to Neumark [Ne_2] for details .

As a special case of the spectral theorem (thm.3.3) we formulate and prove theorem 4.1, below, which refers to a second order Sturm-Liouville expression in $[0,\infty)$, under a Dirichlet-boundary condition at 0, i.e., the condition (4.3) with a=1, b=0 . Or,

$$(4.4) \qquad \mathrm{dom}\, A = \{u \in \mathrm{dom}\, L_1 : u(0) = 0 \;,\; \tilde{u}_\infty \in \mathcal{N}_\infty \} \quad ,$$

where the last condition is void if we have the limit point case. For convenience we introduce the operator L_2 , $L_2 \subset A \subset L_1$, with

$$(4.5) \qquad \mathrm{dom}\, L_2 = \{u \in C_0^\infty([0,\infty)) : u(0) = 0 \} \;.$$

<u>Theorem 4.1.</u> Let $u(x;\lambda)$, $0 \le x < \infty$, $\lambda \in \mathbb{R}$, be the unique real-valued solution of the differential equation $Lu = \lambda u$, satisfying

$$(4.6) \qquad u(0;\lambda) = 0 \;,\; u'(0,\lambda) = 1/p(0) \;.$$

There exists a non-decreasing real-valued function $\rho(\lambda)$, $-\infty<\lambda<\infty$, such that the integral relations

$$(4.7) \qquad \phi(\lambda) = \int_0^\infty u(x;\lambda) f(x)\,dx \quad , \quad f \in \mathrm{dom}\, L_2 \;,$$

and

$$(4.8) \qquad f(x) = \int_{-\infty}^\infty u(x;\lambda)\phi(\lambda)\,d\rho(\lambda) \quad , \quad \phi \in C_0(\mathbb{R}) \;,$$

define a pair of mutually inverse unitary operators $H \leftrightarrow H_\rho$, by continuous extension in H and H_ρ. Moreover, if U denotes the operator of (4.7), then we have

$$(4.9) \qquad UAU^* \phi\,(\lambda) = \lambda\phi(\lambda) \quad , \text{ for all } \phi \in U(\mathrm{dom}\, A) \;.$$

The function $\rho(\lambda)$ is uniquely determined as a left continuous function with $\rho(0) = 0$, with the above properties.

Our proof of theorem 4.1 is based on the abstract spectral theorem for unbounded self-adjoint operators (i.e., I, thm.3.3), and some other preparations are required. The result was first proven by H. Weyl [We_1] in 1910 , who did not use the spectral theorem for his proof. (This was, in fact, earlier than the abstract theorem.) The use of the spectral theorem in the proof became customary with the extension of the result to general order (i.e., thm.3.3).

For $f \in C_0^\infty(\mathbb{R}_+)$, $\mathbb{R}_+ = [0,\infty)$, we introduce the entire function

$$(4.10) \qquad \Phi(\lambda) = \Phi(\lambda;f) = \int_0^\infty u(x;\lambda) f(x)\,dx \quad .$$

<u>Proposition 4.2.</u> We have

$$(4.11)\qquad \Phi(\lambda;Af) = \Phi(\lambda;Lf) = \lambda\Phi(\lambda;f) \quad , \quad f \in \text{dom } L_2 \ .$$

Moreover, if $\Phi(\lambda;f) = 0$ for some $\lambda\in\mathbb{R}$, then there exists $\omega \in$ dom L_2 such that $f = (L_2-\lambda)\omega$.

Proof. (4.11) is a matter of partial integration, using that f=0 for large x, and that u and f both satisfy the Dirichlet condition at 0. Introducing z(x) as the unique solution of $Lz=\lambda z$ satisfying $z=1$ $z'=0$ at 0 , we define

$$(4.12)\qquad \omega(x) = u(x)\int_x^\infty z(y)f(y)dy - z(x)\int_0^x u(y)f(y)dy \ .$$

Note that the 'Wronskian expression' $W = p(uz'-zu')$ is constant (its derivative vanishes), so that $W\equiv 1$, using the values of u and z at 0. Thus one confirms by explicit differentiation that $(L-\lambda)\omega=f$. For large x the first integral in (4.12) vanishes while the second equals $\Phi(\lambda;f)$ (which also vanishes,by assumption),using that $f \in$ dom L_2 vanishes for large x. Hence $\omega = 0$ for large x. Also $\omega(0) = 0$, since $u(0)=0$. Hence $\omega \in$ dom L_2 , q.e.d.

It is easy to find some $\psi \in C_0^\infty(\mathbb{R}_+)$ such that $\Phi(0;\psi)\neq 0$. Then $\Phi(\lambda) = \Phi(\lambda;\psi)$ can have at most countably many zeros λ_1, λ_2, ..., and $\lim |\lambda_j| = \infty$, if the sequence is infinite. Applying proposition 4.2 for $\lambda=\lambda_1$ we get $\omega=\psi_1 \in$ dom L_2 such that $\psi= (L-\lambda_1)\psi_1$ and (4.11) yields $\Phi(\lambda;\psi) = \Phi(\lambda;L\omega)-\Phi(\lambda;\omega) = (\lambda-\lambda_1)\Phi(\lambda,\psi_1)$. Accordingly $\Phi(\lambda;\psi_1)$ has the same roots as $\Phi(\lambda;\psi)$ except that λ_1 has its multiplicity reduced by 1 (or no longer is a zero). It is clear that we may iterate this proceedure to arrive at the lemma, below.

Lemma 4.3. For any compact interval $[a,b] \subset \mathbb{R}$ there exists a function $\psi \in C_0^\infty([0,\infty))$ such that $\Phi(\lambda;\psi)\neq 0$, for all $\lambda \in [a,b]$.

Lemma 4.4. Let $E(\lambda)$ be the left-continuous spectral family of the self-adjoint operator $B = A^{**}$. Let ψ be as in Lemma 4.3 , and let $\Delta = [a,b)$. For any interval $\Delta'=[\mu_1,\mu_2) \subset \Delta$, $E_{\Delta'}=E(\mu_2)-E(\mu_1)$, and any $f \in C_0^\infty(\mathbb{R}^+)$ we have

$$(4.13)\qquad E_{\Delta'}f = \int_{\Delta'}\Phi(\lambda;f)/\Phi(\lambda;\psi)\, dE(\lambda)\psi \ .$$

Proof. It suffices to show, that, for $v \in C_0^\infty(\mathbb{R}_+)$,

$$(4.14)\qquad (E_{\Delta'}f,v) = \int_{\Delta'} \gamma(\lambda)d(E(\lambda)\psi,v) \quad , \ \gamma(\lambda)= \Phi(\lambda;f)/\Phi(\lambda;\psi) \ ,$$

since the existence of the integral (4.13), in norm convergence as a Riemann-Stieltjes integral, is well known. Note that (4.14) amounts to $\kappa(\mu) = 0$, with

$$(4.15) \qquad \kappa(\mu) = \int_a^\mu (d(E(\lambda)f,v) - \gamma(\lambda)d(E(\lambda)\psi,v)) \ , \quad \mu \in \Delta \ .$$

We shall prove that the complex-valued function κ is differentiable and that its derivative vanishes. We have $\kappa(\mu)=(h(\mu),v)$, $h(\mu) = \int_a^\mu (dE(\lambda)f - \gamma(\lambda)dE(\lambda)\psi)$, and want to show that

$$(4.16) \qquad h(\mu+\delta) - h(\mu) = o(\delta) \ , \text{ as } \delta \to 0 \ .$$

For a moment let $\delta > 0$, and let $E = E(\mu+\delta) - E(\mu)$. Write

$$(4.17) \qquad h(\mu+\delta) - h(\mu) = E(f-\gamma(\mu)\psi) - \int_\mu^{\mu+\delta} (\gamma(\lambda)-\gamma(\mu))dE(\lambda)\psi \ .$$

In the second expression, called J_2, we may write $\int_{\mu+0}^{\mu+\delta}$ instead of $\int_\mu^{\mu+\delta}$, since the integrand vanishes at μ . This gives

$$(4.18) \|J_2\|^2 = \int_{\mu+0}^{\mu+\delta} |\gamma(\lambda)-\gamma(\mu)|^2 d\|E(\lambda)\psi\|^2 \leq c\delta^2 \|(E(\mu+\delta)-E(\mu+0))\psi\|^2 = o(\delta^2)$$

For the first expression, J_1, in (4.17), let $w = f - \gamma(\mu)\psi$.
Note that

$$(4.19) \qquad \Phi(\mu;w) = \Phi(\mu;f) - \gamma(\mu)\Phi(\mu;\psi) = 0 \ ,$$

by definition of γ . Thus, by proposition 4.2 again, write $w = (A-\mu)\omega$, with some $\omega \in \operatorname{dom} L_2$. If μ is an eigenvalue of A then one may add a multiple of the corresponding eigenfunction $u(x;\mu)$ as to obtain a revised ω which is orthogonal to $u(x,\mu)$, and still is in dom A . It follows that

$$(4.20) \qquad \|J_1\|^2 = \|E(A-\mu)\omega\|^2 = \int_{\mu+0}^{\mu+\delta} (\lambda-\mu)^2 d\|E(\lambda)\omega\|^2 = o(\delta^2) \ .$$

Similarly for $\delta < 0$, q.e.d.

<u>Lemma 4.5.</u> We may write (4.13) in the form

$$(4.21) \qquad E_{\Delta'}f = \int_{\Delta'} \Phi(\lambda;f) \, dE(\lambda)g \ , \quad g = \int_\Delta dE(\lambda)\psi/\Phi(\lambda;\psi) \ .$$

Here the function $g \in H$ is independent of the specific choice of ψ , subject to the conditions of lemma 4.3.
<u>Proof.</u> It is clear that we may write (4.13) in the form of

(4.21). To show that g is independent of ψ, let $\psi^\sim$ be any function with the properties of ψ. Applying lemma 4.4 with $f = \psi^\sim$ we get

$$E_{\Delta'}\psi^\sim = \int_{\Delta'} \Phi(\lambda;\psi^\sim)/\Phi(\lambda;\psi)dE(\lambda)\psi \ , \text{ hence } g^\sim = \int_\Delta dE(\lambda)\psi^\sim/\Phi(\lambda;\psi) \ .$$

This completes the proof.

We will write $g = g_\Delta$, since this element $g \in H$ evidently is determined by the choice of the interval Δ. One finds that

$$(4.22) \qquad g_{\Delta'} = E_{\Delta'}g_\Delta \ , \text{ as } \Delta' \subset \Delta \ .$$

Introduce a function $g : \mathbb{R} \to H$, by setting

$$(4.23) \quad g(\lambda) = \lim_{\varepsilon\to+0,\mu\to\lambda+0} g_{[\varepsilon,\mu)} \ , \text{as } \lambda>0 \ , \ = -g_{[\lambda,0)} \ , \text{as } \lambda \le 0.$$

Then one confirms that $g_{\Delta'} = g(\mu_2) - g(\mu_1)$. For $f \in C_0^\infty$ we thus get

$$(4.24) \qquad (E(b) - E(a))f = \int_a^b \phi(\lambda)\, dg(\lambda) \ , \ \phi(\lambda) = \Phi(\lambda;f) \ ,$$

for all $a \le 0$, $b \ge 0$, with ϕ as in (4.10). Also (4.23) implies

$$(4.25) \qquad \|E_\Delta f\|^2 = \int_\Delta |\phi(\lambda)|^2 d\rho(\lambda) \quad , \quad \rho(\lambda) = \|g(\lambda)\|^2 \quad .$$

Letting a,b tend to $-\infty,\infty$ we confirm Parseval's relation

$$(4.26) \qquad \|f\|^2 = \int_{-\infty}^\infty |\phi(\lambda)|^2 d\rho(\lambda) \quad , \quad f \in C_0^\infty(\mathbb{R}_+) \quad ,$$

and it is found that

$$(4.27) \qquad f = \int_{-\infty}^\infty \phi(\lambda)dg(\lambda) \ , \ f \in C_0^\infty(\mathbb{R}_+) \ ,$$

with a strongly convergent improper Riemann-Stieltjes integral.

From (4.26) it is clear that the linear map $f \to \phi$ defines an isometry in the dense subspace $C_0^\infty(\mathbb{R}_+)$ of H, which may be continuously extended to a map $H \to H_\rho$. Vice versa, let $\phi \in C_0(\mathbb{R})$ be given, then we may define f by (4.27), and will find that $f \in H$. Then a sequence $f_j \in C_0^\infty(\mathbb{R}_+)$ may be constructed with $f_j \to f$ in H. Let ϕ_j be given by (4.7), with $f = f_j$. We know that (4.26) holds for f_j and ϕ_j. It follows that

$$(4.28) \qquad f-f_j = \int_{-\infty}^\infty (\phi-\phi_j)dg(\lambda) \qquad .$$

Taking norms we get

$$(4.29) \qquad \int_{-\infty}^\infty |\phi-\phi_j|^2 d\rho(\lambda) = \|f - f_j\|^2 \to 0 \quad , \quad j \to \infty \ .$$

This proves that the above isometry maps onto H_ρ, since $C_0(\mathbb{R}_+)$

is dense in H_ρ.

Next we prove the inversion formula (4.8). From (4.25) we get

$$(4.30) \qquad (f,g) = (\phi,\psi)_\rho \ , \quad f,g \in H \ ,$$

with ϕ,ψ defined by (4.7) , with f and g .Choose $f \in H$, $g = \chi_\Delta$, the characteristic function of some interval $\Delta = [t,t+\delta] \subset \mathbb{R}_+$. Calculate that $\psi(\lambda) = \int_\Delta u(x;\lambda)dx$. Therefore (4.30) yields

$$(4.31) \qquad \int_t^{t+\delta} f(x)dx = \int_{-\infty}^{\infty} \phi(\lambda)\Big(\int_t^{t+\delta} u(x;\lambda)dx\Big)d\rho(\lambda) \quad .$$

We assume $\phi \in C_0(\mathbb{R}_+)$, and then may interchange the integrals at right, then divide by δ and let $\delta \to 0$. In the limit we get (4.8).

Finally the uniqueness of ρ is evident, because $\int|\phi|^2 d\rho = \int|\phi|^2 d\sigma$ for a pair of left-continuous nondecreasing real-valued functions ρ , σ , for all $\phi \in C_0^\infty(\mathbb{R}_+)$, implies that $\rho-\sigma$ = const. This completes the proof of thm. 4.1.

CHAPTER 3. SECOND ORDER ELLIPTIC EXPRESSIONS ON MANIFOLDS.

In the present chapter we turn to a discussion of a variety of properties of second order self-adjoint strongly elliptic differential expressions with real coefficients defined on a C^∞-manifold of dimension n. In local coordinates such an expression will be of the form

$$H = -\kappa^{-1}\partial_{x^j}\kappa h^{jk}\partial_{x^k} + q \ , \tag{0.1}$$

with function $\kappa,q>0$, and a positive definite tensor $((h^{jk}))$. It then is self-adjoint with respect to the inner product $(u,v) = \int \bar{u}v d\mu$, with the measure $d\mu=\kappa dx$.

First we again discuss Weyl's lemma for expresssions of this type, and a boundary extension of Weyl's lemma as well, similar to that discussed for ordinary expressions in II,2,3. Instead of the fundamental solution used in II,2 we here use a E.E.Lewy type parametrix of the form (1.9), below. With this preparation we can give a full discussion of the Dirichlet realization , in case of a regular boundary, as an extension of the minimal operator with closure coinciding with the Friedrichs extension.

Next we will require criteria for the Friedrichs extension, hence the closure of the Dirichlet realization, to be of compact resolvent. Such criteria are developed in sec.2.

It is well known that differential expressions of the form (0.1) possess a variety of important special properties, such as the maximum principle, and the so-called Harnack inequality. These facts will be required in the sequel, and will be discussed in detail in sec.4. Also, in sec.3, we will discuss a detailed proof of the existence of a (positive) Green's function for the Friedrichs extension, again as required in the following.

The majority of the results mentioned are local results. However, it is important to see them in this global surrounding, where they will have to be applied.

Finally we will be concerned with two normal forms for expressions of the form (0.1). Both may be achieved by a transformation of dependent variable, but the second only if the minimal operator of the expression H is positive definite. This is discussed in sec.6.

1. 2nd order PDE on manifolds; Weyls lemma; Dirichlet operator.

Let Ω denote a C^∞-manifold without boundary, of dimension $n \geq 1$, not necessarily compact, but paracompact (and connected). In fact, for convenience we assume existence of a countable atlas. On Ω we assume given a positive C^∞-measure $d\mu$, locally of the form $d\mu = \kappa dx$, with κ being C^∞, in the local coordiates x. By H we denote the Hilbert space $L^2(\Omega,d\mu)$, with inner product and norm

(1.1) $$(u,v) = \int_\Omega \bar{u}v\, d\mu \quad , \quad \|u\| = (u,u)^{1/2} \quad , \quad u,v \in H \; .$$

Also we assume given on Ω a second order formally self-adjoint strongly elliptic partial differential expression

(1.2) $$H = -\kappa^{-1}\partial_{x^j} h^{jk}\kappa\partial_{x^k} + q$$

with C^∞-coefficients. In particular h^{jk} denotes a symmetric positive definite contravariant tensor with real C^∞-coefficients:

$$h^{jk} = h^{kj} = \bar{h}^{kj} \; , \; h^{jk}\xi_j\xi_k > 0 \; , \text{ as } \xi \neq 0 \; ,$$

and q denotes a scalar real-valued C^∞-function defined over Ω .

In the next sections we use the <u>summation convention</u> to always sum from 1 to n over a pair of an upper and a lower index, in a tensor, denoted by the same symbol. Also we will write local coordinates with superscript indices $x=(x^1,x^2,\ldots,x^n) \in \mathbb{R}^n$, not subscripts, as for a subdomain $\Omega \subset \mathbb{R}^n$, according to convention.

As in II,1 the expression H induces minimal and maximal differential operators of H . For the minimal operator H_0 with domain $C_0^\infty(\Omega)$ we get

(1.3) $$(u,H_0v) = \int_\Omega (h^{jk}\bar{u}_{|x^j} v_{|x^k} + q\bar{u}v)d\mu = H(u,v) \; , \; u,v \in \text{dom } H_0 \; .$$

Clearly H_0 is hermitian and real, in the sense of I,prop. 2.6, since q and h^{jk} are real. Accordingly there exist self-adjoint extensions of H_0 .

In much of the following we will be interested in the case where the sesqui-linear form H(u,v) of (1.3) is an inner product, called $(u,v)_1$. In fact if we assume $q \geq 1$ in Ω , then we get

(1.4) $$\|u\|_1 \geq \|u\| \text{ , for all } u \in \text{dom } H_0 \text{ ,}$$

with the norm $\|u\|_1 = ((u,u)_1)^{1/2}$, since the tensor h^{jk} is positive definite. Condition (1.4) will be crucial in ch.5f, although the condition $q \geq 1$ will be weakened later on (cf. prop.5.2).

For an expression H with (1.4) the Friedrichs extension of H_0 is well defined as restriction of H_0^* to $(\text{dom } H_0^*) \cap \mathcal{H}_1$, with the completion $\mathcal{H}_1$ of dom $H_0 = C_0^\infty(\Omega)$ under the norm $\|u\|_1$ (cf. the proof of I, thm.2.7). In fact, the norm $\|u\|_1$ of (1.4) precisely coincides with the norm $\|u\|_\sim$ of that proof, and the completion $\mathcal{H}^\sim$ defined there will be denoted by $\mathcal{H}_1$ here. We know that $\mathcal{H}_1$ is naturally imbedded in $\mathcal{H}$.

In ch.5 we will assume that (1.4) holds, and then denote the Friedrichs extension by H again. This will be a self-adjoint operator,satisfying $H \geq 1$. We then shall speak of a <u>comparison operator</u> H . The space $\mathcal{H}_1$ will be called the <u>first Sobolev space</u> of the operator H . Clearly $H^{-1} \in L(\mathcal{H})$ has a unique self-adjoint positive square root $\Lambda = H^{-1/2} = \sum_{k=0}^{\infty}(-1)^k\binom{1/2}{k}(1-H^{-1})^k$, and $0 < \Lambda \leq 1$. We have (cf. I, prop.2.8)

(1.5) $$\mathcal{H}_1 = \text{dom } \Lambda^{-1} \text{ , } \|u\| = \|\Lambda u\|_1 \text{ , } u \in \mathcal{H} \text{ , } \|v\|_1 = \|\Lambda^{-1}v\|, \ v \in \mathcal{H}_1 .$$

That is, Λ is an isometric isomorphism between the Hilbert spaces $\mathcal{H}$ and $\mathcal{H}_1$. Generally the restriction of the self-adjoint operator Λ^{-1} to dom H_0 still is essentially self-adjoint, and Λ^{-1}dom H is dense in $\mathcal{H}$, (while the corresponding is not generally true for H dom H_0 = im H_0).

Note that $\mathcal{H}_1$ may be identified as the space of all functions $u \in \mathcal{H}$ such that a sequence $u^m \in$ dom H_0 exists which is Cauchy in the sense of $\mathcal{H}_1$ and converges to u in $\mathcal{H}$. It follows that there exists a unique covariant tensor (called the gradient of u , and $= (u_{|x^j})$, even though the components are not proper derivatives) which is μ-measurable and satisfies

$$\int_\Omega h^{jk}(\bar{u}_{|x^j} - \bar{u}^m_{|x^j})(u_{|x^k} - u^m_{|x^k})d\mu \to 0 \text{ , } m \to \infty \text{ .}$$

It is clear that $u_{|x^j}$ is the local distribution derivative of $u \in L^2(\Omega,d\mu) \subset \mathcal{D}'(\Omega)$ while the above entitles us to speak of the strong L^2-derivative. One may introduce L^2-norm and inner product of gradients just as for covariant tensors in general by setting

$$(1.6) \qquad (\nabla u, \nabla v) = \int_\Omega h^{jk} \overline{u}_{|x^j} v_{|x^k} d\mu \quad , \; \|\nabla u\| = (\nabla u, \nabla u)^{1/2} .$$

Returning to the general case, where (1.4) needs not to be satisfied, we observe that the expression H is elliptic, since we assumed the tensor h^{jk} positive definite : We have

$$(1.7) \qquad h^{jk}(x)\xi_j\xi_k > 0 \quad , \text{ for all } \xi = (\xi_1,\dots,\xi_n) \neq 0 ,$$

at each point $x \in \Omega$. It was mentioned in II,1 that ellipticity implies hypo-ellipticity, but this will be discussed in general only in $[C_3]$. Since we need (strong) hypo-ellipticity (or at least Weyl's lemma) in the following we shall offer a short independent proof of Weyl's lemma here, which is quite similar to the proof of II, lemma 2.2, in the case of an ODE .

It is clear why a proposition like Weyl's lemma is desirable: Thm.1.3 of ch.II carries over literally, with the same proofs, to the present case of an elliptic operator on a manifold. We express this in thm.1.1, below, the proof of which will be left to the reader, (except for our proof of Weyl's lemma, i.e., thm.1.2.)

<u>Theorem 1.1</u>. The defect spaces $\mathcal{D}_\pm(H_0)$ are subspaces of $\mathcal{H}\cap C^\infty(\Omega)$. Moreover, we have $H_0^* = H_1^{**}$, $H_1^* = H_0^{**}$ (i.e.,"weak" = "strong"), and the direct decomposition

$$(1.8) \qquad \text{dom } H_1 = ((\text{dom } H_0^{**})\cap C^\infty(\Omega)) \oplus \mathcal{D}_+(H_0) \oplus \mathcal{D}_-(H_0) ,$$

which is orthogonal with respect to the inner product of graph H_0^*. The self-adjoint extensions A of H_0 precisely are given as the closures of the 'e.s.a.-realizations' $A^\sim$ obtained from an arbitrary isometry $W: \mathcal{D}_+ \to \mathcal{D}_-$, using formulas II,(1.16) and II,(1.17) .

<u>Theorem 1.2</u>. If $(H_0^*-\lambda)f=g$, for any $f\in \text{dom } H_0^*$, and $g\in \mathcal{H}\cap C^\infty(\Omega)$, then we have $f \in C^\infty(\Omega)$, and $Hf-\lambda f = g$, hence $f \in \text{dom } H_1$.

The proof depends on use of an (E.E.Levy-type) local parametrix of the form

$$(1.9) \qquad \begin{aligned} e(x,y) &= (\rho(x,x-y))^{2-n} \quad , \text{ as } n > 2 , \\ &= \log \rho(x,x-y) \;, \text{ as } n=2 , \end{aligned}$$

with $\rho=\rho(x,z) = (h_{jk}(x)z^j z^k)^{1/2}$. (We may assume $\lambda=0$, without loss of generality.) Note that $e(x,y)$ is defined only locally, in local coordinates, for $x, y \in \Omega' \subset \Omega$, with a chart Ω'. We note the proposition, below, which also will be useful later on.

Proposition 1.3. For a function $\phi \in C_0^\infty(\Omega')$ let

$$(1.10) \qquad v(x) = \int e(x,y)\phi(y)dy \quad .$$

Then we have $v \in C^\infty(\Omega')$, and

$$(1.11) \quad Hv(x) = c(x)\phi(x) + \int_{\Omega'} \gamma(x,y)\phi(y)dy \ , \ x\in\Omega' \ ,$$

with a positive $C^\infty(\Omega')$-function $c(x)$, and with a function $\gamma(x,y)$ of the form

$$(1.12) \qquad \gamma(x,y) = \gamma^0(x,y,x-y) + \partial_{x^j}(\gamma^j(x,y,x-y)) \ .$$

Here the functions $\gamma^j(x,y,z)$, $j=0,1,\dots,n$, are $C^\infty(\Omega'\times\Omega'\times\mathbb{R}^{n*})$, and homogeneous of degree $2-n$ in the variable z , as $n>2$. In the case $n=2$ we have

$$(1.13) \qquad \gamma^j(x,y,z) = h^j(x,y,z)\log \rho(x,x-y) \ ,$$

where the h^j are $C^\infty(\Omega'\times\Omega'\times\mathbb{R}^{2*})$ and homogeneous of degree 0.
Proof. We only consider the case $n>2$. The other case $n=2$ may be treated similarly. Observe that the function $e(x,y)$ may be written in the general form

$$(1.14) \quad e(x,y) = f(x,y,x-y) \ , \ f(x,y,z) \in C^\infty(\Omega'\times\Omega'\times\mathbb{R}^{n*}) \ ,$$

where f is homogeneous in the variable z . (Presently we have

$$(1.15) \qquad f(x,y,z) = (h_{jk}(x)z^j z^k)^{1-n/2} \ ,$$

independent of y and homogeneous in z of degree $2-n$.)

For any function $g(x,y) = f(x,y,x-y)$ of the form (1.14) we have

$$(1.16) \qquad (\partial_{x^j} + \partial_{y^j})g(x,y) = f_j(x,y,x-y) \ ,$$

where f_j again has all the properties of (1.14) , with the same homogeneity degree as f . This follows, because a function $a(x-y)$

is constant in the directions $x_j=y_j$ hence has $(\partial_{x^j}+\partial_{y^j})a=0$.

Now a single derivative ∂_{x^j} may be applied under the integral sign of (1.10), because the differentiated integrand still is integrable. Applying (1.16), and a partial integration, we get

$$(1.17)\qquad v_{|x^j}(x) = \int e(x,y)\phi_{|x^j}(y)dy + \int f_j(x,y,x-y)\phi(y)dy ,$$

where $g_j(x,y) = f_j(x,y,x-y)$ has exactly the properties (1.14) again, (with homogeneity degree 2-n). In particular the partial integration first may be performed on the integral over $|x-y|\geq\varepsilon$, where boundary terms may be explicitly evaluated. These boundary terms tend to zero, as $\varepsilon\to 0$, so that no special terms at $x=y$ from the partial integration appear in (1.17).

Accordingly the process may be iterated, it follows that indeed all partial derivatives $v^{(\alpha)}(x)$ exists, for $x \in \Omega'$, and

$$(1.18)\quad v^{(\alpha)}(x) = \int e(x,y)\phi^{(\alpha)}(y)dy + \sum_{\alpha\leq\beta,\beta\neq\alpha} e_{\alpha,\beta}(x,y)\phi^{(\beta)}(y)dy,$$

where all $e_{\alpha,\beta}$ are of the form (1.14) (degree 2-n). In particular we get $v\in C^{\infty}(\Omega')$. Writing the local expression of H in the form

$$(1.19)\ H=H_w+A_w,\ H_w=-h^{jk}(w)\partial_{x^j}\partial_{x^k},\ A_w=a_{jk}(x,w)\partial_{x^j}\partial_{x^k}+b^j(x)\partial_{x^j}+c(x)$$

where $w\in\Omega'$ is arbitrary, fixed, and $a^{jk}(x,w)=h^{jk}(w)-h^{jk}(x)$, it follows that

$$(1.20)\quad Hv(x)=\int e(x,y)(H_x\phi)(y)dy+\int h^0(x,y)\phi(y)dy+\int h^j(x,y)\phi_{|x^j}(y)dy ,$$

with h^0, h^j again of the form (1.14) (degree 2-n).

In (1.20) the first integral, at right, called I_1, contains the constant coefficient operator H_x (where x is only a parameter) and the function e(x,y), depending on x-y only, as long as x is kept fixed. One may make the linear substitution $\eta=M(y-x)$ in the integral I_1, with $M = ((h_{jk}(x)))^{1/2}$, which brings it onto the form

$$(1.21)\qquad I_1 = -\det((h_{jk}(x)))^{-1/2}\int |\eta|^{2-n}\Delta\psi d\eta ,$$

with the Euclidean Laplace operator Δ, and $\psi(\eta)=\phi(x+M^{-1}\eta)$. Here,

$$(1.22)\quad \int |\eta|^{2-n}\Delta\psi(\eta)d\eta=\lim_{\varepsilon\to 0}\int_{|\eta|\geq\varepsilon}=(2-n)\lim_{\varepsilon\to 0}\int_{|\eta|=\varepsilon}\psi(\eta)|\eta|^{1-n}dS_{\eta},$$

using partial integration, and that $\Delta_{\eta}(|\eta|^{2-n})=0$. Also that

$\lim_{\varepsilon\to 0}\int_{|\eta|=\varepsilon}|\eta|^{2-n}\partial_\nu\psi dS_\eta=0$, where dS_η denotes the surface element on the sphere $|\eta|=\varepsilon$, and $\partial_\nu=-\partial_\rho$, $\rho=|\eta|$, the radial derivative.

From (1.22) we conclude that

$$(1.23)\qquad I_1 = c_n\phi(x)\{\det((h_{jk}(x)))\}^{1/2},\text{ where } c_n = (n-2)\omega_{n-1}\ ,$$

with the area ω_{n-1} of the sphere $|\eta|=1$ in $\mathbb{R}^n$. The above holds for $n>2$. In case of $n=2$ one finds that $c_n = c_2 = 2\pi$. Since the determinant is C^∞ and >0 this term accounts for the first term in (1.11).

On the other hand another partial integration in the third term of (1.20) will remove the derivative from ϕ , and give an integrand $h^j_{|y^j}(x,y)\phi(y)$ instead, confirming (1.12). (In view of (1.16) we again may write $-h_{|y^i} = h_{|x^i}+k$ with $k(x,y,x-y)$ (degree $2-n$), etc.) This completes the proof of prop.1.3.

Now the proof of thm.1.2, in essence, is a repetition of the argument used in the proof of II, 1.2.1. Using our parametrix $e(x,y)$ of (1.9) in place of the fundamental solution $e(x,y)$, we derive a local integral equation of the form II,(2.11) for any pair of functions $f,g\in H$ satisfying II,(2.6), with $L=H$, and with the present Hilbert space $H=L^2(\Omega)$. Then, if $g\in C^\infty$, the right hand side is continuous, using the nature of the kernels. Hence $f(x)$, after correction on a null set, also is continuous. Knowing that f is continuous, one concludes that the right hand side even is C^1 , so that $f\in C^1$, etc. Continuing on, one concludes that $f\in C^\infty$.

Let us get such formula in detail. Let

$$(1.24)\qquad u(x) = \chi(x)\int\phi_\varepsilon(x^\sim-y)e(x,y)dy$$

with a local cut-off function $\chi(x)=1$ near some $x_0\in\Omega'$, $\chi\in C_0^\infty(\Omega')$, and a "regularizing kernel" $\phi_\varepsilon(x)=\varepsilon^{-n}\phi(x/\varepsilon)$, $\varepsilon>0$, where $\phi\in C_0^\infty(\mathbb{R}^n)$. Assume $\chi=1$ in $\Omega''\subset\Omega'$, Ω'' open, and let $x^\sim\in\Omega''$, and ε so small that $\operatorname{supp}\phi_\varepsilon(x^\sim-.)\subset\Omega'$. Then we may apply (1.11) for

$$(1.25)\qquad Hu(x) = c(x)\phi_\varepsilon(x^\sim-x) + \int_{\Omega'}\delta(x,y)\phi_\varepsilon(x^\sim-y)dy\ ,$$

with $c(x)$ of (1.11), and $\delta(x,y)$ of the form (1.12) again. In fact, δ is a sum of $\chi(x)\gamma(x,y)$ and terms of the form $a(x)e(x,y)$, $b^j(x)e_{|x^j}(*x,y)$, with $a,b^j\in C_0^\infty(\Omega')$, where a and b^j vanish outside supp χ . All these terms may be written in the form (1.12).

Substituting u and Hu into the (present) equation II,(2.6), and passing to the limit $\varepsilon \to 0$, we get

$$(1.26)\quad \kappa(x^{\sim})c(x^{\sim})f(x^{\sim}) + \int_{\Omega'} \delta(x,x^{\sim})f(x)d\mu_x = \int d\mu_x g(x)\chi(x)e(x,x^{\sim}),$$

valid for $x^{\sim}\in\Omega''$. This relation is of the general type of II,(2.11) However, the kernel δ requires a somewhat refined proceedure.

If $g \in C^{\infty}$, then the right hand side is a C^{∞}-function $h(x^{\sim})$, by prop.1.3. The integral at left is at least Hoelder continuous, since we find that the function $x^{\sim} \to \delta(.,x^{\sim})$, with values in H is Hoelder continuous: We get

$$(1.27)\quad \|\delta(.,x^{\sim})-\delta(.,x^{\wedge})\|_{L^2}/|x^{\sim}-x^{\wedge}|^{\varepsilon} = O(1) ,$$

for a suitable $\varepsilon>0$ and $x^{\sim}$, $x^{\wedge} \in \Omega''$, by a calculation. Hence (1.26) shows that f is Hoelder-continuous as well. To show that f is even C^1 we write (1.26) as

$$(1.28)\qquad (\kappa cf)(x) = h(x) - \int\delta(y,x)f(y)d\mu_y ,$$

and investigate the function

$$(1.29)\qquad \theta(x) = \int\delta(y,x)f(y)d\mu_y .$$

The difference quotient $\nabla\theta = (\theta(x+\varepsilon\xi)-\theta(x))/\varepsilon$ may be written as

$$(1.30)\quad \nabla\theta = \int\nabla\delta(y)f(y)d\mu_y = \int\nabla\delta(y)(f(y)-f(x))d\mu_y + f(x)\int\nabla\delta(y)d\mu_y,$$

where $\nabla\delta(y) = (\delta(y,x+\varepsilon\xi)-\delta(y,x))/\varepsilon$. Now we use (1.12) for

$$(1.31)\qquad \int\nabla\delta(y)d\mu_y = \int(\nabla\delta^0\kappa-\nabla\delta^j\kappa_{|y^j})dy .$$

The right hand side limit exists under the integral sign, and represents a C^{∞}-function of x, since again the derivatives may be taken over to the function κ , using (1.16) . The other integral in (1.30) has a limit as well, since f is Hoelder continuous, so that the limit of the integrand is L^1 . Thus we get

$$(1.32)\qquad \partial_j\theta(x) = k(x)f(x) + \int\delta_{|x^j}(y,x)(f(y)-f(x))d\mu_y ,$$

confirming that $f\in C^1$. In particular, k(x) is a C^{∞}-function. Knowing that f is C^1 we may integrate by parts in (1.32) , for

$$(1.33)\qquad (\kappa cf)_{|x^j} = h_{|x^j} - kf + \int \delta^{\blacktriangle}(y,x)f_{|x^j}(y)dy ,$$

with a function $\delta^{\blacktriangle}$ of the general for (1.12) again.

It is clear now that this may be iterated, proving that κcf , hence f is C^{∞} , q.e.d.

2. Boundary regularity, for the Dirichlet realization.

It will be of interest to note the following variant of thm.1.2, applying in the case where g is only continuous, not necessarily smooth.

Corollary 2.1. If g of thm.1.2 is only continuous, then we still get $f \in C^1(\Omega)$, and its first derivatives are Hoelder continuous. Moreover, if g is Hoelder continuous, then also $f \in C^2(\Omega)$ follows, and f is a classical solution of $(H-\lambda)f=g$.

Indeed, it is clear that (1.26) is valid for general $f,g \in \mathcal{H}$ satisfying $(H_0^*-\lambda)f=g$. If g is only continuous then the right hand side of (1.26), called $\psi(x)$, still is $C^1(\Omega)$, and $\psi_{|x^j}$ satisfies a Hoelder condition, by the above arguments, using an estimate like (1.27) for the first derivatives of $e(x,x^{\sim})$. Moreover, if g is Hoelder continuous, then the conclusion leading to (1.32) may be applied to integrals representing $\psi_{|x^j}$. It follows then that also the derivatives $\psi_{|x^j x^l}$ exist and are continuous. This shows that, in the first case, f will be C^1 , and that (1.33) still holds. Also the first derivatives of f then still will be Hoelder continuous. For Hoelder continuous g thus we may iterate once more, getting that also $f \in C^2$, q.e.d.

As in II,2 we next try some boundary application of our parametrix $e(x,y)$. Let us investigate the Friedrichs extension H of the minimal operator H_0 in the special case where we have only regular boundary points. Similarly as in the 1-dimensional case of II,2 it may occur that the manifold Ω is a subdomain with smooth boundary $\partial\Omega$ of another manifold $\Omega^{\wedge}$, and that the triple $\{\Omega,d\mu,H\}$ extends to a triple $\{\Omega^{\wedge},d\mu^{\wedge},H^{\wedge}\}$ on $\Omega^{\wedge}$ satisfying our general assumptions. (That is, $\partial\Omega$ is an (n-1)-dimensional C^{∞}-submanifold of $\Omega^{\wedge}$, and we have $d\mu = d\mu^{\wedge}|\Omega$, $H = H^{\wedge}|\Omega$.) In this case we will say that Ω has the regular boundary $\partial\Omega$. Moreover, if even $\Omega\cup\partial\Omega$ is a compact subset of $\Omega^{\wedge}$, then we will speak of an expression H (or a triple) with regular boundary, or we will say that all boundary points of Ω are regular.

In the case of a triple with regular boundary one defines the Dirichlet operator $H_d \supset H_0$ by setting

(2.1) $\quad \operatorname{dom} H_d = \{u \in C^\infty(\Omega \cup \partial\Omega): u=0 \text{ on } \partial\Omega\}, \; H_d u = Hu$, for $u \in \operatorname{dom} H_d$.

Specifically the condition "$u=0$ at $x \in \partial\Omega$" is referred to as the Dirichlet (boundary) condition.

Theorem 2.2. The Dirichlet operator H_d , for an expression H with regular boundary, is an e.s.a.-realization of H , and its closure $H = H_d^{**}$ is the Friedrichs extension of the minimal operator H_0 .

Proof. First we note that H_d indeed is a restriction of the Friedrichs extension H of H_0 . For clearly we have $H_d \subset H_d^*$ (i.e., H_d is hermitian), since a partial integration confirms that

(2.2) $\quad (u, H_d v) = (u,v)_1 = (H_d u, v)$, for all $u,v \in \operatorname{dom} H_d$.

(The boundary terms vanish, since u and v vanish at $\partial\Omega$.) Also, $H_0 \subset H_d$, trivially, hence $H_d^* \subset H_0^*$, so that we get $H_d \subset H_0^*$. For any $u \in \operatorname{dom} H_d$ it is easy to construct a sequence $u_j \in \operatorname{dom} H_0$ with $\|u-u_j\|_1 \to 0$, as $j \to \infty$, since only the first derivatives of u must be modified, while the 1-st Sobolev norm $\|.\|_1$ contains only first derivatives as well. In details, for a boundary chart Ω', thought of as an open set in $\mathbb{R}^n_+ = \{x^n \geq 0\}$, let $\chi(x^n) \in C^\infty(\mathbb{R})$, $\chi \geq 0$, $\chi = 0$ near $x_n = 0$, $\chi = 1$ for $x_n \geq 1$, and let $\chi_j(x_n) = \chi(jx_n)$, $u_j(x) = u(x)\chi_j(x_n)$. Then $u-u_j = u(1-\chi_j)$ has support in $0 \leq x_n \leq 1/j$. If u has compact support in Ω' , then $\|u-u_j\| \to 0$, as $j \to \infty$. The same is true for $\|\nabla(u-u_j)\|$, whenever $u=0$ at $x_n = 0$, since we get $(u-u_j)_{|x^k} = (1-\chi_j)u_{|x^k}$ as $k<n$, while $(u-u_j)_{|x^n} = (1-\chi_j)u_{|x^n} - u\chi_{j|x^n}$, where $v_j = u\chi_{|x^n}$ has support in $0 \leq x^n \leq 1/j$, and is bounded in j and x, due to $u(x)\chi_{j|x^n} = ju(x)\chi'(jx^n)$, where $u = O(x_n) = O(1/j)$ in $\operatorname{supp} \chi'(jx_n)$. Accordingly we conclude that $\|u-u_j\|_1 \to 0$, since the coefficients of H are bounded at the regular boundary. For a general $u \in \operatorname{dom} H_d$ one uses a partition of unity, corresponding to a finite atlas of the compact manifold with boundary $\Omega \cup \partial\Omega$, for a similar result. Therefore indeed we get $\operatorname{dom} H_d \subset \mathcal{H}_1 \cap (\operatorname{dom} H_0^*)$, the domain of the Friedrichs extension, and the above also shows that $H_d u = Hu$ for $u \in \operatorname{dom} H$.

Our theorem is proven if we show that H_d is essentially

self-adjoint. For then it is clear that its (self-adjoint) closure H_d^{**}, as a restriction of the Friedrichs extension, must coincide with the Friedrichs extension.

Since we know that $H_d \geq 1$, it suffices to show that im H_d is dense. Thus, let $f \in H$ satisfy

$$(2.3) \qquad (f,Hu) = 0 \text{ for all } u \in \text{dom } H_d .$$

By thm.1.2 we then get $f \in C^\infty(\Omega)$. We want to show, that, in addition we get $f \in C^\infty(\Omega \cup \partial\Omega)$, and $f=0$ on $\partial\Omega$. If this can be established then (2.2) for $u=v=f$ implies that $\|f\|^2 = \|f\|_1^2 = (f,H_d f) = 0$, hence $f=0$, so that indeed im H_d is dense.

Thus we again focus on a boundary chart Ω', as above. After another change of coordinates one may assume that the co-normal direction at $x^n=0$ councides with the x^n-direction. In other words, this means that we have $h^{nj}(x) = 0$, $j=1,\dots,n-1$, as $x^n=0$.

Let us again write the expression H in the form

$$(2.4) \qquad H = h^{jk}\partial_{x^j}\partial_{x^k} + b^j\partial_{x^j} + c ,$$

for simplicity, and let Ω'' denote the 'doubled' chart Ω' given by

$$(2.5) \qquad \Omega'' = \Omega' \cup \Omega'^- , \quad \Omega'^- = \{x \in \mathbb{R}^n : \bar{x} = (x^1,\dots,x^{n-1},-x^n) \in \Omega'\} .$$

In Ω'' we define the doubled measure $d\mu'' = \kappa'' dx$, with the even extension κ'' of the function κ to Ω'' (i.e., $\kappa''(x) = \kappa(\bar{x})$, $x \in \Omega''$). We drop the notation κ'', $d\mu''$ in favour of the old notation κ, $d\mu$. Let an extension of H be defined by extending h^{nn}, h^{jk}, b^j, c, for $j,k<n$, as even functions, but $h^{nj} = h^{jn}$, $j<n$, and b^n as odd functions. Also let the function f of (2.3) be extended as an odd function, and assume $u \in C_0^\infty(\Omega'')$. We claim that then (2.3) with integrals over Ω'' still holds for f an all such u. Indeed, one may decompose

$$(2.6) \qquad u = u_e + u_o , \quad u_e(x) = (u(x)+u(\bar{x}))/2 , \quad u_o(x) = (u(x)-u(\bar{x}))/2 .$$

Note that the extended H maps even functions into even functions and odd functions into odd functions. Since (the extended) f is odd, we get $(f,Hu_e)_{\Omega''} = \int_{\Omega''} d\mu \bar{f} Hu_e = 0$, as an integral over an odd function with symmetric range. On the other hand, $(f,Hu_o)_{\Omega''} = 2(f,Hu_o)_{\Omega'} = 0$ as well, since (the restriction to Ω' of) u_o

is a function in dom H_d (A continuous odd function vanishes at $x_n=0$). Accordingly, the extended function f satisfies a relation of the form II,(2.6) with the extended expression H .

Note that $u \in C_0^\infty(\Omega'')$ is not required, as long as u_o is continuous at zero, and $u_o|\Omega'$ is $C_0^\infty(\Omega')$.

Thus it appears that the proof of thm.1.2 will give $f \in C^\infty$ on the part of $\partial\Omega$ bordering the chart Ω' . Since f was defined as odd function, it then follows that f=0 on $\partial\Omega$, or, $f \in$ dom H_d .

However, it should be noticed that the assumptions of thm. 1.2 are not satisfied, since the coefficients of the extended H and κ) are not C^∞ accross $x^n = 0$. Instead all leading coefficients h^{jk} are continuous and $C^\infty(\Omega')$, since $h^{jn}=0$ at $x^n=0$. It is easy to make at least $h^{nn} \in C^1(\Omega'')$, by the change of variable

$$(2.7)\qquad \begin{aligned} y^\sim &= (y^1,\dots,y^{n-1}) = (x^1,\dots,x^{n-1}) = x^\sim , \\ y_n &= x_n(1 + x^n . h^{nn}(x^\sim,0)/(4h^{nn}_{|x^n}(x^\sim,0))) , \end{aligned}$$

which takes Ω' onto a similar region in y-space, and makes the normal derivative of h^{nn} zero at $x^n=0$. The lower order coefficients c and b_j , j<n also are continuous and piecewise C^1 , while b^n has a jump accross $x^n = 0$. Furthermore we have $g = \lambda = 0$, in the notation of thm.1.2.

Going over the proof of prop.1.3 we notice that v(x) of (1.10) no longer is $C^\infty(\Omega'')$, but will be $C^\infty(\Omega')$ again, and continuous accross $x^n=0$ .Therefore we again may substitute u of (1.24) into (2.3) . We still get (1.25), as long as $x^\sim$ is not on $\partial\Omega$, for sufficiently small ε , and then (1.26) follows as well, with g = 0 , so that we have

$$(2.8)\qquad f(x) = \int_{\Omega''} \gamma(x,y,x-y)f(y)dy \ , \text{ for } x^n \neq 0 \ , \ x \in \Omega^\sim \ ,$$

with a sufficiently small neighbourhood $\Omega^\sim$ of a boundary point x_0. Here $\gamma(x,y,z)$ is continuous and of compact support in $\Omega''\times\Omega''\times\mathbb{R}^{n*}$, and homogeneous of degree 2-n in z , and its restrictions to $x^n \geq 0$ and to $x^n \leq 0$ are C^∞ . In this way we conclude that indeed f is continuous, and has continuous one-sided derivatives of all orders at $x^n=0$. Thus indeed $f \in$ dom H_d , and the proof is complete.

Note that a small modification of the above proof gives the following.

Corollary 2.3. Under the assumptions of thm.2.2, if $f \in C^\infty(\Omega \cup \partial\Omega)$, relative to the imbedding $\Omega\to\Omega^\wedge$, and if $(H-\lambda)u = (H_d^{**}-\lambda)u = f$, for some $\lambda\in\mathbb{C}$, and $u \in \mathrm{dom}\, H = \mathrm{dom}\, H_d^{**}$, then we have $u\in \mathrm{dom}\, H_d$, and $(H_d-\lambda)u=f$.

Indeed, under these assumptions the interior regularity of u is a consequence of thm.1.2, while, at the boundary, the argument around (2.3) may be repeated, giving that also $u \in C^\infty$ at a boundary point. Thus cor.2.3 follows.

3. Compactness of the resolvent of the Friedrichs extension.

This section discusses various compactness criteria, for the resolvent of a comparison operator, as well as for the inverse square root $\Lambda=H^{-1/2}$, and also, more generally, for $a\Lambda$ with a function $a(x)\in C_0^\infty(\Omega)$. Its main applications will be found in V,3, and V,4,6 , where compactness of commutators, and certain asymptotic expansions are derived and applied. On the other hand, compactness of the resolvent $(H-\lambda)^{-1} = R(\lambda)$ will be essential in the discussions of the present chapter, below, involving the Dirichlet operator H_d .

In each case we consider the (self-adjoint) Friedrichs extension of the minimal operator H_0 , for an expression of the form (1.2). Note that compactness of the resolvent $R(\lambda)$ is an immediate consequence of the compactness of Λ , since we get $R(\lambda) = (H-\lambda)^{-1} = \Lambda^2(1-\lambda\Lambda^2)^{-1}$, for points $\lambda\in\mathrm{Rs}(H)$, where $(1-\lambda\Lambda^2)^{-1} \in L(H)$. Since $H=H_d$ for a manifold Ω with regular boundary, and $\Omega\cup\partial\Omega$ compact, we get compactness of the resolvent of the Dirichlet operator in cor.3.9.

The requirements on a for compactness of $a\Lambda$ will be different near the various 'infinities of Ω' . For a clear formulation of these facts we introduce the concept of subextending triple, below.

For technical reasons, for a while, we consider differential expressions of the (formally) more general form

$$(3.1) \qquad L = q - \frac{\beta}{\kappa}\, \partial_{x^j} \kappa h^{jk} \partial_{x^k} \beta \ , \ d\mu = \kappa\, dx \ ,$$

with a $C^\infty(\Omega)$-function $\beta > 0$, and with all other coefficients q , κ , h^{jk} as in (1.2). Such an expression may be brought into the form (1.2) : One has

$$(3.2) \quad L = -\kappa^{-1}\partial_{x^j}\kappa\beta^2 h^{jk}\partial_{x^k} + q + q^0 \ , \ q^0 = -\frac{\beta}{\kappa}\,(h^{jk}\kappa\beta_{|x^j})_{|x^k} \ .$$

Indeed, we get $(u,Lv) = (u,qv) + (\nabla(\beta u),\nabla(\beta v))$, u , $v \in \mathrm{dom}\, H_0$, where we write $\nabla(\beta u) = \beta\nabla u + u\nabla\beta$, and apply a partial integration to the term $(\beta\nabla\beta,\bar{u}\nabla v+v\nabla\bar{u}) = (\beta\nabla\beta,\nabla(\bar{u}v))$, from the second inner product, to free the term $(\bar{u}v)$ from its gradient. For more detail, cf. the derivation of formula (6.6), below.

If we assume that

$$(3.3) \qquad (u,Lu) \geq (u,u) \ , \ u \in C_0^\infty(\Omega) \ ,$$

then the facts discussed for H in section 1 apply to L as well. We get a well defined Friedrichs extension L of the minimal operator L_0 , and its inverse positive square root $L^{-1/2}$ is a bounded self-adjoint operator of H . The space $L^{-1/2}C_0^\infty(\Omega)$ is dense in H , and $L^{1/2}$ is essentially self-adjoint in $\mathrm{dom}\, H_0$.

Let us adopt the convention to write

$$(3.4) \qquad \lim_{x\to\infty} \gamma(x) = \gamma_0 \quad , \ (\text{in } \Omega\ ) \ ,$$

where $\gamma(x)$ is a function over Ω , (with values in a space X) , and $\gamma_0 \in X$, if for every neighbourhood N of γ_0 a compact set $K \subset \Omega$ can be found such that $\gamma(x) \in N$ whenever $x \in \Omega$ is outside K . We are not excluding the case of a compact manifold Ω . Then condition (3.4) is void: It is true for every function $\gamma(x)$, limit γ_0, and space X whatsoever. In particular condition (3.5), below, is generally true whenever Ω is compact.

Theorem 3.1. Let the expression (3.1) satisfy (3.3) and

$$(3.5) \qquad \lim_{x\to\infty} q(x) = \infty \ , \ (\text{in } \Omega\ ) \ .$$

Then $L^{-1/2}$ is a compact operator of $H = L^2(\Omega,d\mu)$.

The proof is rather technical,and will be prepared by a series of propositions. Let $\omega,\chi \in C_0^\infty(\mathbb{R})$ be such that

$$(3.6) \qquad 0\leq\omega,\chi\leq 1 \ , \ \omega^2+\chi^2=1 \ , \ \omega(t) =1 \ , \ t \leq 0 \ , \ \omega(t)=0 \ , \ t\geq 1.$$

For $0 \leq A < \infty$ define the two functions

$$(3.7) \qquad \lambda_A(x) = \omega(q(x)-A) \ , \ \mu_A(x) = \chi(q(x)-A) \ .$$

It follows from (3.5) that λ_A has compact support.In fact, we have

$$(3.8) \qquad q \leq A+ 1 \ \text{ in supp } \lambda_A \ , \quad q \geq A \ \text{ in supp } \mu_A \ .$$

<u>Proposition 3.2.</u> For every $A \geq 0$ there exists a constant c_A with

(3.9) $$(u,Lu) \geq \frac{A}{2} \|\mu_A u\|^2 + \|\nabla(\lambda_A \beta u)\|^2 - c_A \|\lambda_A \beta u\|^2 \ , \ u \in C_0^\infty(\Omega) \ .$$

<u>Proof</u>. We get

$$(u,Lu) = \int_{q\leq A+1} \lambda_A{}^2 (h^{jk}(\beta\bar{u})_{|x^j}(\beta u)_{|x^k} + q|u|^2)d\mu$$
$$+ \int_{q\geq A} \mu_A{}^2 (h^{jk}(\beta\bar{u})_{|x^j}(\beta u)_{|x^k} + q|u|^2)d\mu \quad .$$

The second term at right is bounded below by $A\|\mu_A u\|^2$. For the first term we use (3.2), with β replaced by λ_A, to obtain, with $\lambda=\lambda_A$, $\mu=\mu_A$,

$$\int \lambda_A{}^2 h^{jk}(\beta\bar{u})_{|x^j}(\beta u)_{|x^k} d\mu = \|\nabla(\lambda_A \beta u)\|^2 + \int \beta^2 \frac{\lambda}{\kappa}(\kappa h^{jk}\lambda_{|x^j})_{|x^k}|u|^2 d\mu.$$

Here we estimate the last term at right by

$$c_A' \|\lambda\beta u\| \, \|u\| \quad , \quad c_A' = \sup_{x\in\Omega} |\tfrac{\beta}{\kappa}(\kappa h^{jk}\lambda_{|x^j})_{|x^k}| \ .$$

Note that $\|u\| = \|(\lambda^2+\mu^2)u\| \leq \|\lambda^2 u\| + \|\mu^2 u\| \leq \|\lambda u\| + \|\mu u\|$, so that

$$(u,Lu) \geq A\|\mu_A u\|^2 + \|\nabla(\lambda_A \beta u)\|^2 + \int (q\lambda^2 + \beta^2 \tfrac{2\lambda}{\kappa}(\kappa h^{jk}\lambda_{|x^j})_{|x^k})|u|^2 d\mu \ ,$$

with the last term estimated below by

$$-c_A''\|\lambda\beta u\|^2 - c_A'''\|\lambda\beta u\|^2 - c_A'\|\lambda\beta u\|\|\mu u\|, \ c_A'' = \sup\{-q/\beta^2 : x\in \operatorname{supp} \lambda_A\},$$

$$c_A''' = c_A' \cdot \sup\{\beta^{-1}(x) : x\in \operatorname{supp} \lambda_A\} \ .$$

Using that $\|\lambda\beta u\|\|\mu u\| \leq \frac{\varepsilon}{2}\|\mu u\|^2 + 1/(2\varepsilon)\|\lambda\beta u\|^2$ we get (3.9), setting $\varepsilon = A/c_A'$, with $c_A = c_A'' + c_A''' + c_A'^2/(2A)$. Q.E.D.

In the following let $\{\Omega_j\}$ be a locally finite atlas of Ω, and let $\omega_j{}^2$ be a corresponding partition of unity, such that $\operatorname{supp} \omega_j \subset \Omega_j$, $\omega_j \in C^\infty(\Omega)$, $\omega_j \geq 0$, $\sum \omega_j{}^2 = 1$, on all of Ω. We assume without loss of generality that each Ω_j is a subset of a larger chart in which it has compact closure.

<u>Proposition 3.3.</u> For every $A \geq 0$ there exists a constant d_A with

(3.10) $$(u,Lu) \geq \frac{A}{2}\|\mu_A u\|^2 + \sum_j \|\nabla(\lambda_A \beta\omega_j u)\|^2 - d_A \|\lambda_A \beta u\|^2, \ u \in C_0^\infty(\Omega).$$

Proof. In view of (3.9) it suffices to show that

$$\|\nabla(\lambda_A \beta u)\|^2 \geq \sum_j \|\nabla(\lambda_A \beta \omega_j u)\|^2 - c_A \|\lambda_A \beta u\|^2 \quad , \; u \in C_0^\infty(\Omega) . \tag{3.11}$$

Write $\|\nabla(\lambda_A \beta u)\|^2 = \sum_j \|\omega_j \nabla(\lambda_A \beta u)\|^2$, and apply the conversion (3.1) → (3.2) , for $\beta = \omega_j$, to bring ω_j under the gradient. Since λ_A has compact support only finitely many of the summands do not vanish, and the corresponding functions q^0 of (3.2) are all bounded. Hence (3.11) follows, q.e.d.

For a moment consider the cube $Q=\{x\in\mathbb{R}^n : 0<x_j<1, j=1,\dots,n\}$, with the Riemannian metric of $\mathbb{R}^n$, (i.e., $|\nabla v|^2 = \sum_{j=1}^n |\partial_{x_j} v|^2$), and the functions $\sin_\alpha \pi x = \Pi_1^n \sin \pi\alpha_j x_j$, $\alpha \in \mathbb{Z}^n$. As a consequence of Parsevals relation for the Fourier sine and Fourier cosine series we have

$$\int_Q |v|^2 dx = \sum |v_\alpha|^2 \; , \; \int_Q |\nabla v|^2 dx = \sum (\alpha,\alpha) |v_\alpha|^2 \; , \; v\in C_0^\infty(\Omega), \tag{3.12}$$

with the sums taken over all multi-indices α with $\alpha_j \geq 1$, and $(\alpha,\alpha) = \sum_j \alpha_j^2$, and with the Fourier coefficients

$$v_\alpha = 2^n \int_Q v(x) \sin_\alpha \pi x \, dx \quad . \tag{3.13}$$

Proposition 3.4. For every $C > 0$ there exists an integer $N = N(C)$, such that $v \in C_0^\infty(Q)$, $v_\alpha = \int_Q \sin_\alpha \pi x \, v(x) dx = 0$ for $|\alpha| \leq N$ implies

$$\int_Q |\nabla v|^2 \, dx \geq C \int_Q |v|^2 \, dx . \tag{3.14}$$

Proof. It is clear that (3.14) is an immediate consequence of (3.12) , setting $N(C) = \mathrm{Max}\,\{|\alpha| : (\alpha,\alpha) \leq C\}$.

Returning to our general case we next prove the following.

Proposition 3.5. For every $B \geq 0$ there exists finitely many functions $z_j \in C_0^\infty(\Omega)$, $j = 1,\dots,M$, $M = M(B)$, such that

$$u \in C_0^\infty(\Omega) \; , \; (z_j,u) = 0 \; , \; j = 1,\dots,M \; , \tag{3.15}$$

implies

$$(u,Lu) \geq B\|u\|^2 . \tag{3.16}$$

Proof. Apply proposition 3.3 , with A = 2B , and assume without loss of generality that each of the compact sets supp ω_j relates to a subset of the cube Q under the coordinates of the corresponding chart Ω_j . In that compact set we also get a positive bound below for the positive definite matrix function $((h^{jk}(x)))$. Thus

$$(3.17) \qquad \|\nabla(\lambda_A\omega_j u)\|^2 \geq (B + d_A)\|\lambda_A\omega_j u\|^2 \quad ,$$

if only

$$(3.18) \qquad \int \lambda_A\omega_j\, u\, \sin_\alpha \pi x\, dx = 0 \;,\; |\alpha| \leq N_j \;.$$

Since only finitely many indices j are involved (i.e., those with supp $\omega_j \cap$ supp $\lambda_A \neq \emptyset$, (3.18) translates into (3.15), and (3.16) follows by combining (3.10) and (3.17), q.e.d.

Proof of theorem 3.1. Setting $u = L^{-1/2}w$ in (3.15) , (3.16) we get

$$(3.19) \; \|L^{-1/2}w\|^2 \leq B^{-1}\|w\|^2, \text{ as } w \in L^{1/2}C_0^\infty(\Omega),\; (L^{-1/2}z_j,w)=0,\; j=1,..,M.$$

Since $L^{1/2}C_0^\infty(\Omega)$ is dense in H, the estimate (3.19) follows for all $w \in H$ orthogonal to $L^{-1/2}z_j$, $j = 1,\dots,M$. Since B may be choosen arbitrarily large this implies compactness of $L^{-1/2}$, by Rellich's criterion ([C_1], A1,4) : We get

$$(3.20) \qquad \|L^{-1/2}|T_B\| \leq B^{-1/2} \;,\; T_B = (L^{-1/2}\{z_1,\dots,z_M\})^\perp$$

where T_B has finite codimension. This completes the proof.

Corollary 3.6. Under the assumptions of theorem 3.1, let $\Omega^\sim \subset \Omega$ be a subdomain of Ω . Then the statement of the theorem holds for the restriction of the differential expression L to $\Omega^\sim$ as well : If $L^\sim$ is the Friedrichs extension of the minimal operator generated by L in $H^\sim = L^2(\Omega^\sim,d\mu)$, then $L^{\sim -1/2} \in K(H^\sim)$. Moreover, the same still holds for any expression

$$(3.21) \qquad L^\sim = q^\sim - (\beta/\kappa)\partial_{x^j}\kappa h^{\sim jk}\partial_{x^k}\,\beta \;,$$

of the form (3.1) with C^∞-coefficients, defined only in $\Omega^\sim$, and satisfying

$$(3.22) \qquad q^\sim \geq q \;,\; ((h^{\sim jk})) \geq ((h^{jk})) \quad \text{in } \Omega^\sim \;.$$

Proof. For $u \in C_0^\infty(\Omega^\sim)$, setting $(.,.)_{\Omega^\sim} = \int_{\Omega^\sim} \bar{.}\,.\,d\mu$, we get

(3.23) $(u,L^{\sim}u)_{\Omega^{\sim}} \geq (u,Lu) \geq B(u,u) = B(u,u)_{\Omega^{\sim}}$,

whenever $(z_j,u) = 0$, as a consequence of prop.3.5. Here we were extending u to Ω by setting it zero outside $\Omega^{\sim}$. Notice that the conditions $(z_j,u)=0$ may be written as $(z^{\sim}_j,u)_{\Omega^{\sim}} = 0$, with $z^{\sim}_j = z_j|\Omega^{\sim} \in H^{\sim}$. In other words, we conclude that prop.3.5 is valid also for $L^{\sim}$ in the subdomain $\Omega^{\sim} \subset \Omega$, except that now the functions $z^{\sim}_j$ no longer are $C_0^{\infty}(\Omega^{\sim})$, but only in $C^{\infty}(\Omega)\cap H^{\sim}$. This weakened condition still allows its use for the argument of the proof of thm.3.1 , because we still get $(z^{\sim}_j,L^{\sim -1/2}w) = (L^{\sim -1/2}z^{\sim}_j,w)$, where $L^{\sim -1/2}z^{\sim}_j \in H^{\sim}$, q.e.d.

We now return to comparison triples. Let triples $\{\Omega,d\mu,H\}$, and $\{\Omega^{\wedge},d\mu^{\wedge},H^{\wedge}\}$ be given, both satisfying the basic assumptions of sec.1. We write

$$\{\Omega,d\mu,H\} <_c) \{\Omega^{\wedge},d\mu^{\wedge},H^{\wedge}\} ,$$

(or $\{\Omega^{\wedge},d\mu^{\wedge},H^{\wedge}\}$ $(_c> \{\Omega,d\mu,H\}$) if

(3.24) (i) Ω is an (open) subdomain of $\Omega^{\wedge}$, (ii) $d\mu = d\mu^{\wedge}|\Omega$,

(iii) $((h^{jk}(x))) \geq ((h^{\wedge jk}(x)))$, $q(x) \geq q^{\wedge}(x)$, for all $x \in \Omega$.

In this case we will say that the triple $\{\Omega^{\wedge},d\mu^{\wedge},H^{\wedge}\}$ <u>sub-extends</u> the triple $\{\Omega,d\mu,H\}$.

The main point for introducing this more complicated notion is the fact that many such singular problems display a very different behaviour near a point $x_0 \in \partial\Omega \subset \Omega^{\wedge}$ than at infinity (of $\Omega^{\wedge}$) . For example, as shown by cor.3.6, we do not have to require $q \to \infty$ near such a point x_0 , to get compactness of $L^{-1/2}$.

This will still become more evident in V,3, where we investigate compactness of commutators of the generators of a comparison algebra C .

Here it may become important that a proper change of dependent and independent variable may be necessary, before $\Omega^{\wedge}$ can be defined (cf. the examples in V,4 and V,5).

<u>Theorem 3.7.</u> For a comparison triple $\{\Omega,d\mu,H\}$ let there exist a triple $\{\Omega^{\wedge},d\mu^{\wedge},H^{\wedge}\}$ $(_c> \{\Omega,d\mu,H\}$ satisfying

(3.25) $q^{\wedge}(x) > 0$ for all $x \in \Omega^{\wedge}\setminus K$,where $K \subset \Omega^{\wedge}$ is compact .

Then, for every function $a \in C(\Omega)$ with $a = o_\Omega\hat{}(\sqrt{q}\hat{})$, i.e., with

(3.26) $a^2 = O(q\hat{})$, and $\lim_{x\to\infty} a^2/q\hat{} = 0$, (in $\Omega\hat{}$) ,

the operator $a\Lambda = aH^{-1/2}$ is compact in $H = L^2(\Omega, d\mu)$.

Proof. Assume first a has a real extension $0<a\hat{} \in C^\infty(\Omega\hat{})$ and set $L=a^{-1}Ha^{-1}$. Consider $L\hat{}= a\hat{}^{-1}H\hat{}a\hat{}^{-1}$, and note that $L\hat{}+\gamma_0$, with a sufficiently large positive constant γ_0 , satisfies the assumptions of thm.3.1. Also the expression $L+\gamma_0$ satisfies the assumptions of cor.3.6. Hence $(L+\gamma_0)^{-1/2}$, $(L+\gamma_0)^{-1}$ are compact, and $L+\gamma_0$ (therefore also L) has discrete spectrum. But $\Lambda^{-1} = aH^{-1}a$ is well defined, hence 0 cannot be eigenvalue of L, and L^{-1} also must be compact. But $L^{-1}= (a\Lambda)(a\Lambda)^*$, so that $a\Lambda$ is compact, since an operator $A \in L(H)$ is compact if and only if AA^* is compact.

For a general $a\in C(\Omega)$ we get $\|a\Lambda u\|\leq\|b\Lambda u\|$, where $b = c|\Omega$ is the restriction to Ω of $0<c\in C^\infty(\Omega\hat{})$ constructed according to lemma A.1, with $g=\sqrt{q}\hat{}$, suitably modified within K . By lemma 3.10, below, it thus suffices to establish compactness of $b\Lambda$. This was just done above. The proof is complete.

Corollary 3.8. Under the assumptions of thm.3.7 we also have $|a|^\varepsilon\Lambda^\varepsilon \in K(H)$, for all $0<\varepsilon\leq 1$.

This corollary is a straight application of I,lemma 5.1.

Corollary 3.9. The operators Λ^ε, $0\leq\varepsilon\leq 1$, are compact whenever Ω is compact. Moreover, then H has discrete spectrum,i.e., its resolvent $(H-\lambda)^{-1}$is compact for all $\lambda\in Rs(H)=\mathbb{C}\setminus Z$, where $Z=Sp(H)$ consists of countably many eigenvalues of finite multiplicity without cluster. The same holds if $\{\Omega,d\mu,H\}<c)\{\Omega\hat{},d\mu\hat{},H\hat{}\}$, with $\Omega\cup\partial\Omega$ compact in $\Omega\hat{}$.

Proof. Under these assumptions the conditions of thm.3.7 hold for $a\equiv 1$, evidently, q.e.d.

Lemma 3.10. Given two bounded operators $A , B \in L(H)$. If A is compact and B satisfies the inequality

(3.27) $\|Bu\| \leq \|Au\|$,for all $u \in H$,

then B also is compact.

Proof. By Rellich's criterion (cf.[C_1],App.AI,4) there exists a closed subspace $S_\varepsilon \subset H$ with $\|A|S_\varepsilon\| \leq \varepsilon$, for every $\varepsilon>0$. But then (3.27) implies that

(3.28) $\|Bu\| \le \|Au\| \le \varepsilon \|u\|$ for all $u \in S_\varepsilon$.

Or, $\|B|S_\varepsilon\| \le \varepsilon$, so that B must be compact, q.e.d.

4. A Greens function for H and H_d, and a mean-value inequality.

For later application it will be useful to obtain a 'clean' fundamental solution for our expression H , and, moreover, a Greens function for our Friedrichs extension H ,as well as for Dirichlet operator H_d of sec.2.

First we work locally, in a given fixed compact subset $\mathcal{B} \subset \Omega'$ of a chart Ω', with nonvoid connected $\mathcal{B}^{int} = \Omega''$, $\Omega''^{clos} = \mathcal{B}$. Let η_0 denote the (positive) distance (in $\mathbb{R}^n$) from $\mathcal{B}$ to $\partial\Omega'$.

It is practical to consider the local expression H of the form (1.2) redefined outside $\mathcal{B}$ in such a way that it equals $1-\Delta$ in Ω'^{compl} , while still satisfying all other conditions of H . For example one may choose a cutoff function $\chi \in C_0^\infty(\Omega')$, $0 \le \chi \le 1$, $\chi=1$ near $\mathcal{B}$, and then define the measure and expression

(4.1) $$d\mu' = \kappa' dx \ , \ H' = q' - \kappa'^{-1} \partial_{x^j} \kappa' h'^{jk} \partial_{x^k} \ ,$$

over $\mathbb{R}^n$, where

(4.2) $$\kappa' = \chi\kappa + (1-\chi) \ , \ q' = \chi q + (1-\chi) \ , \ h'^{jk} = \chi h^{jk} + (1-\chi)\delta^{jk} \ .$$

For a while (until thm.4.4) we will entirely focus on this expression over $\mathbb{R}^n$. Hence we return to our old notation, and will assume that H and $d\mu$ coincide with $1-\Delta$ and dx outside some compact subset $\mathcal{B}$ of $\mathbb{R}^n$.

The 'parametrix function' e(x,y) of (1.9) again will play a central part. Note that it now is defined for $x,y \in \mathbb{R}^n$. Moreover, it is clear that $e(x,y) = |x-y|^{2-n}$ (or $= \log|x-y|$, for n=2) , as $x \in \mathcal{B}^{compl.}$. In fact, we again will depart from (1.26), which will have to be iterated. Let us reinterpret these facts as follows.

<u>Proposition 4.1</u>. For $x,y \in \mathbb{R}^n$ define the function

(4.3) $$g(x,y) = (c_n\sqrt{h}(x))^{-1}(\kappa(y)/\kappa(x))\chi(x-y)e(y,x), \ h=\det((h_{jk})),$$

with the constant c_n of (1.23), and with $\chi(x) \ge 0$ denoting a cutoff function, $\chi \in C_0^\infty(|x|<2)$, $\chi=1$ at $|x| \le 1$. There exist functions $\theta^j \in C^\infty(\mathbb{R}^n \times \mathbb{R}^n \times \mathbb{R}^{n*})$, $j=0,\ldots,n$, with $\theta^j(x,y,z)$ homogeneous of degree $2-n$ in z and support contained in the set $\{(x,y)\in\mathbb{R}^n\times\mathbb{R}^n: |x-y|<2\}\times\mathbb{R}^{n*}$, such that, for every solution u of Hu=v , with $u,v\in C^\infty(\mathbb{R}^n)$, we have the relation

$$(4.4)\qquad u(x) - \int dy\theta(x,y)u(y) = \int g(x,y)v(y)by , \; x \in \mathbb{R}^n ,$$

with

$$(4.5)\qquad \theta(x,y) = \theta^0(x,y,x-y) - (\theta^j(x,y,x-y))_{|y^j} .$$

Here the functions θ^j depend only on the choice of χ , and the coefficients of H in $\mathbb{R}^n$, but not on the functions u, v.

For the proof one must look at the derivation of (1.26) again, and just confirm that the functions θ^j are as stated. A pair of C^∞-functions u,v satisfying Hu=v may be put into the places of f anf g of (1.26), of course, once we interpret II,(2.6) with $\Omega=\Omega'=\mathbb{R}^n$.

<u>Proposition 4.2</u>. Let Θ and G denote the integral operators on $\mathbb{R}^n$, with kernel $\theta(x,y)$ and $g(x,y)$, respectively , so that (4.4) assumes the form $u-\Theta u=Gv$. Then the iterates $G_k=\Theta^k G$, $k=1,\ldots,n$, and Θ^{n+1}, all are well defined integral operators over $\mathbb{R}^n$ with kernels $g_k(x,y)$ and $\theta_{n+1}(x,y)$, respectively. In particular, θ_{n+1}, g_{n-1}, g_n are $C(\mathbb{R}^n\times\mathbb{R}^n)$, while $g_k \in C(\mathbb{R}^n\times\mathbb{R}^n\setminus\{x=y\})$, for $k=0,1,\ldots,n-2$. Moreover, we have $g_k = O(\langle x-y\rangle^{-n+k+2})$, as $0\le k<n-2$, while g_{n-2} is $O(\langle x-y\rangle^{-\eta})$, for all $\eta>0$. If Γ_j denotes the integral operator with kernel $g_{|y^j}(x,y)$, then the kernels of $\Theta^k\Gamma_j$, $j,k=1,\ldots,n$, are $C(\mathbb{R}^n\times\mathbb{R}^n\setminus\{x=y\})$ and $O(\langle x-y\rangle^{-n+k+1})$, as $k<n-1$, and $O(\langle x-y\rangle^{-\eta})$, for all $\eta>0$, as $k=n-1$, and are $C(\mathbb{R}^n\times\mathbb{R}^n)$, for $k=n$. All the above kernels have support in $\{|x-y|\le 2(n+1)\}$. In particular we may iterate (4.4) n-times for (4.6), below, where again $u\in C^\infty(\mathbb{R}^n)$, $v=Hu$:

$$(4.6)\qquad u(x) = (\Theta^{n+1}u)(x) + \textstyle\sum_{k=0}^n (G_k v)(x) , \; \text{for all } x\in\mathcal{B} .$$

The proof is an immediate consequence of prop.4.6, below.

<u>Proposition 4.3</u>. Given two kernels $k_j(x,y)$, continuous over $\mathbb{R}^n\times\mathbb{R}^n\setminus\{x=y\}$, and with support contained in $|x-y|\le c$ for some constant c. Assume that, with constants ν_j , $0\le\nu_j<n$, we have

$$(4.7)\qquad k_j = O(|x-y|^{-\nu_j}) , \quad K_j u(x) = \int k_j(x,y)u(y)dy , \; u\in C_0^\infty(\mathbb{R}^n) .$$

Then the operator $K_3=K_1K_2$ is an integral operator of the same form again, with kernel $k_3 = O(|x-y|^{-\nu_3})$, whenever $\nu_3=-n+\nu_1+\nu_2 > 0$, while the kernel k_3 is continuous at $x=y$ if $\nu_3 < 0$,and k_3 still is $O(|x-y|^{-\eta})$, for every $\eta>0$, if $\nu_3=0$.

Proof. If $\nu_3<0$ then the integral

$$(4.8) \qquad k_3(x,y) = \int k_1(x,z)k_2(z,y)dz$$

exists for all x,y , including $x=y$. It is easily seen that the product $a_{x,y}(z) = k_1(x,z)k_2(z,y)$ defines a continuous family of functions in $L^1(\mathbb{R}^n)$, depending on x,y , hence k_3 also is continuous in x and y . If $\nu_3\geq 0$ the same still is true for $x\neq y$, but no longer at $x=y$. On the other hand, as $\nu_3>0$ one may estimate

$$(4.9) \qquad k_3(x,y) \leq c \int_{R^n} |x-z|^{-\nu_1}|z-y|^{-\nu_2}dz .$$

The right hand side integral $I(x,y)$ exists, since the integrand is $O(|z|^{-\nu_1-\nu_2})$ for large $|z|$, with $\nu_1+\nu_2>n$, while the local singularities at x and y have order $-\nu_j>-n$. Moreover,

$$(4.10) \qquad I(x,y) = \int |z|^{-\nu_1}|z-\sigma e^1|^{-\nu_2}dz , \quad \sigma=|x-y| , \quad e^1=(1,0,\ldots,0),$$

as shown by a substitution of integration variable involving a suitable translation and rotation. Then another substitution $z = \sigma w$ shows that $I(x,y) = \sigma^{n-\nu_1-\nu_2}I(0,e^1)$, where $I(0,e^1)$ is a constant. This confirms the estimate $k_3 = O(|x-y|^{-\nu_3})$. Finally, if $\nu_3=0$ we still get the estimate (4.9) with ν_1 replaced by $\nu_1+\eta$, for sufficiently small $\eta>0$ (or, similarly, for ν_2, if more convenient). Then we get (4.10) with $0=\nu_3$ replaced by $\nu_3+\eta=\eta>0$, q.e.d.

Note that prop.4.2 suggests that the function

$$(4.11) \qquad g^{\sim}(x,y) = \sum_{k=0}^{n} g_k(x,y) = g(x,y) + \sum_{k=1}^{n} g_k(x,y)$$

is an improved parametrix of the expression H , because the corresponding integral operator $G^{\sim}$ satisfies

$$(4.12) \qquad u - \Theta^{n+1}u = G^{\sim}Hu \quad , \text{ for all } u \in C^{\infty}(\mathbb{R}^n) ,$$

which seems better than (4.4) , since the kernel θ_{n+1} of Θ^{n+1} is continuous at $x=y$. In order to find a fundamental solution of H we seek to add another term $f^{\sim}$ to $g^{\sim}$, such that $g^{\wedge}=g^{\sim}+f^{\sim}$ satisfies

$$(4.13) \qquad u = G^{\blacktriangle} Hu \;, \text{ for all } u \in C_0^\infty(\mathbb{R}^n) \;.$$

Subtracting (4.12) from (4.13) we find that $f^\sim$ must satisfy

$$(4.14) \quad \Theta^{n+1}u = \int \theta_{n+1}(x,y)u(y)dy = (G^{\blacktriangle} - G^\sim)Hu = \int f^\sim(x,y)Hu(y)dy, \; u \in C_0^\infty(\mathbb{R}^n).$$

Let us require (4.14) not only for $u \in C_0^\infty$, but even for all $u \in \operatorname{dom} H$ (the Friedrichs extension of H_0) . Also introduce the functions (of y) $b_x(y) = \theta_{n+1}(x,y)/\kappa(y)$, and $a_x(y) = f^\sim(x,y)/\kappa(y)$. Observe that $b_x \in C_0(\mathbb{R}^n) \subset H = L^2(\mathbb{R}^n)$, and require that $a_x \in H$. Under these conditions (4.14) takes the form

$$(4.15) \qquad (b_x, u) = (a_x, Hu) \;, \text{ for all } u \in \operatorname{dom} H \;,$$

so that $a_x \in \operatorname{dom} H$ and $Ha_x = b_x$ follows, since the Friedrichs extension is self-adjoint. In particular, of course, we also get $H_0^* a_x = b_x$, so that cor.2.1 applies, since $b_x \in C_0(\mathbb{R}^n)$. Accordingly a_x at least is C^1, and has its first derivatives Hoelder continuous. Moreover, since $H^{-1} \in L(H)$, and $b_x \in C(\mathbb{R}^n, H)$, as x varies, we also get $a_x \in C(\mathbb{R}^n, H)$.

Now let us carry this back to the function $g^{\blacktriangle} = g^\sim + f^\sim$. Introduce the function $c_x(y) = g^{\blacktriangle}(x,y)/\kappa(y) = g^\sim(x,y)/\kappa(y) + a_x(y)$. Clearly we still get $c_x \in C(\mathbb{R}^n, H)$. Combining (4.12) and (4.15) get

$$(4.16) \qquad u(x) = (c_x, Hu) \;, \text{ for all } u \in \operatorname{dom} H \text{ and all } x \in \mathbb{R}^n \;.$$

Applying (4.16) for $u \in C_0^\infty(\mathbb{R}^n \setminus \{x\})$, for fixed x, one concludes that c_x is a solution of $H_{0,x}^* c_x = 0$, with the minimal operator $H_{0,x}$ of the expression H in the punctured $\mathbb{R}^n$. Thus thm.1.2 implies that we have $c_x \in C^\infty(\mathbb{R}^n \setminus \{x\})$, and $Hc_x = 0$, as $y \neq x$, with H acting on the y-variable.

Finally let us multiply (4.16) by $\bar{v}(x) \in C_0^\infty(\mathbb{R}^n)$, and integrate $d\mu_x$, over $\mathbb{R}^n$. Note that the integration may be interchanged with the inner product, since the family $\{c_x : x \in \mathbb{R}^n\}$ is $C(\mathbb{R}^n, H)$. It follows that

$$(4.17) \qquad (v,u) = \left(\int v(x) c_x d\mu_x, Hu\right) \;, \text{ for all } v \in C_0^\infty \;, \; u \in \operatorname{dom} H \;.$$

Accordingly,

$$(4.18) \quad w = \int v(x) c_x d\mu_x \in \operatorname{dom} H \;, \text{ and } Hw = v \;, \text{ for all } v \in C_0^\infty(\mathbb{R}^n) \;.$$

But the Friedrichs extension H is self-adjoint and ≥ 1, and has a well defined bounded hermitian inverse $H^{-1} \in L(H)$. Applying H^{-1} to (4.18) we conclude that

$$(4.19)\quad H^{-1}v(y)=\int c_x(y)v(x)d\mu_x=\int g^{\blacktriangle}(x,y)(\kappa(x)/\kappa(y))v(x)dx,\ v\in C_0^\infty(\mathbb{R}^n).$$

Then the hermitian symmetry of H^{-1} implies that

$$(4.20)\qquad g^{\blacktriangle}(x,y)\kappa(x) = g^{\blacktriangle}(y,x)\kappa(y)\ ,\ x,y\in \mathbb{R}^n\ ,\ x\neq y\ .$$

We have proven the following result in the special case of a triple $\{\mathbb{R}^n,d\mu,H\}$, coinciding with the Laplace triple outside a compact set. In case of a general manifold Ω we relate the integral operator to the global measure $d\mu=\kappa dx$ again. Also we denote the Green's kernel by $g(x,y)$, not $g^{\blacktriangle}(x,y)$, ignoring the earlier meaning of $g(x,y)$.

Theorem 4.4. For any triple $\{\Omega,d\mu,H\}$ satisfying (1.4) the Friedrichs extension has the form of an integral operator

$$(4.21)\qquad H^{-1}u(x) = \int g(x,y)u(y)d\mu_y\ ,\ u \in H = L^2(\mathbb{R}^n)\ ,$$

with the 'Green's function' $g(x,y) \in C^\infty(\Omega\times\Omega\setminus\{x=y\})$,and with $g(x,.)$, $g(.,x) \in C(\Omega,H)$. Also, $g(x,y) = g(y,x)$, $x,y \in \Omega$, and, in local coordinates, on a chart Ω', g has an expansion of the form

$$(4.22)\quad g(x,y) = (\chi(x-y)/(c_n\sqrt{h(x)})e(x,y) + \textstyle\sum_{k=1}^{n-1}g_k(x,y)\ ,\ x,y\in\Omega',$$

where $e(x,y)$ denotes the parametrix of (1.9) , and $\chi(x)$ denotes an even cutoff function ($\chi(-x)=\chi(x)$, $\chi\in C_0^\infty$, $\chi=1$ near 0 , $\chi\geq 0$). Also, $g_k(x,y) = O(\langle x-y\rangle^{k-n+2})$, $k=1,\dots,n-3$, $g_{n-2}(x,y)=O(\langle x-y\rangle^{-\eta})$, for all $\eta>0$, $g_{n-1} \in C(\Omega'\times\Omega')$, while g_k , $k=1,\dots,n-2$, have support contained in a ball $|x-y|\leq c_0$. Furthermore, the first y-derivatives of the $g_k(x,y)$ exists and are continuous for $y\neq x$. The kernels $g_{k|y^j}(x,y)$ are $O(\langle x-y\rangle^{-n+k+1})$, as $k=1,\dots,n-2$, and $O(\langle x-y\rangle^{-\eta})$ for every $\eta>0$,for $k=n-1$. Finally, for any $x_0\in\Omega$, and a cutoff function $\chi_1\in C^\infty(\Omega)$ $\chi_1=0$ near x_0 , $1-\chi_1\in C_0^\infty(\Omega)$, the function $\psi(y) = \chi_1(y)g(x_0,y)$ is contained in $C^\infty\cap \mathrm{dom}\ H$.

For a proof of thm.4.4 in the general case observe that the function $g(x,y)$ is uniquely determined by the relation

$$(4.23)\qquad u(x) = \int g(x,y)(Hu)(y)d\mu_y\ ,\ \text{for all } u \in \mathrm{dom}\ H\ .$$

For $x\in\Omega'$, a chart of Ω , let B and H' be as initially in this section. Let $g'(x,y)$ be the Greens function of (the Friedrichs extension of) H' in $\mathbb{R}^n$, and let the cutoff function χ have support in Ω' and be $= 1$ in $\mathcal{B}$. Then we also get

$$(4.24) \qquad u(x) = \int g'(x,y)(H(\chi u))(y)d\mu_y \quad , \; u\in C_0^\infty(\Omega) \; .$$

Subtracting (4.23) and (4.24) we get

$$(4.25) \int (g(x,y)-\chi(y)g'(x,y))(Hu)(y)d\mu_y = \int ([\chi,H]g'(x,y))u(y)d\mu_y , u\in C_0^\infty .$$

In particular we note that the commutator $[H,\chi]$ acts on the y-variable, and is a first order expression with support in $\Omega'\setminus\mathcal{B}$, so that the function $[\chi,H]g'(x,y) = q_x(y)$ is $C_0^\infty(\Omega')$, (as a function of y only). Setting $p_x(y) = g(x,y) - \chi(y)g'(x,y)$, (4.25) assumes the form

$$(4.26) \qquad (p_x,Hu) = (q_x,u) \; , \text{ for all } u \in C_0^\infty(\Omega) \; .$$

We even request that (4.26) holds for all $u \in \operatorname{dom} H$. This determines a C^∞-function $p_x \in \operatorname{dom} H$ such that $Hp_x = q_x$ in the sense of ordinary differentiation. Then we define $g(x,y) = p_x(y)+\chi(y)g'(x,y)$. At least for all $u\in \operatorname{dom} H$ with $Hu\in C_0^\infty$ we also get $u\in C^\infty(\Omega)$, and the right hand side of (4.25) may be written as $(g'(x,.)[H,\chi]u)$, so that (4.23) follows.

For the last statement observe that

$$(4.27) \quad (\psi,Hv) = (g_{x_0},H(\chi_1 v)) + ([H,\chi_1]g_{x_0},v) \; , \; v\in C^\infty \cap \operatorname{dom} H \; ,$$

with $g_{x_0}(y) = g(x_0,y)$, where $z = [H,\chi_1]g_{x_0} \in \mathcal{H}\cap C^\infty$, since the commutator $[H,\chi_1]$ vanishes near x_0. Note that the first term at right of (4.27) equals $GH(\chi_1 v)(x_0) = 0$. Hence the selfadjointness of H ,and essential self-adjointness of its restriction to $C^\infty\cap \operatorname{dom} H$ implies that $\psi\in\operatorname{dom} H\cap C^\infty$, as stated. Therefore the proof of thm.4.4 is complete.

Next we will investigate the Greens function $g(x,y)$ near its singularity. Given some fixed $x_0\in\Omega$ let us ask for the surface

$$(4.28) \qquad C_{x_0,\eta} = \{y\in\Omega : \; g(x_0,y) = \delta(x_0)\eta^{2-n}\}, \text{ (or } g=\delta\log\eta, \text{ as } n=2) \; ,$$

with $\delta = \delta(x_0) = c_n\sqrt{h(x_0)}$, as in (4.22), where the constant η

is assumed sufficiently large. More precisely, if g in (4.28) is replaced by g_0 , then the surface (4.28) will be an ellipsoid, in local coordinates. For g one expects a smooth surface $C_{x_0,\eta}$ close to that ellipsoid, and with normal close to the normal of the ellipsoid, as η is small. We leave open the possibility that other points y exist in the set (4.28). These will be ignored, i.e., the statement "y sufficiently close to x_0 " should be added in (4.28).

Let us assume without loss of generality that, at the point x^0, we have $h^{jk}(x_0)=\delta^{jk}=0$, as $j\neq l$, $=1$, as $j=l$, and that $x_0=0$, so that the equation $g(x_0,y)=\delta\eta^{2-n}$ of (4.28) assumes the form

$$(4.29) \qquad |y|^{2-n} + \gamma(y) = \eta^{2-n} ,$$

where we have

$$(4.30) \qquad \gamma(y) = O(|y|^{5/2-n}) , \nabla\gamma(y) = O(|y|^{3/2-n}) .$$

For n=2 the first term in (4.29) must be replaced by $-\log|y|$, and the second term is continuous with derivatives satisfying (4.30). We again focus on the case n>2, with n=2 to be treated similarly.

It follows that for small $|y|$ the first term at left of equation (4.29) is large as compared to the second term. Along each ray ty_0 , with $|y_0|=1$, $0<t<\infty$, the equation (4.29) assumes the form

$$(4.31) \ \phi(t,y_0)=t^{2-n}+\gamma(ty_0)=\eta^{2-n}, \ \partial_t\phi(t,y_0)=(2-n)t^{1-n}+y_0\cdot\nabla\gamma(ty_0)<0,$$

as t is small. In particular, also in the expression for $\partial_t\phi(t,y_0)$ the first term dominates for small t, since $y_0\nabla\gamma(ty_0) = O(t^{3/2-n})$, near t=0 , by (4.30). Let

$$(4.32) \qquad \sigma_0(t)=\sup\{\tau^{n-5/2}|\gamma(\tau y_0)|, \tau^{n-3/2}|y_0\nabla\gamma(\tau y_0)|/(n-2): \tau\leq t\} ,$$

and define $\lambda_\pm(t) = t^{-1/2}\pm\sigma_0(t)$. Then $t^{1/2}\lambda_\pm(t)\to1$, as $t\to0$, and

$$(4.33) \qquad \begin{gathered} t^{5/2-n}\lambda_-(t) \leq \phi(t,y_0) \leq t^{5/2-n}\lambda_+(t) , \\ (n-2)t^{3/2-n}\lambda_+(t) \leq -\partial_t\phi(t,y_0) \leq (n-2)t^{3/2-n}\lambda_-(t) . \end{gathered}$$

Choose a $t_0>0$ such that $|1-t^{1/2}\lambda_\pm(t)|\leq1/2$, for $0<t\leq t_0$. Then the function ϕ decreases with t, for every fixed y_0 . Hence there will be precisely one root $t\in(0,t_0)$ of (4.31) for each y_0 , whenever $\eta^{2-n}>(3/2)t_0^{2-n}$, i.e., $0<\eta<(2/3)^{1/(n-2)}t_0$. Moreover, then

(4.33) implies that $(1/2)\tau^{2-n} \le \phi(t,y_0) = \eta^{2-n} \le (3/2)t^{2-n}$. Or,

$$(4.34) \qquad (1/2)^{1/(n-2)}\eta \le t \le (3/2)^{1/(n-2)}\eta \ .$$

This defines a surface C_η: $t=t_\eta(y_0)$, contained in the spherical shell (4.34) . Combining (4.34) with the second (4.33) we get

$$(4.35) \qquad 2/9(n-2)\eta^{1-n} \le -\partial_t\phi(ty_0) \le 6(n-2)\eta^{1-n} \ .$$

This implies $\nabla\phi \ne 0$ on C_η , so that C_η is smooth. Also we get C_η inside $C_{\eta'}$, as $\eta<\eta'$. Notice that the 'angular derivatives' of the function $\phi(y)$ are $O(\eta^{3/2-n})$, by (4.30), and (4.34), since the first term is constant along the spheres $|y|$ = const. By (4.35) the radial derivative is bounded below by $c\eta^{1-n}$. Accordingly the unit normal of C_η (which is parallel to the gradient) has its radial component of the form

$$(4.36) \qquad n_r(y) = \pm 1 + O(|y|^{1/2}) \ , \text{ as } |y|\to 0 \ .$$

Now we consider the integral

$$(4.37) \qquad I_\eta = \int_{W_\eta} (g(x_0,y)-\delta\eta^{2-n})Hu(y)d\mu_y \ ,$$

for small η , and some C^∞-function u , where W_η denotes the interior of $C_{x_0,\eta}$. With a cutoff function $\psi=1$ near x_0 , $=0$ near C_η , write $u=u\psi+u(1-\psi)=u_1+u_2$, and, correspondingly, $I_\eta=I_{1,\eta}+I_{2,\eta}$. Clearly we get $I_{1,\eta} = u(x_0) - \delta\eta^{2-n}\int qu_1 d\mu$. On the other hand, setting $\omega(y) = g(x_0,y)-\delta\eta^{2-n}$, for a moment,

$$(4.38) \qquad \begin{aligned} I_{2,\eta} &= \int \omega q u_2 d\mu - \int dy \partial_{y^j}(\kappa h^{jk} u_{2|y^k})\omega \\ &= \int H\omega u_2 d\mu + \int_{C_\eta} dS_y \kappa(y) h^{jk}(y)\nu_k(y)\omega_{|y^j} u_2(y) \ . \end{aligned}$$

with the surface element dS and the exterior normal components ν_j. Noting that $H\omega u_2 = -\delta\eta^{2-n}qu_2$, and assuming that $Hu=0$, we conclude that

$$(4.39) \qquad \begin{aligned} &0 = I_\eta = u(x_0) + \int_{C_\eta} dS \, p_\eta u - \delta\eta^{2-n}\int_{W_\eta} qud\mu \ , \\ &\text{with} \quad p_\eta(y)=\kappa(y)h^{jk}(y)g_{|y^j}(x_0,y)\nu_k(y) \ . \end{aligned}$$

Write the surface integral in (4.39) as an integral over the unit sphere $S^{n-1}=\{|y_0|=1\}$, with the surface element dS_1 of S^{n-1}, noting that $dS/dS_1 = |y|^{n-1}/n_r(y)$, with n_r of (4.36). We get

$$(4.40)\quad u(x_0)=\delta\eta^{2-n}\int_{W_\eta} qud\mu-\int_{S^{n-1}}|t_\eta(y)|^{n-1}p_\eta(t_\eta(y))u(t_\eta(y))/n_r dS_1 .$$

Here we introduce the function $\alpha_\eta(z) = |z|^{n-1}p_\eta(z)/n_r(z)$, and observe that $\alpha(z)$ is bounded and bounded below by positive constants independent of η and z. Let us multiply (4.40) by η^{n-1} and integrate this $d\eta$, from 0 to $\eta_0=(2/3)^{1/(n-2)}t_0$:

$$(4.41)\quad \eta_0^n u(x_0) = P - \int_{S^{n-1}} dS_{1,y^0}\int_0^{\eta_0} \eta^{n-1}d\eta(\alpha u)(t_\eta(y_0)) ,$$

with $P = \delta\int_0^{\eta_0} \eta\int_{W_\eta} qud\mu$.

In the inner integral of (4.41) make the substitution $t = t_\eta(y_0)$ of integration variable. The derivative $t'_\eta = \partial_\eta t_\eta(y_0)$ may be obtained by differentiating (4.31) for η. It follows that

$$(4.42)\quad t'_\eta((2-n)t^{1-n} + y_0\nabla\gamma(ty_0)) = (2-n)\eta^{1-n} .$$

Clearly this implies that t'_η is bounded and bounded below by positive constants independent of t,η, and u. Therefore the integral in (4.41) may be written in the form

$$(4.43)\quad \int_{W_{\eta_0}} dy\beta(y)u(y) ,\quad c_1\leq\beta(y)\leq C_1 ,$$

with constants c_1, C_1 independent of η and u. On the other hand the integral P above may be estimated by

$$(4.44)\quad O(\eta_0^2\int_{W_{\eta_0}} dy|u(y)|) .$$

We have proven the following result.

<u>Theorem 4.5</u>. Let the 'local balls' $B(x_0,\eta)$ be defined by

$$(4.45)\quad B(x_0,\eta) = \{x\in\Omega : d(x,x_0)\leq\eta\} ,$$

where $d(x,y)$ denotes the geodesic distance from x to y, with respect to $ds^2=h_{jk}dx^j dx^k$. For every compact subset K of Ω there exist positive constants ε_0,c,C, which generally may depend on the triple $\{\Omega,d\mu,H\}$, and on K, but do not depend on u, such that, for every nonnegative solution of the differential equation $Hu=0$, and every $\varepsilon>0$ with $\varepsilon\leq\varepsilon_0$ we have the inequality

(4.46) $$c\varepsilon^{-n}\int_{B(x,\varepsilon/2)} u\,d\mu \le u(x) \le C\varepsilon^{-n}\int_{B(x,2\varepsilon)} u\,d\mu \ , \ x \in K \ .$$

Indeed, our construction gives such inequality with the two 'balls' of integration replaced by the set W_η above. Moreover, it follows from (4.34) that $B(x,\varepsilon/2) \subset W_{\eta,x} \subset B(x,2\varepsilon)$, for suitably smaller ε_0 , which implies (4.46) . Also, it is evident from our construction that each point $x_0 \in \Omega$ has a neighbourhood N_{x_0} such that the constants c, C, ε_0 may be choosen independent of x, as long as $x \in N_{x_0}$. In particular, our construction was carried out in a special coordinate system, for each fixed x. Hence we must involve a coordinate transform to establish the special coordinates, but the coefficients involved then are locally continuous in x, and all conditions are locally uniform. Since a compact set K may be covered by finitely many such neighbourhoods N_{x_j} , we thus can choose our constants to be valid in K,q.e.d.

5. Harnack inequality; Dirichlet problem; Maximum principle.

In all of this section let H denote a second order strongly elliptic differential expression of the form (1.2), with real C^∞-coefficients, and q(x) of any sign, not necessarily satisfying (1.4). Actually, formal self-adjointness is not required for thm. 5.1 nor thm.5.8. In fact, both results are known to be true under still more general assumptions not important for us (cf. [GT_1]) .

In preparation for some results of sec.11 we next will discuss a proof of the Harnack inequality, using the inequality of thm.4.5.

Theorem 5.1. (Harnack's inequality). For every compact subset $K \subset \Omega$ there exists a constant $c = c_K$ such that, for every nonnegative solution $u \in C^\infty(\Omega)$ of Hu=0 we have

(5.1) $$\sup\{u: x \in K\} \le c \inf\{u: x \in K\} \ .$$

Remark. It is known that the constant c may be choosen to depend only on the dimension n, the ellipticity constant (i.e. the local lower bounds of $h^{jk}\xi_j\xi_k/(\delta^{jk}\xi_j\xi_k)$, in a given fixed countable atlas), on local bounds for all coefficients, and the choice of Ω and K (cf. Gilbarg-Trudinger [GT_1]). We will not require (nor prove) this much stronger result here, however.

Proof. First assume that $K=B(x_0,\varepsilon)$, as in (4.45) for some $x_0\in\Omega$, and that $B(x_0,16\varepsilon)\subset\Omega$ as well, where $8\varepsilon\le\varepsilon_0$, with ε_0 as in thm.4.5. A solution u of Hu=0 assumes its maximum and its minimum at x_M and $x_m \in K$, respectively. Then we use (4.46) for

$$(5.2)\quad u(x_M) \le C\varepsilon^{-n}\int_{B(x_M,2\varepsilon)} u d\mu \le C\varepsilon^{-n}\int_{B(x_m,4\varepsilon)} u d\mu \le 8^n C/c\ u(x_m) ,$$

using that $d(x_m,x_M) \le 2\varepsilon$, hence $B(x_M,2\varepsilon)\subset B(x_m,4\varepsilon)$. Accordingly (5.1) follows with the constant c_K given by $8^n C/c$, where c and C are the constants of thm.4.5.

Now consider a general compact subset $K\subset\Omega$. It is no loss of generality to assume that K^{int} is connected, and $(K^{int})^{clos}=K$, since a larger K with these properties always can be found. The constant c_K will also be valid for the given K, since the sup increases and the inf decreases, as K is enlarged.

Again the solution u assumes its maximum at x_M and its minimum at x_m, and these points may be connected by a path Γ within K^{int} . Then a finite sequence of points $x_j\in\Gamma$, j=1,...,N, may be found such that $x_0=x_M$, $x_N=x_m$, and that the intersections $B(x_j,\varepsilon)\cap B(x_{j+1},\varepsilon)$, j=1,...,N-1 all are nonvoid, while ε is choosen to satisfy the above conditions. Accordingly, letting $B_j= B(x_j,\varepsilon)$, we conclude that

$$(5.3)\quad u(x_M) \le \sup_{B_0} u \le \gamma \inf_{B_0} u \le \gamma \sup_{B_1} u \le \dots \le \gamma^N \inf_{B_r} u \le \gamma^N u(x_m),$$

with the constant $\gamma=8^n C/c$ of (5.2). This completes the proof.

Proposition 5.2. Let $\Omega' \subset \Omega$ be a subdomain (an open connected subset) of Ω . If $u \in C^\infty(\Omega')$ solves $Hu = \lambda u$, $x \in \Omega'$, with $\lambda \in \mathbb{R}$, and u is nonnegative in Ω' then it must be either positive in all of Ω' or identically zero in Ω' .

Proof. Follows from Harnack's inequality.

Next we will make some comments about the Dirichlet problem of the differential expression H in a subdomain $\Omega' \subset \Omega$ with regular boundary, in the sense of sec.2. That is, we assume $\Omega'\cup\partial\Omega'$ compact , and $\partial\Omega'$ a finite union of (connected) compact n-1-dimensional submanifolds of Ω . We also allow the case of $\partial\Omega=\emptyset$. Note that the assumptions of cor.3.9 are satisfied, so that the Friedrichs extension H, (i.e., the closure of the the Dirichlet operator H_d) has compact resolvent. It follows that the spectrum of $H=H_d^{**}$ is discrete. There exists an orthonormal basis

$\{\psi_j, j=1,\dots\}$ of $H'=L^2(\Omega')$ consisting of eigenvectors of H , such that $H\psi_j = \lambda_j\psi_j$, with corresponding eigenvalues

(5.4) $$\lambda_0 \le \lambda_1 \le \dots , \lambda_j \to \infty , \text{ as } j \to \infty .$$

Proposition 5.3. All eigenvectors are in $C^\infty(\Omega'\cup\partial\Omega')$. If $\lambda\in\mathbb{C}$, $\lambda\ne\lambda_l$, $l=0,1,2,\dots$, and if $\partial\Omega'\ne\emptyset$, then the Dirichlet problem

(5.5) $$u \in C^\infty(\Omega'\cup\partial\Omega') , Hu = \lambda u , x \in \Omega'\cup\partial\Omega' , u|\partial\Omega' = u^0 ,$$

is uniquely solvable for all $u^0 \in C_0^\infty(\partial\Omega')$.

Proposition 5.4. For the smallest eigenvalue λ_0 of H_d we have

(5.6) $$\lambda_0 = \text{Min } \{J(u) : u \in H_1' \setminus \{0\} \}, \text{ where } J(u) = \|u\|_1'^2/\|u\|'^2.$$

Here $\|u\|_1'$ and $\|u\|'$ denote the norms $\|u\|_1$ and $\|u\|$, for Ω' .

Proofs. Since H is of compact resolvent we find that λ is not in the spectrum of H , so that the resolvent $(H-\lambda)^{-1}$ exists, as a bounded operator of H'. Hence the equation $(H-\lambda)u = f$ is uniquely solvable, with $u \in \text{dom } H$, for all $f\in H'$. Again, if $f \in C^\infty(\Omega'\cup\partial\Omega')$, then we conclude that $u\in C^\infty(\Omega'\cup\partial\Omega')$, using cor.2.2 . In particular this also implies that the eigenvectors are $C^\infty(\Omega'\cup\partial\Omega')$.

For a given $u^0 \in C^\infty(\partial\Omega')$, if $\partial\Omega'$ is nonvoid, it is easy to construct $v \in C^\infty(\Omega')$ with $v|\partial\Omega' = u^0$. Just extend u^0 a little distance inside of Ω', carrying it to zero,using a partition, etc. then the problem (5.5) is equivalent to

(5.7) $$(H-\lambda)w = -(H-\lambda)v = f ,$$

where $w = u - v$, evidently. Since $(H_d-\lambda)w=f$ is uniquely solvable for $f \in C^\infty(\Omega)$ proposition 5.3 follows.

For prop.5.4 we first notice that, for $u \in C_0^\infty(\Omega')$,

(5.8) $$(u,u)_1' = (u,H_0u)'=(u,Hu)'= \textstyle\sum_{j=0}^\infty \lambda_j|a_j|^2 , a_j = (\psi_j,u),$$

Taking closure in H_1' one confirms equality of the first and fourth term in (5.8) for general $u \in H_1'$. We know that $H_1' \subset H'$. Hence the functional J(u) of (5.6) is well defined over H_1' , and

(5.9) $$J(u) = \textstyle\sum\lambda_j|a_j|^2/\sum|a_j|^2 \ge \lambda_0 = J(\psi_0) .$$

This completes the proof of prop.5.4.

Proposition 5.5. Any $u \in H_1' \setminus \{0\}$ which minimizes the functional

$J(u)$ is in dom H_d ,and is an eigen function of H_d to the minimal eigenvalue λ_0 .

Proof. For a given function u with the assumed properties and any $w \in$ dom H let $\theta(t) = J(u+tw)$ .This is a well defined differentiable function,since $u+tw \in H_1' \setminus \{0\}$ for small t .Using the fact that u minimizes J we conclude that

$$(5.10) \qquad dJ/dt(0) = 2 \operatorname{Re} ((u,w)_1'/(u,u)' - (u,w)'/(u,u)'^2) = 0 .$$

Or, replacing w by $we^{i\upsilon}$ with suitable real υ , it is found that 'Re' in the last equation (5.10) may be omitted. It follows that

$$(5.11) \quad (u,w)_1' = (u,Hw)' = \lambda(u,w)' ,\lambda = J(u) ,\text{for all } w \in \text{dom } H.$$

This implies that λ is an eigenvalue of H ,hence of H_d ,by prop.5.3. It is clear,from prop.5.4 ,that λ is the smallest eigenvalue of H_d , q.e.d.

Proposition 5.6. If u is a real-valued eigenfunction to the smallest eigenvalue λ_0 of H_d ,then either $u \equiv 0$ or $u \neq 0$ in all of Ω'.

Proof. In view of prop.5.2 it is sufficient to show that u cannot assume both positive and negative values. If the latter were true then the function $v(x) = \operatorname{Max} \{u(x),0\}$ vanishes identically in some open subset of Ω' ,but does not vanish on all of Ω' .We claim that v is in H_1' ,and that it minimizes $J(u)$ .This,if proven, implies that $v \in$ dom H_d ,and $Lv = \lambda_0 v$,a contradiction because v then will have to vanish identically,in view of prop.5.2.

We require the following special case of Sard's theorem (cf. [Ml_1] , [GP_1]) .

Proposition 5.7. For any real-valued $C^\infty(\Omega)$-function f the set of all critical values (i.e. values $\eta \in \mathbb{R}$ such that $\eta=f(x)$ at some $x \in \Omega$ with $\nabla f(x)=0$) has Lebesgue measure 0 on the real line $\mathbb{R}$.

Using prop.5.7 we conclude the existence of a decreasing sequence $\{\varepsilon_j\}$,$\varepsilon_j \to 0$,such that the level sets $u(x) = \varepsilon_j$ do not contain a critical point ,for each j . Indeed the nulset of all critical values of $u(x)$ cannot contain any interval $(0,\varepsilon)$, $\varepsilon>0$. It follows from the implicit function theorem that the sets $\partial\Omega'_{\varepsilon_j} = \{x \in \Omega' : u(x) = \varepsilon_j\}$ are finite unions of smooth compact n-1-dimensional submanifolds of Ω'. In particular, $\partial\Omega'_{\varepsilon_j} \cap \partial\Omega' = \emptyset$, since $u=0\neq\varepsilon_j$ on $\partial\Omega'$. (In case of infinitely many components there would have to be a limit point in the interior of the compact set $\Omega' \cup \partial\Omega'$, which would make ε_j a critical value.) It follows that the set $\Omega'_{\varepsilon_j} \cup \partial\Omega'_{\varepsilon_j}$, with $\Omega'_{\varepsilon_j} = \{x \in \Omega' : u(x) > \varepsilon_j\}$ is a finite disjoint union of compact manifolds with boundary. Accordingly we may integrate by parts in these sets to obtain

$$(5.12)\qquad \int_{\Omega''} h^{jk} \bar{u}_{|x^j} u_{|x^k} d\mu = -\int_{\Omega''} (\bar{u}-\varepsilon_1)\kappa^{-1}(\kappa h^{jk} u_{|x^j})_{|x^k} d\mu$$
$$= \int_{\Omega''} (u-\varepsilon_1)(\lambda_0 - q) u \, d\mu \qquad , \ \Omega'' = \Omega'_{\varepsilon_1} \ .$$

Or, setting $\Omega_+ = \cup\Omega'_{\varepsilon_1}$, we get

$$(5.13)\qquad (v,v)'_1 = \int_{\Omega_+} (h^{jk} \bar{u}_{|x^j} u_{|x^k} + q|u|^2) d\mu = \lambda_0 (v,v)' < \infty \ .$$

This indeed shows that $v \in H'_1$ minimizes J, q.e.d.

Finally, in this section we formulate and prove a maximum principle for second order elliptic expressions, needed below. Again we will not require the strongest known results (cf. GT_1]).

<u>Theorem 5.8</u>. (Maximum principle) Assume a differential expression of the form (1.2), with $q(x) \geq 0$. Let $u \in C^2(\Omega)$ be a solution of the differential inequality

$$(5.14)\qquad Hu \leq 0 \ , \ x \in \Omega \ .$$

Then, if u assumes a positive maximum at some point of Ω, or if $q(x) \equiv 0$, we have $u = \text{const}$.

<u>Proof</u>. Assume that $u_M = u(x_0) \geq u(x)$ for all $x \in \Omega$, and some $x_0 \in \Omega$. Let $u_M > 0$. Since we have $q \geq 0$, and since $\nabla u = 0$ for $x = x_0$, it then follows from (5.14) that

$$(5.15)\qquad h^{jk} u_{|x^j x^k} \geq 0 \text{ at } x = x_0 \ .$$

At a maximum we also have the 'Hessian matrix' $((u_{|x^j x^k})) = M$ negative semi-definite, as well known. On the other hand, it is no loss of generality to assume that $h^{jk}(x_0) - \delta^{jk}$, by virtue of a suitable linear coordinate transform. Then (5.15) amounts to the condition that the trace of the matrix M is nonnegative. One thus concludes that the Hessian matrix must be zero at x_0. A contradiction results if we even have $\Delta_\mu u > 0$ at x_0, since then it is implied that the trace of a negative semi-definite matrix is positive.

A proof of the maximum principle results by proving that a suitable modification of u indeed results in a function w with $Hw>0$ at its maximum. The construction, below, has been adopted from Protter-Weinberger [PW_1] , p61f.

Suppose we have $u(x_1)<u_M$ at some $x_1\in\Omega$. Connect x_1 and x_0 by a path Γ , and let x_2 be the first point, coming from x_1 , where $u=u_M$. Choose a chart for x_2, and let B be a ball with center x_2 , and radius d, in that chart. On the part between x_1 and x_2 of Γ choose a point x_3 with distance from x_2 less than d/2. Let B_1 be the largest open ball with center at x_3 such that $u<u_M$ in B_1 . Clearly B_1 has radius $<d/2$, hence is entirely in the given chart. Also there must be a point ξ at the boundary of B_1 where $u=u_M$. Finally let K be a ball inside of B_1 tangent to B_1 at ξ , and with smaller radius, and let y be its center.

As a result of this construction we have a ball K (of radius r and center y) such that $u=u_M$ at precisely one point ξ of ∂K, while $u<u_M$ in all of $K\cup\partial K\setminus\{\xi\}$. Also $K\cup\partial K$ is contained in a ball around y , twice its radius, contained in a chart (Fig.5.1). Let $0<r_1<r$, and let K_1 denote the ball around ξ with radius r_1. Since $u_M>0$ we can choose r_1 so small that $u\geq 0$ in K_1 , so that $\Delta_\mu u\geq 0$ in K_1. The boundary ∂K_1 is the union of a closed arc C_1' in the set $K\cup\partial K$, and an open arc C_1'' outside K . Note that $\max\{u:x\in C_1'\} = u_M-\zeta<u_M$, since C_1' is compact, and $u<u_M$ on C_1'

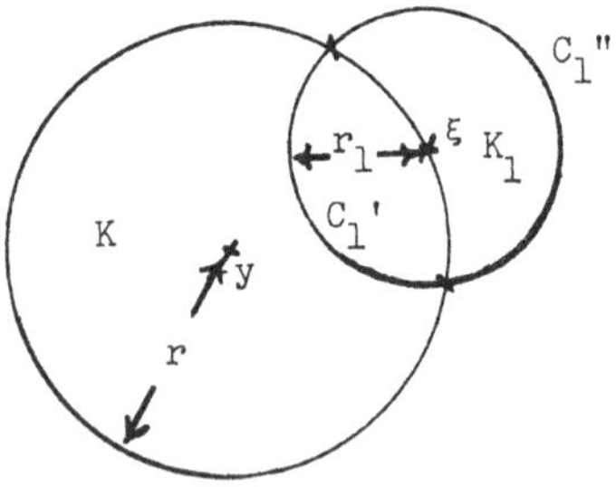

Fig.5.1

Now introduce the auxiliary function

$$z(x) = e^{-\alpha|x-y|^2} - e^{-\alpha r^2} , \tag{5.16}$$

with $\alpha>0$ to be determined. Confirm that

$$\Delta_\mu z = e^{-\alpha|x-y|^2}\{4\alpha^2 h^{jk}(x-y)_j(x-y)_k-2\alpha(h^{jj}+b^j(x_j-y_j))\}, \tag{5.17}$$

where, for a moment we set $\Delta_\mu = h^{jk}\partial_{x^j}\partial_{x^k}+b^j\partial_{x^j}$, and where the coefficients h^{jk} , b^j all are taken at x . By choosing α large we can arrange for $\Delta_\mu z>0$ in $K_1\cup\partial K_1$, using ellipticity of H .

Finally define $w = u + \varepsilon z$, with $0<\varepsilon<\zeta(1-e^{-\alpha r^2})^{-1}$. Confirm that w has the following properties.

(i) $w<u_M$ on C_1', due to $\varepsilon z < \varepsilon(1-e^{-\alpha r^2}) < \zeta$.

(ii) $w<u_M$ on C_1'' , since $z<0$ there, and $u\le u_M$ everywhere.

(iii) $w=u_M$ at ξ , since $z=0$ there, but $u=u_M$.

It follows that there must be a maximum of w inside the ball K_1, while we conclude that $\Delta_\mu w = \Delta_\mu u + \varepsilon\Delta_\mu z>0$ in K_1. Thus we get the contradiction mentioned above, and the theorem is established.

6. Change of dependent variable; Normal forms; Positivity of g.

Let us return to the configuration of section 1, of a triple $\{\Omega,d\mu,H\}$ of a measure $d\mu = \kappa dx$, and an elliptic self-adjoint expression H of the form (1.2) on a manifold Ω . The presence of the tensor h^{jk} on Ω suggests introduction of the Riemannian metric

$$ds^2 = h_{jk}dx^j dx^k \quad , \quad ((h_{jk})) = ((h^{jk}))^{-1} \tag{6.1}$$

on Ω . Then one also has the corresponding surface measure

$$dS = \sqrt{h}\, dx \quad , \quad h = \det\, ((h_{jk})) = (\det\, ((h^{jk}))\,)^{-1} \quad , \tag{6.2}$$

and the 'Beltrami-Laplace expression'

$$\Delta = \sqrt{h}^{-1}\partial_{x^j}\sqrt{h}\, h^{jk}\partial_{x^k} \quad , \tag{6.3}$$

which is self-adjoint with respect to the measure dS . The choice of $H = -\Delta + 1$, $d\mu = dS$, for a general Riemannian space Ω with positive definite metric (6.1) will give a configuration meeting (1.4). Then the Friedrichs extension H will be called the Laplace comparison operator.

Two triples $\{\Omega,d\mu,H\}$ will be called equivalent if they can be transformed into each other by a "transformation of dependent variable". To explain this in more detail, if we set $u=\gamma u^\sim$, $v=\gamma v^\sim$, with a positive $C^\infty(\Omega)$-function γ then the inner product (1.1) assumes the form $(u,v) = \int_\Omega \bar{u}^\sim v^\sim \gamma^2 d\mu = (u^\sim,v^\sim)^\sim$. Also, the equation $Hu=v$ may be written as $\gamma^{-1}H\gamma u^\sim = v^\sim$. Accordingly we introduce

$$d\mu^\sim = \gamma^2 d\mu \quad , \quad H^\sim = \gamma^{-1}H\gamma \quad , \tag{6.4}$$

as a new measure and second order differential operator, and the triple $(\Omega,d\mu^\sim,H^\sim)$ will be called equivalent to $(\Omega,d\mu,H)$. (We then write $(\Omega,d\mu,H)\sim(\Omega,d\mu^{\sim},H^{\sim})$.) It is clear then that the linear map $u^{\sim}\to\gamma u^{\sim}$ defines an isometry $L^2(\Omega,d\mu^{\sim})\to L^2(\Omega,d\mu)$. Moreover, we get

$$(6.5)\quad (u,v)_1=(u,Hv)=(u^{\sim},\gamma^2(\gamma^{-1}H\gamma)v^{\sim})=\int_\Omega \bar{u}^{\sim}H^{\sim}v^{\sim}d\mu^{\sim}=(u^{\sim},v^{\sim})_1^{\sim},$$

for $u,v\in C_0^\infty(\Omega)$, where $(.,.)_1^{\sim}$ is the inner product (1.3) for $H^{\sim}$.

Since $(\Lambda^{-1}|C_0^\infty)^{**}=\Lambda^{-1}$, and the corresponding for $\Lambda^{\sim}=H^{\sim-1/2}$ is true, (6.5) implies that $u^{\sim}\to\gamma u^{\sim}$ also defines an isometry $\tilde{H}_1\to H_1$.

Furthermore $H^{\sim}=\gamma^{-1}H\gamma$, of course, is a second order differential operator with principal part $-h^{jl}\partial_{x^j}\partial_{x^l}$. We get $(H^{\sim}u,v)^{\sim}$ $=(\gamma^{-1}H\gamma u,\gamma^2v)=(\gamma^2u,\gamma^{-1}H\gamma v)=(u,H^{\sim}v)^{\sim}$, for $u,v\in C_0^\infty(\Omega)$, so that $H^{\sim}$ is formally self-adjoint,with respect to the new inner product. In fact, $H^{\sim}$ again is of the form (1.2), with

$$(6.6)\qquad \kappa^{\sim}=\gamma^2\kappa\ ,\quad q^{\sim}=H\gamma/\gamma=q-(\gamma\kappa)^{-1}(\kappa h^{jk}\gamma_{|x^j})_{|x^k}\ .$$

To verify this we write, for $u,v\in C_0^\infty$,

$$(u,H^{\sim}v)^{\sim}=(\gamma^2u,\gamma^{-1}H\gamma v)=(\gamma u,H\gamma v)=\int(h^{jk}(\gamma\bar{u})_{|x^j}(\gamma v)_{|x^k}+q\gamma^2\bar{u}v)d\mu$$

$$=\int h^{jk}\bar{u}_{|x^j}v_{|x^k}\gamma^2d\mu+\int(h^{jk}\gamma_{|x^j}\gamma_{|x^k}+q\gamma^2)\bar{u}vd\mu+\int h^{jk}\gamma\gamma_{|x^j}(\bar{u}v)_{|x^k}d\mu$$

A partial integration transforms the last integral into

$-\int\kappa^{-1}(\gamma\kappa h^{jk}\gamma_{|x^j})_{|x^k}\bar{u}v\,d\mu$. Accordingly, by a calculation,

$$(6.7)\qquad (u,H^{\sim}v)^{\sim}=\int(h^{jk}\bar{u}_{|x^j}v_{|x^k}+q^{\sim}\bar{u}v)d\mu^{\sim}\quad ,\ u,v\in C_0^\infty(\Omega)\ ,$$

with $q^{\sim}$ of (6.6) , and $d\mu^{\sim}=\kappa^{\sim}dx$, $\kappa^{\sim}$ of (6.6) .Comparing this with (1.3) one concludes that $H^{\sim}=\gamma^{-1}H\gamma$ indeed must have the desired form.

It is natural to use the equivalence defined for introducing a suitable normal form for comparison triples. We shall find two specific normal forms convenient.

Proposition 6.1. Every comparison triple is equivalent to a unique triple of the form

$$(6.8)\qquad (\Omega\ ,dS,-\Delta+q^{\sim})$$

with surface measure dS and Laplace operator Δ of the given metric ds^2 , and with a $C^\infty(\Omega)$-function $q^{\sim}$, explicitly given by

(6.9) $$q^{\sim} = H\gamma/\gamma \quad , \quad \gamma = \{\kappa^2/h\}^{-1/4} \quad .$$

Proof. Note that γ , as defined in (6.9) indeed is a C^∞-function on Ω because the transformation properties of the measure $d\mu$ and the metric h^{jk} imply coordinate independence of $q^{\sim}$ and γ of (6.9), as easily checked. Then the proposition follows from the general facts about the transformation $u = \gamma u^{\sim}$ derived above, noting that

$\kappa^{\sim} = \gamma^2\kappa = \sqrt{h}$,q.e.d.

A triple (6.8) will be said to be in Sturm-Liouville normal form.

We note that the condition $q \geq 1$ imposed on comparison triples so far has only one purpose : It will guarantee that H is semi-bounded below, and, moreover, that $H \geq 1$, implying (1.4). On the other hand, prop. 6.2, below, shows that for a triple with $H \geq 1$ always another normal form can be introduced -i.e. an equivalent triple - such that $q^{\sim}$ is a constant ≥ 1 . In fact, if Ω is noncompact, we always can arrange for $q^{\sim}=1$.

In the following we often no longer require $q \geq 1$, but only that (1.4) holds.

Proposition 6.2. Let $(\Omega, d\mu, H)$ satisfy all assumptions of early sec. 1, especially (1.4), i.e.,(6.10), below, but not necessarily $q \geq 1$.

(6.10) $$(u,Hu) \geq (u,u) \quad , \; u \in C_0^\infty(\Omega) \; .$$

Then, if Ω is noncompact, there exists a positive $C^\infty(\Omega)$-function γ, solving the differential equation $H\gamma = \gamma$ in all of Ω .

Using γ in a transformation of dependent variable $u = \gamma u^{\sim}$, as described, one arrives at an equivalent triple of the form $\{\Omega, d\mu^{\sim}, 1-\Delta_{\mu^{\sim}}\}$ with

(6.11) $$\Delta_{\mu^{\sim}} = \kappa^{\sim -1}\partial_{x^j}\kappa^{\sim}h^{jk}\partial_{x^k} \quad , \; d\mu = \kappa^{\sim}dx \; , \; \kappa^{\sim} = \gamma^2\kappa \quad ,$$

in local coordinates.

On the other hand, if Ω is compact then the equation $Hu=\lambda u$, for some real constant $\lambda \geq 1$ admits a positive solution γ which may be used as above to obtain an equivalent triple with $q^{\sim}=\lambda \geq 1$.

This proposition is an immediate consequence of the facts derived on transformations of dependent variable,as soon as we establish existence of a positive solution γ of the equation $H\gamma=\gamma$. Such a solution will be constructed below. In fact it will become evident that the condition $H \geq 1$ is necessary and sufficient for existence of a positive solution.

The result, below is required for prop.6.2, and perhaps of independent interest. It was shown by the author for an ODE and a compact interval in [CEd], p.65 (in 1957) (also [CS],IV,1.9.), and for elliptic operators of varying generality by Allegretto [Al_1], [Al_2], [Al_3], Piepenbrink [Pp_1], [Pp_2], Moss-Piepenbrink [MP_1], Agmon [Ag_1]. We are indebted to M.Meier for the suggestion to use the Harnack principle, which independently also was used by Agmon [Ag_1] , for the same result.

Theorem 6.3. Given a positive measure $d\mu$ and differential expression H as in (1.2) on a manifold Ω, with real symmetric positive definite h^{jk} , and real-valued q , all C^∞ , but not necessarily $q \geq 1$.
(a) If Ω is noncompact, and the expression H-1 is positive, i.e.,

$$(6.12) \qquad \int (h^{jk}\bar{u}_{|x^j} u_{|x^k} + (q-1)|u|^2)d\mu \geq 0 \ , \text{ for all } u \in C_0^\infty(\Omega) \ ,$$

then there exists a positive solution $u = \gamma \in C^\infty(\Omega)$ of the partial differential equation $Hu = u$.
(b) If Ω is compact then the same is true under the additional assumption that

$$(6.12a) \qquad \inf \{(Hu,u)/(u,u) : u \in C_0^\infty(\Omega) \ , \ u \neq 0\} = 1 \ .$$

(c) Vice versa, regardless whether Ω is compact or not, if a positive C^∞-solution of $Hu = u$, defined over all of Ω , can be found then H-1 is positive (i.e. , (6.12) holds. If Ω is compact then also (6.12a) follows.

Remark. Notice that a positive solution γ of $H\gamma = \gamma$ may be used for a transformation of dependent variable, as in section 1, to take the triple $\{\Omega,d\mu,H\}$ to an equivalent one $\{\Omega,d\mu^\sim,H^\sim\}$, with $q^\sim = H\gamma/\gamma = 1$, using (6.6). Since the new triple satisfies $q^\sim = 1 \geq 1$, we trivially get (1.4), coinciding with (6.12). Moreover, if Ω is compact, then a positive solution of $H\gamma = \gamma$ trivially represents an eigenfunction of H to the eigenvalue 1. Since we know that (6.12) holds we then conclude that

$$(6.12b) \quad 1 = (H\gamma,\gamma)/(\gamma,\gamma) \leq (Hu,u)/(u,u), \text{ for all } u \in C_0^\infty(\Omega) = C^\infty(\Omega).$$

This evidently implies (6.12a) . Accordingly we already have proven (c), using the above transformation of dependent variable.

Proposition 6.4. On a manifold Ω satisfying our general assumptions there exists a $C^\infty(\Omega)$-function ϕ taking positive values on Ω such that $\lim_{x\to\infty}\phi(x) = \infty$, in the sense of sec.3 (or app.A).
Proof. If Ω is compact then the function $\phi \equiv 1$ will satisfy all assumptions. If Ω is non-compact then set $\phi(x) = \sum_{j=1}^{\infty} j\omega_j(x)$, with the partition of unity of app.A. Q.E.D.

Proposition 6.5. If Ω is noncompact then there exists an infinite sequence

$$(6.13)\qquad \begin{gathered}\Omega_1 \subset \Omega_2 \subset \dots \subset \Omega \ , \ \cup\, \Omega_j = \Omega \ , \ \Omega_j \neq 0 \ , \\ \Omega_j = \cup_l\, \Omega_{jl} \ , \ \Omega_{jl} \cap \Omega_{jm} = \emptyset \ , \ l \neq m \ ,\end{gathered}$$

where each Ω_{jl} is a nonvoid open subdomain of Ω with compact closure and nonvoid smooth non-self-intersecting boundary (i.e., $\partial\Omega_{jl}$ is an n-1-dimensional submanifold of Ω). Every sum in (6.13) is finite,and every inclusion in (6.13) is proper. Moreover, each Ω_{jl}^{clos} is a proper subset of some unique $\Omega_{j+1,m}$.
Proof. Using proposition 5.7 on the function ϕ of proposition 6.4 we conclude the existence of a strictly increasing sequence $1 < \eta_1 < \eta_2 < \dots < \eta_j < \dots$, $\eta_j \to \infty$ such that the "level set" $\partial\Omega_j = \{x \in \Omega : \phi(x) = \eta_j\}$ does not contain any point x with $\nabla f(x) = 0$ .Indeed the set of all critical values cannot contain any interval (η,∞) ,so that a sequence $\{\eta_j\}$ must exist. Then define

$$(6.14)\qquad \Omega_j = \{x \in \Omega : \phi(x) < \eta_j\} \ ,$$

so that $\partial\Omega_j$ is the set defined above, containing no critical points. By the implicit function theorem we conclude that $\partial\Omega$ indeed is a compact n-1-dimensional manifold,not necessarily connected. Each Ω_j is nonvoid and each $\Omega_j \cup \partial\Omega_j$ is compact, by construction of ϕ. Thus each connected component is the interior of a manifold with compact closure and smooth (compact boundary). There can be at most finitely many components. (Otherwise a subsequence of components must converge to a point $x^0 \in \partial\Omega_j$, near which $\partial\Omega_j$ is homeomorphic to a connected open piece of $\mathbb{R}^{n-1}$ while it is supposed to contain an infinity of nonconnected boundaries of components.) Clearly each component Ω_{jl} has a nonvoid boundary, since we assume Ω connected. Clearly we get $\partial\Omega_j \subset \Omega_{j+1}$, which implies the remaining statements, q.e.d.

Let us observe next, that prop.5.6 establishes theorem 6.3 for the case of a compact manifold Ω .

Proposition 6.6. Let Ω be noncompact, and let Ω_j be defined as in (6.13) Then , if λ_0^j is the smallest eigenvalue of H_d in Ω_j , the sequence $\{\lambda_0^j\}$ is strictly decreasing , and $\lim \lambda_0^j = \lambda^0 \geq 1$.

Remark: The operator H_d for Ω_j is defined as the direct sum of the corresponding operators for the components Ω_{jl} .

Proof. Notice that the minimal eigenvalue ν of H_d is minimal eigen value for at least one of the connected components Ω_{jl}. But $\Omega_{jl} \subset \Omega_{j+1,m}$, for some m , by prop.6.5, and this is a proper inclusion. If $\psi \not\equiv 0$ is a corresponding (positve) eigen function then define

$$(6.15) \qquad w = \psi \text{ in } \Omega_{jl} \ , = 0 \text{ elsewehere in } \Omega_{j+1,m} \ .$$

Then the conclusion of prop.5.5 may be repeated to show that $w \in H_1$ - for $\Omega_{j+1,m}$, and $J(u) = \nu$, so that the minimal eigenvalue ν' of H_d in Ω_{j+1} cannot be larger than ν . If $\nu=\nu'$ then w will have to be an eigen function of H_d in $\Omega_{j+1,m}$ which is impossible since w as defined in (6.15) vanishes in some open set, by prop.5.2, thus will have to vanish identically. This amounts to a contradiction unless $\nu' > \nu$. Thus we indeed get a strictly decreasing sequence, as j increases. Evidently there is the lower bound 1 as well , q.e.d.

For any j and some $\omega_j \in C^\infty(\partial\Omega_j)$, $\omega_j > 0$ we now solve the Dirichlet problem

$$(6.16) \qquad u \in C^\infty(\Omega_j) \ , \ Hu = u \text{ in } \Omega_j \cup \partial\Omega_j \ , \ u|\partial\Omega_j = \omega_j \ .$$

This is possible, since 1 is less than the smallest eigenvalue of H_d for Ω_j (due to prop.6.6 and prop.5.3). The solution will be called u_j .

Proposition 6.7. The function u_j is positive in all of $\Omega_j \cup \partial\Omega_j$.

This proposition follows exactly as prop.5.6 .

Proof of theorem 6.3. To complete the proof in the noncompact case let us now consider the sequence $\{v_j = \alpha_j u_j : j=1,2,\dots\}$, with the positive reals α_j chosen to normalize v_j according to

$$(6.17) \qquad \sup \{v_j(x) : x \in \Omega_1\} = 1 \ , \ j = 1,2,\dots \ .$$

Using the Harnack inequality (i.e., thm.5.1) we get

$$\text{(6.18)}\qquad \begin{aligned} &\sup\{v_j(x) : x\in\Omega_3\} \le C \inf\{v_j(x) : x\in\Omega_3\} \\ &\le C\inf\{v_j(x) : x\in\Omega_1\} \le C\sup\{v_j(x) : x\in\Omega_1\} = C\ , \end{aligned}$$

where the constant C is independent of j , j = 4,5,... .

Now (6.18) at once implies a corresponding estimate

$$\text{(6.19)}\qquad \int_{\Omega_2} d\mu(q|v_j|^2 + h^{kl}\bar{v}_{j|x^k} v_{j|x^l}) \le C'\ ,\ C' \text{ independent of } j\ .$$

Indeed, with a cutoff function $\chi\in C_0^\infty(\Omega_3)$, $\chi=1$ in Ω_2 , we get

$$\text{(6.20)}\qquad (\chi v,(q-1)v)_{\Omega_3} = (\chi v,\Delta_\mu v)_{\Omega_3} = -\int h^{kl}(\chi v)_{|x^k} v_{|x^l} d\mu\ ,$$

where we set $v=v_j$, for a moment. In the right hand side write $(\chi v)_{|x^k} = \chi v_{|x^k} + v\chi_{|x^k}$, and, for the integral with the second term, integrate by parts again, using that $vv_{|x^l} = (v^2)_{|x^l}/2$.

It follows that

$$\text{(6.21)}\qquad \begin{aligned} &\int_{\Omega_2} h^{kl} v_{|x^k} v_{|x^l} d\mu \le \int_{\Omega_3} \chi h^{kl} v_{|x^k} v_{|x^l} d\mu \\ &= -(\chi v,(q-1)v)_{\Omega_3} - (1/2)(\Delta_\mu\chi, v^2)_{\Omega_3} \le \operatorname{Max}\{|q-1|\chi + |\Delta_\mu\chi|/2 : x\in\Omega_3\}\ . \end{aligned}$$

Clearly this proves (6.19).

In turn, (6.19) implies existence of a subsequence $\{v_{j_k}\}$ converging in $L^2(\Omega_1)$ to a positive solution v of Hv=v. Indeed, with a new cutoff function $\chi_1\in C_0^\infty(\Omega_2)$, $\chi_1=1$ in Ω_1 we conclude that $w_j = \chi_2 v_j \in H_1(\Omega_2)$, and that the norm $\|w_j\|_1$ (with respect to Ω_2) is uniformly bounded in j . By cor.3.9 the imbedding $H_1\to H$ (for Ω_2) is compact. Thus w_j must have a convergent subsequence in $H(\Omega_2)$. Then the restrictions to Ω_1 define a convergent subsequence of v_j in $H(\Omega_1)$. Again we get $(H-1)v_{j_k}=0$ for x with $\chi_1=1$, i.e.,in a fixed neighbourhood Ω'' of Ω_1 . It follows that $(v,(H-1)\phi)_{\Omega''} = \lim(v_{j_k},(H-1)\phi)_{\Omega''}=0$, for all $\phi\in C_0^\infty(\Omega'')$, i.e. v is a weak, hence a strong nonnegative solution of Hv=v .Also Harnacks inequality gives $v_j\ge 1/C$ in Ω_1, since $\sup v_j = 1$. Accordingly v=0 is impossible.

A similar argument may be carried out for each Ω_k , instead of Ω_1 .Using the Cantor diagonal scheme one arrives at a subsequence of v_j converging in every Ω_j . The limit v cannot vanish identically anywhere and will be the desired positive solution of Hv = v . This completes the proof of thm.6.3.

Finally, in this section we will show that (1.4) also implies positivity of the Greens function of the Friedrichs extension H .

Theorem 6.8. Assume that condition (1.4) holds. then the Greens function g(x,y) of the Friedrichs extension H , of the minimal operator H_0 , as constructed in thm.4.4, is positive for all $x,y\in\Omega$, $x\neq y$.

Proof. First we note that it is sufficient to consider the case of q=const.$\geq$1, by virtue of thm.6.3, since we have (1.4). Indeed, the Greens function of the transformed operator $\gamma^{-1}H\gamma=H^{\sim}$ is given by

$$(6.22) \qquad g^{\sim}(x,y) = (\gamma(y)/\gamma(x))g(x,y) \ ,$$

which is positive if and only if g is positive.

Notice that we certainly have g(x,y)>0 as x and y are sufficiently close together, by virtue of the expansion (4.22) . Indeed, the first term in (4.22) is positive and is large in comparison to the other terms. Accordingly the positivity of g follows from the maximum principle (thm.5.8), whenever Ω is compact.

If Ω is noncompact, then we apply prop.6.5, constructing an increasing sequence Ω_j with smooth boundary, and let $H_{d,j}$ be the Dirichlet operator of Ω_j . Let $g_j(x,y)$, $x,y\in\Omega_j$ be the Greens function of $H_{d,j}$. From thm.4.4 we know that $\chi(.)g_j(x_0,.)\in\mathrm{dom}\, H^{**}_{d,j}$,

as $x_0\in\Omega_j$, so that cor.2.2 gives $g_j(x_0,.)\in C^{\infty}(\Omega_j\cup\partial\Omega_j\setminus\{x_0\})$, and $g_j(x_0,y)=0$ as $y\in\partial\Omega_j$. Accordingly we again may use the maximum principle to conclude that $g_j(x,y)\geq 0$ as $x,y\in\Omega_j$, and $x\neq y$, and then Harnack's inequality to show that even $g_j>0$.

Now we conclude the same for g by showing a type of weak convergence $g_j\to g$. Define the sequence of operators $G_j^{\sim}= G_j\chi_{\Omega_j}$, with the characteristic function χ_{Ω_j} of Ω_j , where it is understood that the function $G_j^{\sim}u$, defined in Ω_j is to be extended zero outside Ω_j. Let $v_j=G_j^{\sim}u-Gu$. Note that $v_j\in H_1$, and that even $\|v_j\|_1\le c$, with c independent of j . Indeed, we get $Gu\in\mathrm{dom}\,H$, hence

$$(6.23)\qquad \|Gu\|_1^2 = (Gu,HGu) = (Gu,u) = \|\Lambda u\|^2 \le \|u\|^2 .$$

Similarly one confirms that, with $u_j= u|\Omega_j$,

$$(6.24)\qquad \|G_j^{\sim}u\|_1^2 = \|G_ju_j\|_1^2 \le \|u_j\|^2 \le \|u\|^2 ,$$

so that $\|v_j\|_1\le 2\|u\|=c$.

Next we note that, for $u\in H\cap C^\infty$, we have $Hv_j= HG_j^{\sim}u-HGu = 0$, as $x \in \Omega_j$, and $v_j\in C^\infty(\Omega_j)$ as well. This implies that

$$(6.25)\qquad (v_j,\phi)_1 = (v_j,H\phi) = 0 \text{ , as } \phi\in C_0^\infty(\Omega_j) ,$$

hence we also get

$$(6.26)\qquad \lim_{j\to\infty}(v_j,\phi)_1 = 0 \text{ , for all } \phi\in H_1 .$$

(This follows first for all $\phi\in C_0^\infty$, and then for H_1, since $\|v_j\|_1$ was seen bounded, and C_0^∞ is dense in H_1 .)

Proposition 6.9. We have

$$(6.27)\qquad G_j^{\sim}u \to Gu \text{ in weak operator convergence of } L(H,H_1) .$$

Moreover, we have strong operator convergence (in $L(H,L^2(K))$) of the restrictions $u \to (G_j^{\sim}u)|K$ (to $(Gu)|K$) , for each compact subset $K \subset \Omega$.

The first part of prop.6.9 was derived above. The second part follows since the restriction map $H_1 \to L^2(K)$ is compact, by cor.3.9 , so that a weakly convergent sequence is turned into a strongly convergent sequence,q.e.d.

We will apply only the weak convergence (6.27): Suppose we have $g(x_0,y_0)<0$, for some $x_0\ne y_0$. Then the same holds for $x\in N_{x_0}$, $y\in N_{y_0}$, with neighbourhoods N_{x_0} , N_{y_0} of x_0 and y_0 . We may assume that $N_{x_0}\cap N_{y_0} = \emptyset$. For any $\phi\subset C_0^\infty(N_{x_0})$ $\phi\ge 0$, we conclude $(G\phi)(y)\le 0$, as $y\in N_{y_0}$, and we get "<" whenever ϕ is not $\equiv 0$. Accordingly we have $(\phi,Gu) <0$ whenever $u\in C_0^\infty(N_{y_0})$, $u\ge 0$, and both ϕ, u are not $\equiv 0$. On the other hand it is clear that $(\phi,G_j^{\sim}u)\ge 0$. Since the weak convergence in H_1 implies weak convergence in H, we get a contradiction. Accordingly we must have $g(x,y)\geq 0$, and Harnack's inequality then implies $g>0$. This completes the proof of thm.6.8.

CHAPTER 4. ESSENTIAL SELF-ADJOINTNESS OF THE MINIMAL OPERATOR.

In this section we will discuss a variety of criteria, all establishing essential self-adjointness of the minimal operator H_0 , for an expression H of the form III,(0.1), or one of its powers. In other words, we seek to establish the definite case, in the language of Carleman. This is a classical subject, with a large list of contributions, too numerous to be discussed in detail (cf. the bibliography in [F_1]).

Our selection, below, is guided by the requirements in chapter's 5f. In particular we only consider the case of an expression H bounded below.

All proofs offered are elementary. However we will require the statement of Weyl's lemma for expressions H as well as for their powers H^m , m=1,2,.... . While the first was discussed in III,1, the corresponding property for powers can be easily derived with the same proof, using the explicit Green's function we have established for the operators H . Clearly H^m has the m-th iterate of the integral kernel g(x,y) as its Green's function, and one easily analyzes the character of its singularity. Then the proof of III,thm.1.2 may be repeated.

Actually, the latter argument is not required if distribution calculus is used: It is a trivial fact that products of of hypo-elliptic operators are hypo-elliptic again. The proof of III,thm.1.2 is easily rewritten to show that the expression H there is hypo-elliptic. Hence H^m , for m = 1,2,3,... , all all must be hypo-elliptic. Hence Weyls Lemma holds for H^m . The only objection to this argument is that distribution calculus must be used, which we have not introduced here (cf.[C_1]).

Also, since the result is local, one may use a result in $\mathbb{R}^n$, for an expression H equal to $1-\Delta$ outside a compact set. Such a result is easily derived, even for general elliptic expressions of arbitrary order, using a Green inverse of the form introduced in [C_1],IV,3. This indicates a third alternative for proving Weyl's lemma for H^m. Accordingly, and since the result is generally well known and contained in standard texts on elliptic partial differential equations, we will not discuss a detailed proof.

1. Essential self-adjointness of powers of H_0.

In this section we turn to the question about criteria to insure Carleman's definite case for an expression H of the form III,(1.2). We will not only look for H itself, but also for its powers which will be needed in chapter 5. It was shown by Gaffney [Gf_1], and Roelcke [Ro_1] that the Laplace operator Δ on a complete Riemannian manifold always is in the definite case.

This was generalized to all powers Δ^m, m=1,2,..., by the author [CWS], together with a more general result on expressions H of the form (1.2) (cf. thm.1.1, below). Chernoff [Ch] gave a very elegant proof for the powers Δ^m on a complete manifold, using as an essential ingredient that the first order hyperbolic system $\partial u/\partial t = (d+\delta)u$, with the operators d and δ acting on the differential form u, has finite propagation speed, so that a solution of the Cauchy problem with C_0^∞-initial-values will be C_0^∞ for all t. We shall discuss Chernoff's proof in [C_3], in connection with theory of hyperbolic equations.

On the other hand, for the special case of $\Omega=\mathbb{R}^n$, $d\mu=dx$, m=1, there is a multitude of results. We mainly will be interested in the case of a semi-bounded H_0, and, moreover, $q\geq 0$. The most general set of results perhaps are those of J.Frehse [F_1], some of which we will discuss in sec.4 (cf. thm.1.9). Notice that the case of n=1, under these assumptions, does not require any additional assumptions (cf.thm.1.10). For many other contributions to this subject we refer to the bibliography in [F_1], c.f. also Devinatz [Dv_1] .

As in III,1 we work with the expression

$$(1.1) \qquad H = -\Delta_\mu + q \ , \ \Delta_\mu = 1/\kappa \partial_{x^j} \kappa h^{jk} \partial_{x^k} \ , \ d\mu = \kappa dx \ ,$$

Theorem 1.1. Let us assume that there exists a non-negative Lipschitz continuous function σ, defined over Ω, and satisfying the following conditions

(i) $\sigma(x_0) = 0$ for some given fixed $x_0 \in \Omega$;

(ii) $\lim_{x\to\infty} \sigma(x)=\infty$, (i.e., for N>0 there exists a compact set K_N such that $\sigma(x) > N$, as $x \in \Omega\setminus K_N$);

(iii) $|\nabla\sigma|^2 \le q$, $x \in \Omega \setminus K$, with some compact set $K \subset \Omega$.

$(\text{iv})_\eta$ $|\nabla\sigma| \le \zeta e^{\eta\sigma}$, $x \in \Omega$, with constants $\eta,\zeta \ge 0$.

Then there exists a nonincreasing sequence η_m , m=1,2,... , such that (i),(ii),(iii),and $(\text{iv})_{\eta_m}$ implies essential self-adjointness of $H_0{}^m = (H^m)_0$ (i.e., the definite case for the expression H^m.) Moreover, η_1 may be choosen as any number with $0 < \eta_1 < 1$.

Remark 1.2. Notice that the conditions (i), (ii), (iii), $(\text{iv})_\eta$ of thm.1.1 are equivalent to the following alternate conditions:

There exists a positive Lipschitz-continuous function $\tau(x)$, $x \in \Omega$, such that

(i') $\tau(x_0)=1$ for some fixed $x_0 \in \Omega$;

(ii') $\lim_{x\to\infty}\tau(x) = 0$;

(iii') $\nabla\tau(x)$ is bounded on Ω ;

$(\text{iv}')_\eta$ $q(x) \ge \eta^{-2}|\nabla\tau(x)|^2/\tau^2(x)$, for all $x \in \Omega\setminus K$, with some compact set K .

Indeed, (i), (ii), (iii), $(\text{iv})_\eta$ translate into (i'), (ii'), (iii'), $(\text{iv}')_\eta$, if we set $\tau(x)=e^{-\eta\sigma(x)}$, as shown by a calculation.

Remark 1.3. Notice that the conditions of the thm. imply semiboundedness of H_0 , because we will have $q \ge c_0$ in the compact set K , and hence q bounded below in all of Ω , since (iii) implies $q \ge 0$ outside K .

Our proof of thm.1.1 is elementary, but quite technical. It will be discussed in the following sections. Presently let us discuss the theorem and some of its consequences.

Proposition 1.4. For a semi-bounded H_0 , if $(H_0+c_0) \ge 1$, and $(H_0+c_0)^m$ is e.s.a., for some real c_0, then $(H_0+c)^k$ is e.s.a., for all $c \in \mathbb{R}$, and k=1,...,m .

Proof. Let $C' = (H_0+c_0)^m$, $B'=H_0+c_0$, for a moment, and let C be the unique self-adjoint extension of C'. Clearly $B' \ge 1$ implies $\|u\|^2 \le (u,B'u) \le \|u\|\|B'u\|$, or, $\|B'u\| \ge \|u\|$, $u \in C_0^\infty$, so that $\|u\|^2 \le \|B'u\|^2 = (u,B'^2u) \le (B'u,B'^2u) = (u,B'^3u) \le \|B'^2u\|^2 = (u,B'^4u) \le \dots \le (u,C'u)$, so that $C' \ge 1$. It follows that C must be the Friedrichs extension of C', so we get $C \ge 1$ as well. As in the proof of I, prop.2.7 we find that C^{-1} is bounded, and $0 \le C \le 1$, implying $\|C^{-1}\| \le 1$, im C $=H$. Let B be the Friedrichs extension of B' , then B^m is self-adjoint, since it is hermitian, and one easily concludes im $B^m = H$. Also, $B^m \supset B'^m = C'$, so that $C = B^m$. Now for $f \in H$ there exists $f_j = C'u_j = B'(B'^{m-1}u_j) \to f$, with $u_j \in C_0^\infty$ = dom B'. Note that $v_j = B'^{m-1}u_j \in C_0^\infty$, and $v_j \to v$, due to $\|B'z\| \geq \|z\|$. It follows that im B' is dense in H, and that $B'^c = B$, so that B' is e.s.a. Similarly we may write $f_j = B'^k(B'^{m-k}u_j)$,above,and conclude that B'^k is e.s.a.

Now we are finished if we also can derive that $(B'+\eta)^k$ is e.s.a., for every real η , because clearly $H_0 + c = B' + \eta$, $\eta = c - \eta$. For this we note that, for a given η there exist positive constants λ, γ_1, γ_2 such that

$$\gamma_1 \|B'^k u\|^2 \leq \|(B'+\eta)^k u\|^2 + \lambda \|u\|^2 \leq \gamma_2 \|B'^k u\|^2, \text{ for } u \in \text{dom } (B')^k \ . \tag{1.2}$$

Indeed, the second inequality is evident, and the first follows from

$$B'^j \leq \varepsilon B'^k + c_\varepsilon \ , \ j=1,\ldots,k-1 \ , \tag{1.3}$$

valid for every $\varepsilon > 0$, and sufficiently large $c_\varepsilon > 0$, as easily derived. Now (1.2) at once implies that $(B'+\eta)^{k**} = (B+\eta)^k$, which is self-adjoint, q.e.d.

Theorem 1.5. Let Ω be the interior of a complete Riemannian manifold Ω_0 with compact boundary $\partial\Omega$, and assume that

$$q(x) \geq \gamma/(\text{distance}(x,\partial\Omega))^2 \ , \ x \in \Omega \setminus K \ . \tag{1.4}$$

There exists a nondecreasing sequence γ_m , $m=1,2,\ldots.$, such that for $\gamma \geq \gamma_m$, and $H = -\Delta + q$, we have $H_0^m = (H^m)_0$ essentially self-adjoint. Here condition (1.4) will be considered void (i.e., always satisfied) if the boundary $\partial\Omega$ is void. Also, Δ is the Laplace operator of the Riemannian metric of Ω .

Proof. Let $\rho(x)$ be the Riemannian distance from x to $\partial\Omega$, and let $\kappa(x)$ be the distance from any given fixed point $x^0 \in \Omega$. Then

$$\tau(x) = \text{Max}\ \{ e^{-\eta\kappa(x)} \ , \rho(x)\} \ , \tag{1.5}$$

where we set $\rho \equiv \infty$ if $\partial\Omega = \emptyset$, so that then $\tau = \kappa$, may be seen to satisfy all conditions ('). Indeed, the first three conditions are trivially satisfied, using that $\nabla\kappa$ and $\nabla\rho$ are bounded, and that ρ and κ tend to ∞ , of the same order, as $x \to \infty$ on Ω , in view of the completeness of Ω_0 (cf. [KG_1]).

In view of remark 1.3 and prop.1.4 we may consider $H + c_0$, for large c_0 , instead of H . Note that $\tau = e^{-hk}$ near $\partial\Omega$, and $\tau = \rho$ near ∞ (of Ω_0). We get

(1.6) $q+c_0 \geq \nabla\kappa$ near ∞ of Ω_0 , and $q+c_0 \geq \gamma/\rho^2 \geq \eta^{-2}\nabla\rho^2/\rho^2$ near $\partial\Omega$,

for sufficiently large γ, depending on η. Thus the theorem follows

Corollary 1.6. Under the assumptions of thm.1.5, if we have

(1.7) $$\lim_{x\to\partial\Omega}(q(x)(\mathrm{dist}(x,\partial\Omega))^2 = +\infty ,$$

then all powers H_0^m are e.s.a.

Corollary 1.7. For a complete Riemannian manifold without boundary all powers of the Laplace operator have their minimal operator essentially self-adjoint.

The proofs are evident, in view of thm.1.5 and prop.1.4.

We shall discuss Chernoff's proof of cor.1.7 in [C_3], since it depends on the basic properties of the solution operator of the hyperbolic equation $\partial_t^2 u = \Delta u$, to preserve the compact support of its initial values, as t propagates (i.e.,finite propagation speed). We note that his proof also applies to the case of the Laplacian acting on differential forms, which we do not consider here.

The theorem,below,is am immediate consequence of thm.1.5, and III,thm.6.3, regarding a normal form of $H \geq 1$. It appears as an interesting generalization of a conjecture by F.Rellich [Re_1] for Schroedinger type operators in $\mathbb{R}^n$, first proven by Wienholtz [Wi_1] (cf. also [Ch], sec.4).

Theorem 1.8. Let Ω be a complete Riemannian manifold (with countable atlas), and let a differential expressions H of the form (1.1) be given, with the metric tensor h^{jk} of the space, but a general positive measure $d\mu$. Then, if the minimal operator H_0 is semi-bounded below, all its powers H_0^m are essentially self-adjoint in the Hilbert space $\mathcal{H} = L^2(\Omega,d\mu)$.

Indeed, we conclude from III, prop.6.2, that the dependent variable may be changed such that $q=\gamma$ = constant. In view of I, prop.2.4 we may consider $H+c_0$ instead of H ,and thus may assume that $\gamma=1$. Then thm.1.5 may be applied with $\sigma(x)=\mathrm{dist}(x,x^0)$ again, with a fixed point x_0 . Thus thm.1.8 follows.

Finally let us mention some results of J.Frehse [F_1] concerning the first power of a second order operator of the form (1.1) on $\Omega = \mathbb{R}^n$. Actually we are stating weaker results, since we will get restricted to C^∞-coefficients, as we always do here. Also, in [F_1] one finds two more applications of refined nonlinear techniques to problems of essential self-adjointness of the minimal operator which we do not include.

On $\mathbb{R}^n$ we of course can get restricted to a fixed global coordinate system $x = (x_1,\dots,x_n)$ (written with subscripts again) Then, with fixed coordinates, let $\Lambda(x)$ and $\lambda(x)$ be the largest, and the smallest eigenvalue of the n×n-matrix $((h^{jk}))_{j,k=1,\dots,n}$. Let $\mathcal{B}_r(x)$ be the closed ball with center x and radius r , and let $B_r = B_r(0)$. Also we get restricted to the case $\kappa \equiv 1$.

Theorem 1.9. Given the differential expression

$$H = -\sum \partial_{x_j} h^{jk} \partial_{x_k} + q \quad , \ x \in \mathbb{R}^n \ , \ d\mu = dx \ . \tag{1.8}$$

where h^{jk} , $q \in C^\infty(\mathbb{R}^n)$, and the matrix $((h^{jk}(x)))$ is real, symmetric and positive definite. Then each of the following conditions is sufficient for essential self-adjointness of the minimal operator H_0 .

1) q bounded, below, and

$$\sup\{\Lambda(x):x\in B_{2R}\setminus B_R\}/\inf\{\lambda(x):x\in B_{2R}\setminus B_r\} = O(R^2+1) \ , \text{ as } R\to\infty \ . \tag{1.9}$$

2) q bounded, below, and

$$\sup\{\Lambda(x):x\in B_r(y)\}/\inf\{\lambda(x):x\in B_r(y)\} = O(1), \text{ with some } r>0, \tag{1.10}$$

The proof will be discussed in sec.4.

In the case of n=1, and a q bounded below it turns out that (1.9) and (1.10) are superfluous:

Theorem 1.10. For the singular Sturm-Liouville expression

$$H = -\partial x p(x) \partial_x + q(x) \ , \ -\infty<x<\infty \ , \ q(x)\geq 0 \ , \tag{1.11}$$

in $\mathcal{H}=L^2(\mathbb{R})$ the minimal operator H_0 is essentially self-adjoint.

Proof. We may assume that $q \geq 0$, since H_0 may be replaced by H_0+c. Since H is hypo-elliptic, it suffices to show that Hu=0 $u\in C_0^\infty(\mathbb{R})\cap\mathcal{H}$ implies $u\equiv 0$. Now, since $u\in L^2(\mathbb{R})$ there must exist sequences $\xi_j\to-\infty$, $\zeta_j\to+\infty$, such that $u(\xi_j)\to 0$, $u(\zeta_j)\to 0$, as $j\to\infty$. However, since $q\geq 0$ we have the maximum principle (III, thm.5.8) valid, so that $\mathrm{Max}\{|u(x)|: \xi_j\leq x\leq\zeta_j\} \to 0$, i.e., $u\equiv 0$, q.e.d.

2. Essential self-adjointness of H_0.

It is practical to discuss the proof of thm.1.1 for m=1 first. This will be a matter of an established elementary technique, applied to manifolds by W. Roelcke. The lemma, below, is crucial.

Let $\Omega, d\mu$, $H=-\Delta_m+q$ be as in (1.1). For any compact set $\Sigma\subset\Omega$ we define the 'local inner products'

$$(2.1)\quad (u,v)_\Sigma = \int_\Sigma \overline{u}(x)v(x)d\mu \quad , \quad (\nabla u,\nabla v)_\Sigma = \int_\Sigma h^{jk}\overline{u}_{|x^j}u_{|x^k}d\mu \ .$$

Lemma 2.1. Let $\sigma(x)$ be a nonnegative Lipschitz continuous function over Ω, satisfying $\sigma(x_0) = 0$,for some given x_0. For some $R>0$ let the set $\Sigma_R = \{x\in\Omega : \sigma(x)\leq R\}$ be compact. Then we have

$$(2.2)\quad \int_0^R ((\Delta_\mu u,v)_{\Sigma_\rho} + (\nabla u,\nabla v)_{\Sigma_\rho})d\rho \leq \int_{\Sigma_R} |\nabla\sigma||\nabla u||v|d\mu \ .$$

Proof. For given fixed $\varepsilon,\rho > 0$, $0 < \rho \leq R$ introduce the continuous, piecewise linear function $\eta = \eta_{\rho,\varepsilon}$ on $(-\infty,\infty)$ with $\eta(t)=0$ on $[\rho,\infty)$, $\eta(t)=1$ on $(-\infty,\rho-\varepsilon]$. This function is Lipschitz continuous, hence the composition $\zeta(x) = \eta(\sigma(x))$ defines a Lipschitz continuous function over Ω , and we get $\nabla\zeta = \eta'\nabla\sigma$, almost everywhere. For $u,v \in C^\infty(\Omega)$ we get ζv absolutely continuous, and $\nabla(\zeta v)$ is locally bounded. Thus Green's formula may be applied for

$$(2.3)\qquad 0 = \int_{\Sigma_R} (\Delta_\mu \overline{u}\, v\zeta + h^{jk}\overline{u}_{|x^j}(\zeta v)_{|x^k})d\mu \ .$$

It follows that

$$(2.4)\quad \int_{\Sigma_R} (\Delta_\mu \overline{u}\, v + h^{jk}\overline{u}_{|x^j}v_{|x^k})\zeta d\mu \leq \int_{\Sigma_R} |\eta'||\nabla\sigma||\nabla u||v|d\mu \ .$$

Note that (2.4) holds for all ρ , $0 \leq \rho \leq R$, and fixed $\varepsilon > 0$. We may integrate (2.4) $d\rho$ from 0 to R , observing that the only term at right depending on ρ is the term $\eta' = d\eta_{\rho,\varepsilon}/dt(\sigma(x))$. Hence we get

$$(2.5)\quad \int_0^R d\rho\int_{\Sigma_\rho} \zeta d\mu(\Delta_\mu \overline{u}\, v + h^{jk}\overline{u}_{|x^j}v_{|x^k}) \leq \int_{\Sigma_R} d\mu|\nabla\sigma||\nabla u||v|\int d\rho|\eta'| \ .$$

As $\varepsilon \to 0$ the function ζ converges to the characteristic function of Σ_ρ , boundedly, and almost everywhere, while $\int_0^R|\eta'|d\rho$ converges to 1 , as easily calculated. Therefore (2.2) follows, q.e.d.

We now can prove thm.1.1 for m= 1 , as follows. Suppose a function σ exists with the properties of the thm. since $\sigma\to\infty$, as $x\to\infty$, all the sets Σ_R of lemma 2.1, formed with this function σ, are compact. Now the operator H_0 is ≥ 1 , (if necessary we may replace H by $H+c_0$, using rem.1.3). Thus it is sufficient to prove that im H_0 is dense (I, prop.2.7). Or, we may show that $(f,H_0u)=0$ for all $u \in \text{dom } H_0 = C_0^\infty$ implies u=0. In view of Weyls lemma (i.e., III, prop.1.2) any such f must be $C^\infty(\Omega)$, and then must solve Hf=0. Thus we are finished if we can prove the lemma, below.

Lemma 2.2. Under the assumptions of thm.1.1 , for m=1 , if $f \in H\cap C^\infty(\Omega)$ solves Hf=0 , then we get $f\equiv 0$.

Proof. Define the functions $\phi(\rho)$, $\psi(\rho)$, $\lambda(\rho)$, by setting

$$(2.6)\qquad \lambda=\phi+\psi\ ,\ \phi(R) = \int_0^R \|f|\nabla\sigma|\|^2_{\Sigma_\rho}\, d\rho\ ,\ \psi(R) = \int_0^R \|\nabla f\|^2_{\Sigma_\rho}\, d\rho,\ 0<R<\infty\ .$$

Note that ϕ , ψ are integrals of nondecreasing functions over $[0,\infty)$ each. They are continuous and have a left and right first derivativative at each point. Moreover, the first derivative exists at all points except at the (at most countably many) jumps of the integrand in (2.6). Moreover, ϕ and ψ are convex. The same is true for λ . We apply lemma 2.1, now valid for all R > 0 , and use that Hf=0, i.e., $\Delta_\mu f = qf$, to get

$$(2.7)\qquad \int_0^R d\rho(\|\sqrt{q}f\|^2_{\Sigma_\rho} + \|\nabla f\|^2_{\Sigma_\rho}) \leq \||\nabla\sigma|f\|_{\Sigma_R}\|\nabla f\|_{\Sigma_R}\ ,\ 0 < R < \infty\ .$$

With condition (iii) of thm.1.1 and the above functions we have

$$(2.8)\qquad \lambda(R) \leq (\phi'(R+0)\psi'(R+0))^{1/2} \leq \lambda'(R+0)/2\ ,\ 0 < R < \infty\ .$$

Notice that λ,λ' are ≥ 0, by (2.6) . If λ does not vanish identically then we get $\lambda > 0$, as $\rho \geq \rho_1$, for some $\rho_1 > 0$. Clearly λ' is absolutely continuous. (2.8) may be integrated in $[\rho_1,\infty)$, for $\lambda(\rho) \geq c\, e^{2\rho}$, $c= \lambda(\rho_1)e^{-2\rho_1}$. Using $(iv)_\eta$ of thm.1.1 we get

$$(2.9)\ |\nabla\sigma| \leq \zeta e^{\eta\sigma} \leq \zeta e^{\eta\rho} \leq \zeta(\lambda/c)^{\eta/2} = c_1\lambda^{\eta/2},\ \text{as } x \in \Sigma_\rho,\ \rho \geq \rho_1.$$

Now we go back to (2.6), and conclude that $\psi' \leq \lambda'$, $\phi' \leq c_2\lambda^\eta\|f\|^2 = c_3\lambda^\eta$, since $\|f\| < \infty$. This may be taken into (2.8), for $\lambda \leq c_4\sqrt{\lambda'}\lambda^{\eta/2}$. Or ,

$$(2.10)\qquad \lambda' \geq c_5\lambda^{2-\eta}\ ,\ \text{as } \rho \geq \rho_1\ ,$$

with a positive constant c_5 . Now it is well known (and easily derived) that the differential inequality (2.10) cannot have a positive solution in an interval of infinite length, whenever $\eta<1$. Therefore, for $\eta < 1$ we get a contradiction, and $\lambda \equiv 0$, i.e. $f=0$ follows, q.e.d.

3. Proof of theorem 1.1.

We now will discuss the proof of thm.1.1 for general $m=2, 3, \ldots$ .The basic tool will be lemma 2.1 again, which will yield a system of differential inequalities instead of the single inequality (2.8). It then is a rather complicated matter to show that the system obtained admits only the identically zero solution.

We start by arguing as before that the proof is accomplished if we can show that the only solution $f \in H \cap C^{\infty}(\Omega)$ of the differential equation $H^m f = 0$ is the function $f \equiv 0$ (Here we first have added a constant to q to insure that $q \geq 1$, and $H_0{}^m \geq 1$, which may be done, in view of prop.1.4). This works just like in sec.2 , and will not be discussed again.

For an f with above properties define $g_j = H^{m-j}f$, $j=0,\pm1,\pm2,\ldots$ .Clearly $g_j = 0$ for $j\leq 0$, and we interpret H^k as a differential operator applying to a C^{∞}-function, for $j=0,\ldots,m$. The negative powers H^{m-j} , $j > m$, will be interpreted as the (inverse) powers of the Friedrichs extension H , a self-adjoint operator (and the closure of H_0 , since our assumptions (and $\eta < 1$) already give the e.s.a.-property for H_0 itself.

It follows that $g_j \in H$, for all $j \in \mathbb{Z}$. Using Weyl's lemma a finite number of times one also concludes that all g_j are $C^{\infty}(\Omega)$, and we generally have $H^1 g_j = g_{j-1}$, with derivatives in the ordinary sense.

Similarly as in sec.2. we now define the functions

$$(3.1) \quad \phi_j(R) = \int_0^R \| |\nabla\sigma| g_j \|_r^2 dr \ , \ \psi_j(R) = \int_0^R \|\nabla g_j\|_r^2 dr, \ j=0,\pm1,\ldots,$$

where we use the abbreviation $\|.\|_r^2 = \|.\|_{\Sigma_r}^2 = (.,.)_{\Sigma_r}$ (cf.(2.1)).

Clearly ϕ_j, ψ_j have similar properties as ϕ, ψ of (2.6). Specifically they are absolutely continuous and the left and right derivatives exist at each point and coincide, except possibly at countably many jumps. The first derivatives exist except perhaps at the jumps. In the estimates below we write $\phi'(R)$ for the right derivative of ϕ , i.e., $\phi'(R) = \lim_{h\to0,h>0}(\phi(R+h)-\phi(R))/h$, and similarly for $\psi'(R)$. Clearly than $\phi_j, \phi_j', \psi_j, \psi_j'$ are nondecreasing and nonnegative, and ϕ_j , ψ_j are convex. We have the fundamental theorem of calculus valid in the form $\phi(R)-\phi(S) = \int_S^R \phi'(\rho)d\rho$ with a well existing Riemann integral. For $j \leq 0$ both ϕ_j and ψ_j vanish identically. Moreover, using (iii) of thm.1.1 we get

$$(3.2) \quad \| |\nabla\sigma| g_j\|^2 + \|\nabla g_j\|^2 \leq \|\sqrt{q} g_j\|^2 + \|\nabla g_j\|^2 = (g_j, Hg_j) = (g_j, g_{j-1}) < \infty ,$$

where the partial integration without boundary terms is justified by the essential self-adjointness of H_0, already proven (One may carry it out for a sequence g_j^1 in C_0^∞ with $g_j^1 \to g_j$, $Hg_j^1 \to Hg_j$) . Notice that (3.2) yields boundedness of the derivatives ϕ_j', ψ_j' , for all j .

Proposition 3.1. The functions ϕ_j , ψ_j satisfy the system (3.3) of differential inequalities, where the series at right represent finite sums, since ϕ_j , ψ_j vanish for $j \leq 0$, and where the argument at right is R+0 , not R (just as in (2.8)).

$$\phi_j + \psi_j \leq (\phi_j' \psi_j')^{1/2} + \sum_{l=0}^{\infty}((\phi_{j+l+1}' \psi_{j-l-1}')^{1/2} + (\psi_{j+l+1}' \phi_{j-l-1}')^{1/2}),$$

$$(3.3) \qquad \phi_j \leq \zeta e^{\eta r} \sum_{l=0}^{\infty}((\phi_{j+l+1}' \psi_{j-l}')^{1/2} + (\psi_{j+l+1}' \phi_{j-l}')^{1/2}) ,$$

$$j = 1,2,\ldots, \quad 0 \leq r < \infty .$$

Proof. This follows by repeated application of lemma 2.1 to the inequalities

$$(3.4) \qquad \begin{aligned} \phi_j + \psi_j &\leq \int_0^R (\|\sqrt{q} g_j\|_r^2 + \|\nabla g_j\|_r^2) dr , \\ \phi_j &\leq \zeta e^{\eta R} \int_0^R \|g_j\|_r^2 dr , \end{aligned}$$

which follow from (iii) and $(iv)_\eta$ of thm.1.5. For example, the first inequality (3.4) implies

$$\begin{aligned} \phi_j + \psi_j &\leq \int_0^R (g_j, (q-\Delta_\mu) g_j)_r dr + \int_0^R ((g_j, \Delta_\mu g_j)_r + (\nabla g_j, \nabla g_j)_r) dr \\ &\leq \int_0^R (g_j, g_{j-1})_r dr + \| |\nabla\sigma| g_j\|_R \|\nabla g_j\|_R \\ &= \int_0^R ((q-\Delta_\mu) g_{j+1}, g_{j-1})_r dr + \sqrt{\phi_j' \psi_j'} . \end{aligned}$$

The first integral may be "integrated by parts" again, using lemma 2.1, for

$$= \sqrt{\phi_j'\psi_j'} + \int_0^R ((g_{j+1},(q-\Delta_\mu)g_{j-1})_r dr$$

$$- \int_0^R ((\nabla g_{j+1},\nabla g_{j-1})_r + (\Delta_\mu g_{j+1},g_{j-1})_r)dr$$

$$+ \int_0^R ((\nabla g_{j+1},\nabla g_{j-1})_r + (g_{j+1},\Delta_\mu g_{j-1})_r)dr$$

$$\leq \sqrt{\phi_j'\psi_j'} + \sqrt{\phi_{j+1}'\psi_{j-1}'} + \sqrt{\psi_{j+1}'\phi_{j-1}'} + \int_0^R ((q-\Delta_\mu)g_{j+2},g_{j-2})_r dr \ .$$

It is clear now that this process may be iterated until an index $j-1\leq 0$ is reached, at which time the last term vanishes. This results in the first formula (3.3). For the second inequality one starts with

$$\int_0^R \|g_j\|_r^2 dr = \int_0^R ((q-\Delta_\mu)g_{j+1},g_j)_r dr \ ,$$

and again uses lemma 2.1 to increase the index of the first factor, and decrease the index of the second factor, until again the inner product vanishes. Then the second (3.4) implies (3.3), completing the proof of prop.3.1.

Only the first 2m inequalities (3.3) will be used, but we leave the system infinite for technical reasons.

We now reduce the system (3.3) as follows. First introduce new variables by setting

$$\lambda_j = \phi_j + \psi_j \ , \ \mu_j = \phi_j - \psi_j \ . \tag{3.5}$$

Using that $\sqrt{\phi_k'\psi_m'} + \sqrt{\phi_m'\psi_k'} \leq \sqrt{\lambda_k'\lambda_m'}$, by Schwarz' inequality, we get

$$\begin{aligned} &\lambda_j \leq \tfrac{1}{2}\sqrt{\lambda_j'^2-\mu_j'^2} + \textstyle\sum_{l=0}^{\infty}\sqrt{\lambda_{j+l+1}'\lambda_{j-l-1}'} \ , \\ &\tfrac{1}{2}(\lambda_j+\mu_j) \leq \zeta e^{\eta r}\textstyle\sum_{l=0}^{\infty}\sqrt{\lambda_{j+l+1}'\lambda_{j-l}'} \ . \end{aligned} \tag{3.6}$$

Note that λ_j , μ_j satisfy the following conditions.

(Λ): Both λ_j , μ_j are absolutely continuous and their right derivatives λ_j', μ_j' exist everywhere. The functions λ_j are convex and λ_j, λ_j' are nondecreasing and nonnegative. The λ_j' are bounded the μ_j' of bounded variation. Furthermore, $\lambda_j(0) = \mu_j(0) = 0$, and

$$|\mu_j| \leq \lambda_j \ , \ |\mu_j'| \leq \lambda_j', \text{ as } 0 \leq r < \infty \ . \tag{3.7}$$

Proposition 3.2. The inequalities (3.6) and conditions (Λ) imply

$$(3.8)\quad \lambda_j^3 \le 3\sqrt{2}\,\lambda_j'^{3/2}\sum_1(\sqrt{\zeta}e^{\eta r/2}\lambda_j\sqrt[4]{\lambda_{j+1+1}\lambda_{j-1}} + \sqrt{2r\lambda_j\lambda_{j+1+1}\lambda_{j-1-1}}),$$

for $j=1,2,\ldots,$ and $0\le r<\infty$.

Proof. If $\lambda_j'\equiv 0$ the corresponding j-th estimate is trivial, since $\lambda_j(0)=0$ then implies also $\lambda_j\equiv 0$. Since λ_j' is nondecreasing, we may have $\lambda_j'\equiv 0$ only in some initial interval $[0,\rho]$, but then will get $\lambda_j'>0$ in (ρ,∞), (unless $\lambda_j=0$ for all r). We derive the j-th estimate (3.8) from the j-th pair of estimates (3.6). Note that all j-th estimates involve the λ_k , $k\ne j$ only in form of their derivatives. We thus may translate left by ρ and replace λ_k by $\lambda_k-\lambda_k(\rho)$, without changing the j-th pair (3.6). If the j-th (3.8) can be derived for these new λ_j then the same follows for the old λ_j . Thus, we may assume that $\lambda_j'>0$ for $r>0$.

From (3.6) we obtain

$$(3.9)\qquad \lambda_j \le \tfrac{1}{2}\sqrt{2}\sqrt{\lambda_j'(\lambda_j'+\mu_j')} + \sum_1\sqrt{\lambda_{j+1+1}'\lambda_{j-1-1}'}\ .$$

We plan to integrate (3.9). Note that

$$(3.10)\qquad 2\int_\varepsilon^R \lambda_j dr = \int(\lambda_j')^{-1}d(\lambda_j^2) = \lambda_j^2/\lambda_j'\Big|_\varepsilon^R + \int_\varepsilon^R(\lambda_j/\lambda_j')^2 d(\lambda_j')\ ,$$

with well existing Stieltjes integrals. In (3.10) we may let $\varepsilon\to 0$, since $\int_0^R\lambda_j$ exists, and the integral at right (called I_ε) increases, as ε decreases. If $I_\varepsilon\to\infty$, then $\lambda_j^2/\lambda_j'\to\infty$, hence $\lambda_j'/\lambda_j^2\le C$ in $[0,1]$. Integrate this to get

$$1/\lambda_j(\delta) - 1/\lambda_j(r) \le C(r-\delta)\ ,\ 0<\delta\le r\ .$$

As $\delta\to 0$ the left hand side goes to ∞ , since $\lambda_j(0)=0$. Thus a contradiction results. Not only must the integral I_0 exists but also we must have $\lambda_j^2/\lambda_j'\to 0$, as $r\to 0$. It follows that

$$(3.11)\qquad \lambda_j^2/(2\lambda_j') \le \int_0^R\lambda_j(r)dr\ .$$

Here we substitute λ_j from (3.6), and use the inequality

$$(3.12)\qquad \int_0^R\sqrt{a'b'}dr \le \sqrt{ab}\ ,\ \text{as } a,b\in C^1([0,R])\ ,\ a,a',b,b'\ge 0\ .$$

[From Schwarz' inequality we get

$$(3.13)\qquad \int_0^R\sqrt{a'b'}\,dr \le (\int a'dr\int b'dr)^{1/2}\le((a(R)-a(0))(b(R)-b(0)))^{1/2},$$

confirming (3.12).] Thus we get

$$(3.14)\qquad \lambda_j^2/(2\lambda_j') \le \tfrac{1}{2}\sqrt{2}\ \sqrt{\lambda_j}\ \sqrt{\lambda_j+\mu_j} + \textstyle\sum_1 \sqrt{\lambda_{j+1+1}\lambda_{j-1-1}}\ .$$

Here we substitute $(\lambda_j+\mu_j)/2$ from the second (3.6), for

$$(3.15)\qquad \tfrac{1}{2}\lambda_j^2/\lambda_j' \le \textstyle\sum_1 (\sqrt{\zeta}e^{\eta r}\lambda_j \sqrt[4]{\lambda_{j+1+1}'\lambda_{j-1}'} + \sqrt{\lambda_{j+1+1}\lambda_{j-1-1}}\ )\ .$$

A similar proceedure is now applied again to (3.15). Multiply by $\sqrt{\lambda_j'}$, and integrate from 0 to R . Clearly (3.11) implies $\lambda_j^2/\sqrt{\lambda_j'} \to 0$, as $r\to 0$, so that the integral at left exists. We apply a partial integration as before:

$$(3.16)\ 3\int_\varepsilon^R \lambda_j^2/\sqrt{\lambda_j'}dr = \int_\varepsilon^R (\lambda_j')^{-3/2}d(\lambda_j^3) = \lambda_j^3\lambda_j'^{-3/2}\Big|_\varepsilon^R - \int_\varepsilon^R \lambda_j^3 d(\lambda_j'^{-3/2}).$$

As $\varepsilon\to 0$, the last integral, J_ε, is monotone, hence either converges or diverges to $-\infty$, as $\varepsilon\to 0$. The latter proves impossible, because then $\lambda_j^3/\lambda_j'^{3/2} = (\lambda_j^2/\lambda_j')^{3/2} \to \infty$, while we showed above that it tends to 0 as $\varepsilon\to 0$. It follows that

$$(3.17)\qquad \tfrac{1}{3}\lambda_j^3\lambda_j'^{-3/2}(R) \le \int_0^R \lambda_j^2/\sqrt{\lambda_j'}\ dr\ .$$

Combining (3.17) and the integrated (3.14) we get

$$(3.18)\qquad \begin{aligned}\tfrac{1}{6}\lambda_j^3\lambda_j'^{-3/2}(R) \le \sum_1 \{\sqrt{\zeta}e^{\eta R/2}\int_0^R \sqrt{\lambda_j\lambda_j'}\sqrt[4]{\lambda_{j+1+1}'\lambda_{j-1}'}\ dr \\ + \int_0^R \sqrt{\lambda_j'}\sqrt{\lambda_{j+1+1}\lambda_{j-1-1}}\ dr\}\ .\end{aligned}$$

Using (3.12) twice, the first integral at right of (3.18) is estimated

$$(3.19)\qquad \le (\lambda_j^2/2\int\sqrt{\lambda_{j+1+1}'\lambda_{j-1-1}'}dr)^{1/2} \le \tfrac{1}{2}\sqrt{2}\lambda_j(\lambda_{j+1+1}\lambda_{j-1}\quad)^{1/4}\ .$$

Similarly the second integral is estimated by

$$(3.20)\qquad \le (R\lambda_j\lambda_{j+1+1}\lambda_{j-1-1})^{1/2}\ ,$$

using the monotony of λ_k .

Now we may substitute (3.19) and (3.20) into (3.18) to obtain the desired estimate (3.8), q.e.d.

For the derivation of thm.1.1 from prop.3.2 we introduce the following two order relations, defined for the set of nonnegative functions a, b, defined over $[0,\infty)$. We write a $<_p)$ b , if $a(r) \le cb(r)$, with a constant c, and we write a $<_q)$ b if $a(r) \le c(\log(2+K(r))^s b(r)$, where $K(r)=\int_{j=1}^m \lambda_j(r)$, with above functions

λ_j . Clearly both relations are transitive and reflexive, and are compatible with addition and multiplication of functions, and with taking positive powers. Moreover, a $<_p)$ b implies a $<_q)$ b .

Notice that K,K' are ≥ 0 and nondecreasing, and

(3.21) $\lambda_j \leq K$, $\lambda_j' \leq K'$, $j=1,\ldots,m$, $\lambda_j <_p) r$, $\lambda_j' <_p) 1$, as $j>m$, as a consequence of cdn.(Λ) , above.

Proposition 3.3. If K is not $\equiv 0$, then we have $e^{\varepsilon r} <_p) K+1$, for some $\varepsilon>0$.

Proof. We sum over j in the first estimate (3.6), replacing μ_j by 0 and using (3.21). Thus it follows that $K <_p) K' + \sqrt{K'} <_p) K' + \gamma$, for every $\gamma>0$. This amounts to the differential inequality $\varepsilon\sigma \leq \sigma'$, for $\sigma=K=\gamma$ an some $\varepsilon>0$. Suppose K is not $\equiv 0$, then $K(\rho)>0$ for some $\rho>0$. For small $\gamma>0$ we get $\sigma(r)>0$, as $r>\rho$. Then we get $\varepsilon \geq \sigma'/\sigma = (\log \sigma)$. This may be integrated for $\sigma(r) \geq \sigma(\rho)e^{\varepsilon(r-\rho)}$, or, $K(r) \geq ce^{\varepsilon r}$, as $r \geq \rho$, with $c>0$, implying the statement.

Corollary 3.4. We have either $K\equiv 0$ or $r <_q) 1$.

Proof. Take logarithm's.

For the reminder of this section we assume that K is not $\equiv 0$. This will lead us into a contradiction, thus completing the proof of thm.1.1. Define the functions $\sigma_{m+1} \equiv 0$, $\sigma_j = \sum_{k=j}^{m} \lambda_k$ $j=1,\ldots,m$.

Proposition 3.5. We have

$$\lambda_j <_q) K'^{1/2}\{K^{5/12}\sigma_{j+1}^{1/12} + K^{1/3}\sigma_{j+1}^{1/6} + K^{5/12} + K^{1/3} \tag{3.22}$$

$$+ K^{5/12+\eta/6\varepsilon}\sigma_{j+1}^{1/12} + K^{5/12+\eta/6\varepsilon}\} , \quad j=1,\ldots,m .$$

Proof. Use prop's 3.2 and 3.3. Note that $\sqrt{r} <_q) 1$, by cor.3.4. and that $\lambda_j' <_q) K'$. For the two terms S_j and T_j in (3.8) we get

$$S_j e^{-\eta r/2} <_q) K^{5/4}(\textstyle\sum_{k=0}^{j-1}\lambda_{j+k+1})^{1/4} <_q) K^{5/4}(\sigma_j+1)^{1/4} , \tag{3.23}$$

$$T_j <_q) K (\textstyle\sum_{k=1}^{j-2}\lambda_{j+k-1})^{1/2} <_q) K(\sigma_j+1)^{1/2} .$$

Taking the cube root we thus get (3.22) , also using prop.3.4. The proof is complete.

Proposition 3.6. If $\eta \leq \eta_m$ is sufficiently small, then, for every

$j=1,2,\ldots,m$, there exist finitely many positive numbers $\alpha^{j,k}$, $\beta^{j,k}$ $\gamma^{j,k}$, $\delta^{j,k}$ satisfying $\alpha^{j,k}+\beta^{j,k}<1$, $\gamma^{j,k}+\delta^{j,k}<1$, such that

$$(3.24)\qquad \sigma_j <q) \sum_k K'^{\alpha^{j,k}} K^{\beta^{j,k}} , \quad \lambda_j <q) \sum_k K'^{\gamma^{j,k}} K^{\delta^{j,k}} .$$

Proof. We use induction and first note that $\sigma_{m+1}=0$, $\sigma_m=\lambda_m$, so that (3.22) give the assertion for $j=m$. Assume it true for a certain $j\leq m$. Then substitute σ_j from (3.24) into (3.22) for $j-1$. It is clear then that

$$(3.25)\qquad K^{5/12}\sigma_j^{1/12} <q) \sum\{K^{5/12+\beta^{j,k}/12}\ K'^{1/2+\alpha^{j,k}/12}\} ,$$

which again is a sum of terms of the form $K'^{\xi}K^{\zeta}$, in view of $5/12 +1/2+(\beta^{j,k}+\alpha^{jk})/12 = 11/12+(\beta^{j,k}+\alpha^{j,k})/12<1$. Similarly for the other terms in (3.22). This shows that λ_{j-1} is dominated in the described way, and we have $\sigma_{j-1}=\lambda_{j-1}+\sigma_j$, of course, q.e.d.

Note that $K=\sigma_1$ now is dominated by a sum of the form (3.24). Also we get $K(R)\leq \int_0^R K'dr \leq RK'(R)$, by monotony of K. Thus we get $K\leq rK'$, and cor.3.4 implies $K <q) K'$. This may be substituted into the right hand side of (3.24), for $j=1$. It follows that

$$(3.26)\qquad K <q) K'^{\xi} , \text{ for some } 0<\xi<1 .$$

Finally we observe that (3.26) implies

$$(3.27)\qquad K <p) K'^{(1+\xi)/2} ,$$

because we have

$$(3.28)\qquad (\log(2+K))^s <p) K^{\varepsilon} , \text{ for every } \varepsilon>0 ,$$

since we know that $K\to\infty$, as $r\to\infty$, in view of prop.3.4.

Now it is well know that the differential inequality (3.27) (or in details

$$(3.29)\qquad K\leq c\,K'^{\gamma} ,$$

with constants c, $\gamma>0$, $\gamma<1$, does not have a nonnegative solution, defined in $[0,\infty)$, except $K(r)\equiv 0$. Indeed, write (3.29) in the form

$$(3.30)\qquad -(1/K^{\tau})' = \tau K'/K^{\tau+1} \geq c^{\sim} , \ \tau = 1/\gamma - 1 > 0 , \ c^{\sim}>0 .$$

where r must be choosen large enough so that $K \neq 0$. Integrating (3.30) we get

$$(3.31) \qquad K^{-\tau}(\rho) \geq K^{-\tau}(\rho) - K^{-\tau}(r) \geq c^{\sim}(r-\rho) \ , \text{ as } r \geq \rho \ ,$$

where the right hand side tends to ∞ , as $r \to \infty$, while the left hand side remains constant. This results in a contradiction so that we must have $K = 0$ for all R. This, in turn, implies that $f \equiv 0$.

This completes the proof of thm.1.1.

4. Proof of Frehse's theorem.

In this section we finally want to discuss a proof of thm. 1.9. As in the discussion of thm.1.1 we only must show that every C^{∞}-solution of the differential equation $Hu=0$ in $H=L^2(\mathbb{R}^n)$ vanishes identically, under the assumptions of the thm., such as (1.9). Here we may assume that us is real-valued, since the coefficients of H are real-valued. Let us assume $q \geq 0$ again, without loss of generality.

In the following we consider open balls B_r with center x_0 kept fixed for a while, with radius r. For $u \in C^{\infty}(\mathbb{R}^n)$ solving $Hu=0$, and $\phi \in C_0^{\infty}(B_r)$ we have

$$(4.1) \qquad \int_{B_r} (h^{jk} u_{|j} \phi_{|k} + qu\phi) dx = 0 \ .$$

In (4.1) we introduce $\phi = \chi^2 |u|^p \mathrm{sgn}(u)$, with sgn $u = \pm 1$ and 0, as $u>0$, $u<0$, and $u=0$, respectively, and with a suitable Lipschitz continuous cut-off function $\chi \in C_0(B_r)$, to be chosen later on.

Clearly such ϕ is not C^{∞} , but still Lipschitz-continuous. Nevertheless the relation remains valid, by an approximation argument, as long as $p \geq 1$. (For example one may introduce a mollified $\phi_{\varepsilon} = \psi_{\varepsilon} * \phi$, as in the proof of II,lemma.2.1, and then pass to the limit $\varepsilon \to 0$.) From (4.1) we conclude that

$$(4.2) \qquad \int_{B_r} h^{jk} u_{|j} (\chi^2 |u|^p \mathrm{sgn}\ u)_{|k} dx \leq 0 \ ,$$

since the second term is nonnegative.

Let us write the integrand J in (4.2) in the form

$$(4.3) \qquad J = p\chi^2 |u|^{p-1} h^{jk} u_{|j} u_{|k} + 2\chi |u|^p (\mathrm{sgn}\ u) h^{jk} u_{|j} \chi_{|k} \ .$$

In (4.3) introduce

$$u_{|j}\chi|u|^{(p-1)/2} = 2/(p+1)\ (\zeta_{|j}-\eta_j)\ , \tag{4.4}$$

$$\zeta = \chi|u|^{(p+1)/2}\operatorname{sgn} u\ ,\ \eta_j = \chi_{|j}|u|^{(p+1)/2}\operatorname{sgn} u\ .$$

A calculation yields

$$J = 4(p+1)^{-2}(ph^{jk}\zeta_{|j}\zeta_{|k} - h^{jk}\eta_j\eta_k - (p-1)h^{jk}\zeta_{|j}\eta_k)\ . \tag{4.5}$$

Integrating (4.5) and applying (4.2) we thus get

$$0 \geq \int_{B_r} h^{jk}\zeta_{|j}\zeta_{|k}dx - p\int_{B_r} h^{jk}\eta_j\eta_k dx\ . \tag{4.6}$$

Or, with $\Lambda_+=\Lambda_+(r)= \sup\{\Lambda(x):x\in B_r\}$, $\lambda_-=\lambda_-(r)= \inf\{\lambda(x):x\in B_r\}$, with the largest and smallest eigenvalue $\Lambda(x)$ and $\lambda(x)$ of the matrix $((h_{jk}(x)))$, respectively,

$$\int_{B_r} dx|\nabla\zeta|^2 \leq p\Lambda_+/\lambda_-\int_{B_r} dx|\nabla\chi|^2|u|^{p+1}\ . \tag{4.7}$$

We now require a special case of the Sobolev estimate, as stated in prop.4.1, below.

Proposition 4.1. For every Lipschitz continuous $u\in C_0(\mathbb{R}^n)$, and every p satisfying $1<p<n$ we have the estimate

$$\|u\|_{np/(n-p)} \leq n^{-1/2}p(n-1)/(n-p)\ \|\nabla u\|_p\ , \tag{4.8}$$

with the $L^\rho(\mathbb{R}^n)$-norms ($|\nabla u|=(\sum|u_{|j}^2)^{1/2}$, integrals over $\mathbb{R}^n$)

$$\|u\|_\rho = \{\int|u|^\rho d\rho\}^{1/\rho}\ ,\ \|\nabla u\|_\rho = \{\int|\nabla u|^\rho\}^{1/\rho}\ . \tag{4.9}$$

The proof of prop. 4.1. will be discussed at the end of this section.

Let us assume $n>2$, so that the assumptions of prop.4.1 hold for $p=2$. (The case of $n=2$ will be discussed later on.) Then we may combine (4.8), for $u=\zeta$, and (4.7) to obtain

$$\|\chi|u|^{(p+1)/2}\|_{2n/(n-2)} \leq C_1\sqrt{(p+1)}\|\,|\nabla\chi|\,|u|^{(p+1)/2}\|_2\ , \tag{4.10}$$

with

$$C_1 = 2n^{-1/2}(n-1)/(n-2)(\Lambda_+/\lambda_-)^{1/2}\ . \tag{4.11}$$

Finally we replace p by p-1 in (4.10), assuming $p\geq 2$, and take the 2/p-th power, and arrange for L^ρ-norms of u :

(4.12) $\|\chi^{2/p}u\|_{pn/(n-2)} \le (pC_1^2)^{1/p}\||\nabla\chi|^{2/p}u\|_p$, for all $2\le p<\infty$.

The above is correct for every χ meeting the above requirements, and it should be noted that the constant C_1 depends on the choice of χ only insofar as Λ_+ and λ_- might change, according to the support of χ . Also C_1 is independent of p.

Next let us choose a sequence of cutoff functions χ_j, j=1,2, ... , such that $\chi_j\in C_0(B_R)$, and that $\chi_j\ge 0$, and $\nabla\chi_j = \chi_{j-1}\nabla\chi_j$, for all j. We explicitly choose $\chi_j(x)=\psi_j(|x-x_0|/R)$ with $t_j=(2j)^{-1}+1/2$,

(4.13) $\psi_j(r)$ =1 in $r\le t_{j+1}$, =0 in $r\ge t_j$, $=(r-t_j)/(t_{j+1}-t_j)$ else .

Then clearly χ_j is Lipschitz continuous, and we get

(4.14) $$|\nabla\chi_j| \le 2j(j+1)/R = \delta_{j-1} .$$

In (4.12) let $p=p_j=2\tau^j$, with $\tau=n/(n-2)$, and let $\chi = \chi_{j+1}$. Using (4.14) and $\nabla\chi_j\chi_{j-1}=\nabla\chi_j$, it follows that

(4.15) $\|\chi_{j+1}^{2/p_j}u\|_{p_{j+1}} \le (\delta_j^2 p_j C_1^2)^{1/p_j}\|\chi_j^{2/p_{j-1}}u\|_{p_j}$, j=0,1, ...

This relation may be iterated for

(4.16) $$\|\chi_{j+1}^{1/p_j}u\|_{p_{j+1}} \le \Pi_{k=0}^{j}(\delta_k^2 p_k C_1^2)^{1/p_k}\|u\|_{p_0} .$$

Notice that the product in (4.16) is bounded; we get

(4.17) $\Pi(\delta_k^2 p_k C_1^2)^{1/p_k} \le e^{2\Sigma\log\delta_\kappa/p_k}((\tau)^{1/p_0})^{\Sigma k/\tau^k}((p_0C_1^2)^{1/p_0})^{\Sigma\tau^{-k}}$,

with infinite series, at right, taken from 0 to ∞ . All 3 series converge, since $\tau>1$, and $(\log\delta_k)/p_k \le c\tau^{-k/2}$, for large k . Accordingly we may replace the product in (4.16) by the constant C_2 at right of (4.17), and then pass to the limit $j\to\infty$. Note that

(4.18) $\|u\|_{L^\infty(B_{R/2})} = \lim_{j\to\infty}\|u\|_{L^{p_j}(B_{R/2})} \le \lim_{j\to\infty}\|\chi_j^{2/p_{j-1}} u\|_{p_j}$.

Therefore, expressing the L^∞-norm at left by the sup, we get

(4.19) $$\sup\{|u(x)|: x\in B_{R/2}\} \le C_2\|u\|_{L^2(B_R)} .$$

In particular we used that $p_0=2$. The dependence of the constant C on R, Λ_+ and λ_- is easily evaluated: Only the first infinite series in (4.17) involves the radius R. Using (4.14) we get

(4.20) $2\sum(\log\delta_k)/p_k=\sum\tau^{-k}\log 2k(k+1) -\sum\tau^{-k}\log R = C_3-n/2\log R$, with C_3 independent of R . Similarly only the third term at right of (4.17) involves the quotient $\Lambda_+/\lambda_-=\sigma$. With $\sum\tau^{-k}=(1-(n-2)/n)^{-1}$ $=n/2$, as used above, and with $p_0=2$, the term becomes $C_1^{n/2}= C_4\sigma^{n/4}$ with C_4 independent of σ . In conclusion we get

$$(4.21)\qquad C_2(R,\Lambda_+,\lambda_-) = C_5\ R^{-n/2}(\Lambda_+(R)/\lambda_-(R))^{n/4}\ .$$

with C_5 depending on n only.

The above iteration technique was first applied by J.Moser [Ms_1], cf. also [GT_1].

We have proven the following result in the case n>2.

Proposition 4.2. If n>2, and $q\geq 0$, then every $C^\infty(\mathbb{R}^n)$-solution u of Hu=0 satisfies the estimate (4.18) with C_2 of the form (4.19) where C_5 depends only on n , while the center and radius of the ball B_R are completely arbitrary.

Exactly the same proof works in the case of n=2, with the following modification. Hoelder's inequality implies that

$$(4.22)\qquad \|u\|_\rho \leq V^{(2-\rho)/2\rho}\|u\|_2\ ,\ \text{with } V = \int_{B_R} dx = \pi R^2\ ,\ 1<\rho<2\ .$$

Here we set $u=|\nabla\zeta|$, and then apply prop.4.1 with p = r , for

$$(4.23)\qquad \|\zeta\|_{2\rho/(2-\rho)} \leq (\pi R^2)^{(2-\rho)/2\rho}\|\nabla\zeta\|_2\quad ,\ \rho < 2\ .$$

Now we combine (4.23) with (4.7), replacing p with p-1:

$$(4.24)\qquad \|\chi|u|^{p/2}\|_{2\rho/(2-\rho)} \leq c_\rho R^{(2-\rho)/\rho}\sqrt{p}\sigma^{1/2}\|\,|\nabla\chi|\,|u|^{p/2}\|_2\ .$$

with $c_\rho=c'\gamma\pi^{(2\gamma)^{-1}}$, $\gamma=\gamma(\rho)=\rho/(2-\rho)$, c' a constant independent of R,σ,ρ . Or,

$$(4.25)\qquad \|\chi^{2/p}u\|_{\gamma p} \leq (c_\rho R^{1/\gamma}(p\sigma)^{1/2})^{2/p}\|\,|\nabla\chi|^{2/p}u\|_p\ ,\ p\geq 2\ .$$

Notice that $\sum_0^\infty\gamma^{-k}=\rho/2(\rho-1)$. Thus, for $p_j=2.\gamma^j$ the iteration leading to (4.16) yields (4.16) again, but now with the constant

$$(4.26)\qquad c''R^{-1}(\sqrt{\sigma})^{\gamma(1-\gamma)} = c''R^{-1}(\Lambda_+/\lambda_-)^{\delta/4}\ ,$$

where c" depends on n only, and $\delta=\delta(\rho)=\rho/(\rho-1) > 2=n$, $\delta(\rho)\to 2$, as $\rho\to 2$. It is clear that this is somewhat weaker than required in thm.1.9, at least under estimate (1.9), but will be enough for an estimate (1.9) with '$O(1+R^2)$' replaced by '$O(1+R^{2-\varepsilon})$', $\varepsilon>0$. However, for n=2, a simpler proceedure, not using the above technique, still will give the same result as for general n, as will not be discussed here.

Proof of theorem 1.9. In view of thm.1.10 we may assume that $n\geq 2$, so that prop.4.2 holds. Let us first focus on assumption (1.10). For $R=r/2$, with the $r>0$ of (1.10) we conclude that

$$(4.27)\qquad \sup\{|u(x)|:|x|=\eta\} \leq K\int_{\eta-r\leq|x|\leq\eta+r}|u(x)|^2dx \to 0,\ \text{as } \eta\to\infty,$$

since $u\in L^2(\mathbb{R}^n)$, where K denotes a constant independent of η . Indeed, for each x_0 with $|x_0|=\eta$ we get (4.19) with the ball B_R around x_0 and a constant independent of x_0 and η . However the maximum principle (III, thm.5.8) holds, since $q\geq 0$. Therefore $\mu_\eta=\mathrm{Max}\{|u(x)|:|x|\leq\eta\}$ is assumed on $|x|=\eta$, and we indeed must have $\lim_{\eta\to\infty}\mu_\eta = 0$, or, $u=0$ in all of $\mathbb{R}^n$, showing that H_0 is essentially self-adjoint.

Next let us assume (1.9). Notice then that the sphere $\partial B_{3\eta/2}=\{x:|x|=3\eta/2\}$, for large η, may be covered by balls $B_{\eta/2}$ with center on $\partial B_{3\eta/2}$, and contained in the spherical shell $B_{2\eta}\setminus B_\eta$. The corresponding concentric balls $B_{\eta/4}$ still cover $\partial B_{3\eta/2}$, and the constants $\eta^{-n/2}(\Lambda_+/\lambda_-)^{n/4}$ are uniformly bounded, in view of (1.9). Accordingly, again using the maximum principle, we get

$$(4.28)\qquad |u(x)| \leq c\int_{\eta\leq|x|\leq 2\eta}|u(x)|^2dx \to 0\ ,\ \text{as } \eta\to\infty\ .$$

This again implies $u=0$, hence H_0 again is essentially self-adjoint, and theorem 1.9 is established.

Finally, in this section, we discuss a proof of prop.4.1. (cf.[GT_1]). We shall require Hoelder's inequality for n functions:

$$(4.29)\qquad \int u_1u_2\ldots u_k dx \leq \Pi_{j=1}^k\|u\|_{p_j}\ ,\ u\in L^{p_j}(\mathbb{R}^n)\ ,$$

valid if $p_1, \ldots , p_k \in \mathbb{R}$ satisfy the relation

$$(4.30)\qquad 1/p_1 + \ldots + 1/p_k = 1\ .$$

Also we use that, for k nonnegative numbers $a_1, \ldots, a_k$ the 'geometric mean' $(\Pi_{j=1}^k a_j)^{1/k}$ is never larger than the arithmetic mean $k^{-1}\Sigma_{j=1}^k a_k$. The latter is well known. Also, (4.29) is easily derived from the well known case of k=2, by induction.

For $u\in C_0^\infty(\mathbb{R}^n)$ write

$$(4.31)\qquad |u(x)| \le \int_{-\infty}^{x_j} |u_{|j}|dx_j \le \int_{-\infty}^{\infty} |u_{|j}|dx_j .$$

Taking the product of these n inequalities, and the power $1/(n-1)$, we get

$$(4.32)\qquad |u(x)|^{n/(n-1)} \le \Pi_{j=1}^n(\int |u_{|j}|dx_j)^{1(n-1)} .$$

The inequality (4.32) may be integrated with respect to each x_j, $j=1,2,\dots,n$. After each integration Hoelder's inequality is applied, with $k=n-1$, $p_1=p_2=\dots=p_{n-1}=n-1$. For example, the first factor, at right of (4.32) is independent of x_1. Hence there are n-1 unknown functions under the integral $\int dx_1$. We get

$$(4.33)\qquad \int dx_1 |u(x)|^{n/(n-1)} \le \{\int dx_1|u_{|1}| \ \Pi_{j=1}^n\int dx_1dx_j|u_{|j}|\}^{1/(n-1)} .$$

For each subsequent integration one of the n factors at right is a constant, and (4.29) may be applied with n-1 factors. We get

$$(4.34)\qquad \|u\|_1 \le (\Pi_{j=1}^n\|u_{|j}\|)^{1/n} \le n^{-1}\sum\|u_{|j}\|_1 \le n^{-1/2}\||\nabla u|\|_1 .$$

Notice that (4.34) amounts to (4.8) for $p=1$. In particular the inequality may be obtained for Lipschitz continuous $u \in C_0(\mathbb{R}^n)$ by an approximation argument.(For example one may use a mollifier as in the proof of III, thm 1.2).

The general inequality (4.8) then follows by applying (4.34) to the function $v=|u|^\gamma$, with $\gamma = p(n-1)/(n-p)$. We get $|\nabla|u|^\gamma| = \gamma|u|^{\gamma-1}|\nabla u|$. Substituting this into (4.34) and using Hoelder's inequality we conclude (4.8) for general p.

5. More criteria for essential self-adjointness.

In this section we discuss an extension of thm.1.5, useful in ch.8. Let us assume that the expression H , defined over Ω , has an essentially self-adjoint minimal operator H_0 . Let $U \subset \Omega$ be an open subdomain of Ω with compact smooth boundary ∂U, where ∂U is assumed to be an n-1-dimensional submanifold of Ω . Assume the potential q of H bounded below on U , and let $q'\in C^\infty(U)$ denote a nonnegative potential satisfying (1.4) in some neighbourhood of ∂U , with a suitable constant γ . Assume that $q'=0$ outside another neighbourhood of ∂U .

Theorem 5.1. Under the above assumptions the minimal operator H'_0 of the expression $H'=H+q'$, in the Hilbert space $L^2(U,d\mu)$, is essentially self-adjoint, whenever $\gamma\geq\gamma_1$, where γ_1 is a positive constant.

Proof. One may repeat the proof of thm.1.5 in the following amended form. Let $\tau(x)=\mathrm{dist}(x,\partial U)$, and define

$$(5.1) \qquad \sigma(x) = \eta^{-1}\log(\varepsilon/\tau) \text{ , as } \tau(x)\leq\varepsilon \text{ , } = 0 \text{ , as } \tau(x)>\varepsilon \text{ ,}$$

for suitable constants $\varepsilon,\eta>0$. Then use this redefined function $\sigma(x)$ as in the proofs of lemma 2.1 and lemma 2.2, to get essential self-adjointness of H' .

In particular the 'spheres' Σ_R again are defined by the inequality $\sigma(x)\leq R$. These, in general, are noncompact sets, so that (2.3), (2.4), (2.5) are correct only for $u,v\in C_0^\infty(\Omega)$. Now, if $f\in C^\infty(U)\cap L^2(U,d\mu)$ satisfies $H'f=0$ in U , then we get $\omega f \in \mathrm{dom}\, H$, for a suitable cutoff function, zero near ∂U , =1 outside a neighbourhood of ∂U in $U\cup\partial U$. Since H_0 (in Ω) is essentially self-adjoint it follows that there exists a sequence $z_k\in C_0^\infty(U)$, such that $z_k\to\omega f$, $Hz_k\to H\omega f$, $k\to\infty$, in $L^2(\Omega,d\mu)$. Using this, with a suitable limit $k\to\infty$ in (2.3), (2.4), (2.5) , one shows that the 'partial integration' , taking $\int H'\bar{f}f d\mu$ into $\int(|\nabla f|^2+q|f|^2)d\mu$ still will give no boundary terms at infinity of U , although there will be such terms at the boundary of Σ_R , of course.

In other words, one finds that inequality (2.7), for $q+q'$ instead of q , remains intact. Therefore the remainder of the proof in sec.2 may be repeated, and one finds that $f=0$. This proves thm.5.1.

Remark 5.2. It appears that a corresponding statement, regarding essential selfadjointness of powers ${H'_0}^m$, can be proven by repeating the chain of arguments in sec.3, with similar amendments. However, we have not checked this in detail.

CHAPTER 5. C^*-COMPARISON ALGEBRAS.

In this chapter we open the discussion of one of our main objectives: Comparison algebras are certain C^*-algebras of linear operators on an L^2-space. In essence their operators may be explicitly represented either as Cauchy-type singular integral operators or else as pseudo-differential operators of order 0. However, we never use (or even discuss) such explicit representations.

Except for a short excursion into Fredholm results for differential operators on compact manifolds in VI,3 and VI,4, in ch's V,VI and VII we will entirely focus on the discussion of the structure of such C^*-algebras. The reader should keep in mind, however, that the results obtained will be used later on (ch.X) to discuss the properties of differential operators. For certain differential tial expressions (said to be 'within reach' of a comparison algebra C) a realization will be defined (cf. def.6.2 below) In ch.X we will focus on the spectral (and Fredholm) properties of such realizations.

While we have discussed abstract spectral theory in ch.I, we refer to [C_1],app.A2, regarding an abstract theory of Fredholm operators. Also we will use some results of abstract spectral theory not discussed in detail in earlier chapters, such as the discussion of the essential spectrum of a differential operator. (Note that at least a survey of some such abstract results is made in ch.X.)

We work in the Hilbert space $\mathcal{H}= L^2(\Omega,d\mu)$, on a C^∞-manifold Ω , with a positive C^∞-measure $d\mu$, as introduced in ch.3. On Ω we also have a <u>comparison operator</u> H defined, where H is the (self-adjoint) Friedrichs extension of the minimal operator of a differential expression H, as in III,(0.1) . In [C_1] we considered the special noncompact manifold $\mathbb{R}^n$. There we introduced a class of C^*-algebras generated by multiplication operators and convolution operators. Especially the singular integral operators $S_j = D_j\Lambda$, $\Lambda=(1-\Delta)^{-1/2}$, were used as such convolution generators.

In the present approach, on a general compact or non-compact manifold Ω, we replace the generators S_j by operators of the form $D\Lambda = DH^{-1/2}$, with certain first order differential operators D defined on Ω , and with $\Lambda = H^{-1/2}$, with the above self-adjoint positive definite operator H .

Accordingly, a comparison algebra C is a C^*-algebra generated as norm closure from generators

$$(0.1) \qquad a\ ,\ D\Lambda\ ,\ a \in A^{\#}\ ,\ D \in \mathcal{D}^{\#}\ ,$$

with certain classes $A^{\#}$, $\mathcal{D}^{\#}$, of bounded C^∞-functions and first order linear partial differential expressions (folpde's), respectively.

We work with general assumptions on $A^{\#}$, $\mathcal{D}^{\#}$ which always imply that $C \supset K(H)$, with the ideal $K(H)$ of all compact operators on H . The commutator ideal E of C may or may not be $K(H)$ as well. At any rate, for all examples considered there is a chain

$$(0.2) \qquad C \supset E \supset K(H)\ ,$$

where the second inclusion may be proper or improper, but such that both C/E, and $E/K(H)$ are isometrically isomorphic to an algebra of continuous functions. Then the Fredholm property of an operator $A \in C$ is governed by one or two symbol functions, arising naturally from the above ideal chain. For more details and simple explicit examples we then refer to [C_1], where both kinds of ideal chains are obtained.

It should be noted that algebras with two-link ideal chains also have been studied elsewhere (cf. Dynin [Dy_1], Upmeyer, [Up_1]) although, as it seems, not in connection with noncompact manifolds In applications it is found that the two-inversion symbol calculus has its disadvantages, insofar as the second symbol usually is available only after constructing an explicit E-inverse, not available from the first symbol alone. In some later examples we have remedied this by extending the second symbol to the entire algebra C . Then usually the second symbol takes values in some algebra of singular integral operators over a lower dimensonal space. This will be discussed in [CMe_1], [CPo] .

In the present chapter we only discuss some basic facts. In particular it is proven that every comparison algebra C contains the compact ideal, and that $K(H)$ even is contained in the commutator ideal E of C (lemma 1.1). We explicitly discuss L^2-continuity of $D\Lambda$, and $\Lambda M\Lambda$, for certain first and second order expressions D and M, and derive explicit formulas for commutators (sec.2). In sec.3 we derive sufficient conditions for compactness of commutators in a comparison algebra. In sec.4 and 5 we discuss a variety of examples. Specifically sec.5 considers the one-dimensional case.

Finally, in sec.6, we prepare the later application of our theory by introducing the concept of an expression within reach of a comparison algebra. Specifically, a Taylor-type expansion (with integral remainder) is discussed for "H-compatible expressions", having all H-commutators within reach. Such expansions prove useful as an efficient substitute for the calculus of pseudodifferential operators, which we do not assume here.

1. Comparison operators and comparison algebras.

Let Ω denote a paracompact C^∞-manifold without boundary, with a countable atlas, as in III,1, and again consider a measure $d\mu=\kappa dx$, and a second order elliptic expression

$$(1.1) \qquad H = -\kappa^{-1}\partial_{x^j} h^{jk}\kappa\partial_{x^k} + q \quad ,$$

which is self-adjoint with respect to the inner product of the Hilbert space $H = L^2(\Omega,d\mu)$. As in III,1 we assume that H has real C^∞-coefficients, that $((h^{jk}))>0$, and that $H_0 \geq 1$, i.e.

$$(1.2) \qquad (u,Hu) \geq (u,u) \ , \text{ for all } u \in C_0^\infty(\Omega) \ .$$

Then we introduce the first Sobolev norm and inner product by

$$(1.3) \qquad (u,H_0 v) = \int_\Omega (h^{jk}\bar{u}_{|x^j} v_{|x^k} + q\bar{u}v)d\mu = (u,v)_1 \ , \ u,v \in \text{dom } H \ .$$

The Friedrichs extension H is a well defined self-adjoint realization of H. We get $H \geq 1$. Clearly $H^{-1} \in L(H)$ has a self-adjoint positive square root $\Lambda = H^{-1/2} = \sum_{k=0}^\infty (-1)^k \binom{1/2}{k}(1-H^{-1})^k$, and $0<\Lambda\leq 1$. Λ is an isometric isomorphism between H and H_1. We get

$$(1.4) \quad H_1 = \text{dom } \Lambda^{-1} \ , \ \|u\| = \|\Lambda u\|_1 \ , \ u \in H \ , \ \|v\|_1 = \|\Lambda^{-1}v\| \ , \ v \in H_1 .$$

The restriction $\Lambda^{-1}|\text{dom } H_0$ is essentially self-adjoint, and

$\Lambda^{-1}\mathrm{dom}\, H_0$ is dense in H , while the corresponding is not always true for H dom H_0 = im H_0. For details cf.III,1. In fact, we already discussed this extensively. The operator H_0 will have to be essentially self-adjoint for this.

We also note the possibility of transforming the dependent variable,to get from the triple $\{\Omega,d\mu,H\}$ to an 'equivalent triple' $\{\Omega,d\mu^{\sim},H^{\sim}\}$ with possibly better properties,such as a normal form for H , etc. (cf.III,6).

If Ω,$d\mu$,and H are given as described we shall speak of the comparison triple $\{\Omega,d\mu,H\}$, and of the comparison operator H on Ω. Also H_1 will be called the corresponding first Sobolev space.

Let a comparison triple be given, and let

$$(1.5)\qquad D = b^j\partial_{x^j} + p$$

denote a first order differential expression on Ω , where (b^j) and p denote a contra-variant tensor field and a complex-valued function on Ω , both assumed C^∞ . Then the linear operator $D\Lambda: C_0^\infty(\Omega)\to L^2_{loc}(\Omega)$ is well defined, since the square root Λ of H^{-1} maps onto the first Sobolev space H_1 (cf.III,(1.6)). In fact, assuming that

$$(1.6)\qquad \|D\|_H = \|(|b|^2 + |p|^2/q)^{1/2}\|_{L^\infty(\Omega)} < \infty ,$$

(with $|b|=\{h_{jk}\bar{b}^j b^k\}^{1/2}$) , it follows that

$$(1.7)\qquad \|Du\|^2 \le \|D\|_H^2(u,Hu) = \|D\|_H^2\|u\|_1^2 ,$$

provided that $q \ge 0$ on Ω .Using (1.4) , it then follows that

$$(1.8)\qquad \|D\Lambda v\| \le \|D\|_H\|v\| \ , \ v\in \mathrm{im}(\Lambda^{-1}|C_0^\infty(\Omega)) ,$$

where the space $\mathrm{im}(\Lambda^{-1}|C_0^\infty(\Omega))$ is dense in H. Accordingly, by continuous extension of $D_0\Lambda$, we get a bounded operator $A\in L(H)$, and it is clear that $A = D\Lambda$, with Λ as a map $H\to H_1$, and a differential operator $D:H_1\to H$, defined with the strong L^2-gradient of III,(1.6) .

The adjoint operator A^* of $A\in L(H)$ is in $L(H)$ again, of course, and we have

$$(1.9)\qquad A^*u = \Lambda D^*u \ , \ u\in C_0^\infty(\Omega) \ , \ D^* = -\kappa^{-1}\partial_{x^j}\kappa\bar{b}^j + \bar{p} .$$

In chapter X we will study the Fredholm properties of certain C^*-subalgebras of the algebra $L(H)$, called H-comparison algebras. Here by an H-comparison algebra we mean a C^*-algebra $C \subset L(H)$ generated as operator norm closure of the smallest algebra C^0 containing all multiplication operators $u \to au$, and all operators A, A^* of the above form $A = D\Lambda$, where $a \in A^\#$, $D \in \mathcal{D}^\#$. Here $A^\#$, $\mathcal{D}^\#$ denote a class of bounded continuous complex-valued functions, and a class of first order linear partial differential expressions (folpde's) on Ω, of the form (1.5), with $\|D\|_H < \infty$. We will write $C = C(A^\#,\mathcal{D}^\#)$, $C^0 = C^0(A^\#,\mathcal{D}^\#)$, omitting the arguments if no confusion can arise.

We require the following general assumptions:

Condition (a_0): $A^\#$ contains $C_0^\infty(\Omega)$, and its complex conjugates.

Condition (a_1): $A^\#$ contains the constant functions.

Condition (d_0): $\mathcal{D}^\#$ contains its complex conjugates $\bar{D}=\bar{b}^j\partial_{x^j}+\bar{p}$, and contains all C^∞-folpde's with compact support.

Here (a_0) and (d_0) are always imposed, and (a_1) is required unless explicitly stated otherwise. Especially there are two exceptions to this last rule: The minimal algebra J_0, defined below, and the class $A^\#_{U,0}$ of VIII,3 normally will not contain all constant functions.

In these cases we have the following condition instead.

Condition (a_1'): For every $D \in \mathcal{D}^\#$ there exists an $a \in A^\#$ such that ∇a has compact support, and $aD = Da = D$.

For every comparison triple we have a minimal comparison algebra, denoted by J, obtained by choosing $A^\#$ as the class of functions $a=a_0+c$, $a_0 \in C_0^\infty(\Omega)$, $c \in \mathbb{C}$, and $\mathcal{D}^\#$ as the class of folpdes $D = b^j\partial_{x^j}+p$ with $b^j \in C_0^\infty$, $p \in C_0^\infty(\Omega)$. The C^*-algebra (not necessarily with unit) generated by norm closing the algebra generated by a and $D\Lambda$, for all $C_0^\infty(\Omega)$-functions a and C^∞-folpde's with compact support, will be denoted by J_0. Clearly J is obtained from J_0 by adjoining the identity 1. We shall see in VI,1 that J and J_0 have compact commutators. J_0 is unital if and only if Ω is compact. Both J and J_0 will be called minimal comparison algebras.

Lemma 1.1. The minimal algebra J_0, and hence every comparison algebra C, always contains the entire ideal $K(H)$ of compact operators. Moreover, $K(H)$ even is contained in the commutator ideal $E = E(C)$ of any comparison algebra C, and we have $E(J_0)=K(H)$.

In lemma 1.1. the commutator ideal $E(C)$ of C denotes the smallest closed 2-sided ideal containing all the commutators of operators in C .

Proof. As in $[C_1]$,IV,thm.1.1, we use the fact that an irreducible C^*-subalgebra A of $L(H)$, containing at least one nontrivial compact operator, must contain the entire compact ideal (cf.Dixmier $[Dx_1]$, cor.4.1.10). Here 'irreducible' means that the space Au is dense in H , for every $0 \neq u \in H$.

It is easy to construct nontrivial compact operators in a comparison algebra. For example, for any function $a \in C_0^\infty(\Omega)$, not vanishing identically we have $a\Lambda = aH^{-1/2} \in K(H)$, by III,thm.3.7, below, while trivially $a\Lambda \neq 0$. In fact, also the commutator $[a,b\Lambda]$, for suitable $a,b \in C_0^\infty(\Omega)$ is compact and $\neq 0$. Hence it is sufficient to show that Cu is dense in H , for every $0 \neq u \in H$. (Clearly then the construction of $[C_1]$,IV,lemma 3.3 may be repeated to show that $K \subset J_0$, and, starting with the given compact commutator $\neq 0$, that $K \subset E$ holds as well. Indeed, one first will construct an operator of rank 1 in C or E , of the form $\phi\rangle\langle\psi$, with $\phi,\psi \in H$, $\phi,\psi \neq 0$. Then the operator $A(\phi\rangle\langle\psi)B = (A\phi)\rangle\langle(B^*\psi)$ comes arbitrarily close to every dyad $f\rangle\langle g$, $f,g \in H$, as $A,B \in L(H)$. Accordingly C and E contain all such dyads, and hence $K(H)$, the closure of the class of operators of finite rank.)

Note that the operator H^{-1} may be written as an integral operator

$$H^{-1}u(x) = \int_\Omega G(x,x')u(x')d\mu \tag{1.10}$$

where the 'Green's function' $G(x,x')$ is positive on all of Ω . Indeed, from III,thm.6.2 we know that $H \sim H^\sim = 1 - \Delta_{\mu^\sim}$, after a suitable change of dependent variable. Such an operator satisfies the conditions of the maximum principle, which insures that the Green's function is positive, since G solves $HG(.,x')=0$, and is $\sim c|x-x'|^{2-n}$ (or $c \log|x-x'|$) near its singularity $x=x'$, and is limit of the sequence of Green's functions G_k of the Dirichlet problem on an exhausting sequence Ω_k of subdomains of Ω with smooth boundary .Clearly the G_k are positive, since their boundary values are.Hence G is positive. For details cf. III, thm.6.8.

For any $0 \neq u \in H$ we have $0 \neq H^{-n}cu$ continuous for a suitable $C_0^\infty(\Omega)$-function c, by the Sobolev imbedding lemma, (VI,1.4.2), or since the iterated Green's function has the corresponding smoothening property). Hence we may choose $a \in C_0^\infty(\Omega) \subset A^\#$ such that $aH^{-n}cu = a\Lambda^{2n}cu$ does not change sign, and is not always zero. Assume it ≥ 0, without loss of generality. Then $w = \Lambda^2 a\Lambda^{2n}cu$ is positive on Ω, and every $v \in C_0^\infty$ can be written as $v = b\Lambda^2 a\Lambda^{2n} cu$, where $b = v/w \in C_0(\Omega)$. Choose $b_k \in C_0^\infty(\Omega)$ with $\|b-b_k\|_{L^\infty} \to 0$, then let $A_k = b_k\Lambda^2 a\Lambda^{2n} c$, and note that $\|v - A_k u\| \to 0$. Furthermore, if $0 \le \phi_k \le 1$, $\phi_k \in C_0^\infty(\Omega)$, $\phi_k(x) \to 1$ for all $x \in \Omega$, then $\Lambda\phi_j \to \Lambda$, hence $(\Lambda\phi_j)^2 \to \Lambda^2$, $(\Lambda\phi_j)^{2n} \to \Lambda^{2n}$, in strong operator convergence. Thus we also get $\|v - B_k u\| \to 0$, for $B_k = b_k(\Lambda\phi_{j_k})^2 a(\Lambda\phi_{j_k})^{2n} c \in J_0$, with a suitable sequence $\{j_k\}$ of integers , $j_k \to \infty$. This shows that indeed $J_0 u$ is dense in H , q.e.d.

For a comparison algebra C we focus on the commutator ideal E , i.e., the closed 2-sided ideal generated by the commutators. Note that C normally has a unit (the function $\equiv 1$). The quotient C/E, as a unital C^*-algebra, is isometrically isomorphic to a function algebra $C(\mathbb{M})$, with a certain compact space $\mathbb{M}$, called the symbol space, by the Gelfand-Naimark theorem. The homomorphism $\sigma: C \to C/E \to C(\mathbb{M})$ assigns to $A \in C$ a function σ_A, called symbol of A. In sec. 5 we will prove that $\mathbb{M}$ always contains a subset $\mathbb{W}$ homeomorphic to the bundle of unit spheres in the cotangent space $T^*\Omega$ of Ω . We call $\mathbb{W}$ the wave front space (of the algebra C). The continuous function $\sigma_A : \mathbb{M} \to \mathbb{C}$ corresponding to the coset mod E of an operator $A \in C$ coincides with the principal symbol of D on $\mathbb{W}$ if $A = D\Lambda$, with a differential operator D .

In the simplest cases we will get $E = K(H)$ equal to the ideal of compact operators on H . We already mentionned that this is true for the minimal algebras J and J_0 . The space $\mathbb{W}$ mentionned above coincides with the maximal ideal space of J_0/K . On the other hand the set $\mathbb{M} \setminus \mathbb{W}$ can be described as a certain set located "over ∞ of the manifold Ω ". Then an operator $A \in C$ will be Fredholm if and only if $\sigma_A \neq 0$ on all of $\mathbb{M}$. However, in some more general cases we have a proper inclusion $E \supset K(H)$. Then we shall tend to obtain a similar "function structure" for the quotient E/K. The Fredholm theory then will be somewhat more complicated.

The techniques to be described in the following can be extended to other types of differential operators as comparison operators, not necessarily of second order, or elliptic, or self-adjoint. However, the fact that so many properties of these 'Schroedinger operators' are known makes the use of them particularly attractive. In particular the e.s.a.-properties discussed in ch.5 will be important, in this connection.

For many parts of this chapter it will be crucial to have Carleman's definite case of II,1 , for the operator H_0 , in the Hilbert space H . In other words, no boundary conditions at infinity will be required ; the Friedrichs extension is the only self-adjoint extension of H_0 , and it coincides with H_0^{**}, the closure of H_0 .

We shall use this condition in weaker or stronger form, as follows:

Condition (w) : H_0 is essentially self-adjoint.

Condition (s) : For all m = 1,2,.., the minimal operator $(H^m)_0 = H_0{}^m$ of the m-th power H^m is essentially self-adjoint.

Condition (s_k): (where k = 1,2,...,∞) : $(H^k)_0 = H_0{}^k$ is essentially self-adjoint.

It follows easily that (s_k) implies (s_l), for $1 \le l \le k$. Accordingly (s_1) equals (w) and (s_∞) equals (s) (cf.IV, prop.1.4).

Various results asserting the validity of (w) or (s) have been discussed in ch.4. In particular (s) holds whenever the manifold Ω is complete under the metric (1.7). On the other hand we never will have (w) nor (s) if Ω is the interior of a manifold with (nonvoid) boundary onto which the triple may be extended.

2. Differential expressions of order not larger than two.

For a comparison triple {Ω,dμ,H} , as in section 1, let us consider more general differential expressions on the manifold Ω . We speak of a differential expression L of order N on Ω if a linear operator $L: C^\infty(\Omega) \to C^\infty(\Omega)$ is defined such that locally(within a chart with coordinates x) Lu is given by a differential expression of the form II,(1.1). Using the measure $d\mu = \kappa dx$ a general ($\le$) second order differential operator M on Ω always may be written in the form

$$(2.1) \qquad M = -\frac{1}{\kappa} \partial_{x^j} \kappa m^{jk} \partial_{x^k} + b^j \partial_{x^j} + p \quad ,$$

with contravariant C^∞-tensor fields m^{jk} , b^j , of second and first degree, and with a C^∞-function p over Ω . This follows because for an N-th order differential operator on Ω the local principal part defines a contravariant tensor of N indices. For N = 2 , and any tensor field a^{jk} one finds that $-\kappa^{-1} \partial_{x^j} \kappa a^{jk} \partial_{x^k} = A$ defines a global second order differential operator with principal part $-a^{jk}$. Indeed, for u,v with support in a chart one gets $(u,Av) = \int a^{jk}\bar{u}_{|x^j} v_{|x^k} d\mu$ which is coordinate invariant, by the well known transformation properties of tensors. For a general second order operator M one may subtract this operator, formed with its principal part $a^{jk} = m^{jk}$ to obtain an operator of order one or zero. But every first order global differential operator is of the form (1.5). Therefore (2.1) gives the general 2-nd order operator. For additional symmetry and later convenience let us replace (2.1) by

$$(2.2) \qquad M = -\tfrac{1}{\kappa}\partial_{x^j}\kappa m^{jk}\partial_{x^k} + b^j\partial_{x^j} - \kappa^{-1}\partial_{x^j}\kappa c^j + p \quad ,$$

which is only formally more general, of course.

Now the comparison operator H , or rather its inverse square root Λ of (1.4) may be used to convert a general first or second order operator into an L^2-bounded operator. For D of the form (1.5) and M of the form (2.1) we formally consider the operators

$$(2.3) \qquad A = D\Lambda \quad , \quad B = \Lambda M \Lambda \quad .$$

It is not hard to set up conditions insuring that A and B of (2.3) define bounded operators of H . In section 1 we already found that

$$(2.4) \qquad \|A\| = \|D\Lambda\| \le \|D\|_H \ ,$$

where $\Lambda : H \to H_1$, and $D : H_1 \to H$, with the strong L^2-gradient in D , and with $\|D\|_H$ of (1.6). Similarly one may look at

$$(2.5) \quad (u,Mv) = \int(m^{jk}\bar{u}_{|x^j}v_{|x^k} + \bar{u}b^j v_{|x^j} + \bar{c}^j\bar{u}_{|x^j}v + p\bar{u}v)d\mu, \ u,v \in C_0^\infty(\Omega) \ ,$$

by a partial integration. For the integrand m of (2.5) one gets

$$(2.6) \quad |m(u,v)|^2 \le c_1^2\, h(u)h(v) \ , \ h(u) = h^{jk}\bar{u}_{|x^j}u_{|x^k} + q|u|^2 \ ,$$

assuming finiteness of (2.7) , with the supremum over all $x \in \Omega$:

$$(2.7) \quad c_1 = \sup\{(h_{ij}h_{kl}\bar{m}^{ik}m^{jl} + h_{jk}\bar{b}^j b^k/|q| + h_{jk}\bar{c}^j c^k/|q| + |p|^2/|q|^2)^{1/2}\},$$

by a calculation. From (2.5),(2.6),(2.7), and (1.4) we get

$$(2.8) \qquad |(u,Mv)| \le c_1\|u\|_1\|v\|_1 = c_1\|\Lambda^{-1}u\|\|\Lambda^{-1}v\| \ ,$$

which yields

$$(2.9) \qquad |(u,Bv)| \le c_1\|u\|\|v\| \quad , \ u,v \in \Lambda^{-1}\mathrm{dom}\, H_0 \quad .$$

Recall that $\Lambda^{-1}\mathrm{dom}\, H_0$ is dense in H, so that (2.9) implies

(2.10) $$\|Bu\| \le \|M\|_H \|u\| \ , \ \|M\|_H = c_1 \ (\text{with } c_1 \text{ of } (2.7)) \ ,$$

for all u of a dense subspace. It follows that B admits a continuous extension to H, with the operator bound of (2.10).

Finally it may be observed that we may improve on the estimates (2.4) and (2.10) by replacing $\|D\|_H$ and $\|M\|_H$ with

(2.11) $$\inf\{\|D^{\sim}\|_{H^{\sim}}\} \ , \ \inf\{\|M^{\sim}\|_{H^{\sim}}\} \ ,$$

where the infimum is taken over all operators $\gamma^{-1}D\gamma = D^{\sim}$, $\gamma^{-1}M\gamma = M^{\sim}$, $\gamma^{-1}H\gamma = H^{\sim}$, $\gamma \in C^{\infty}(\Omega)$, $\gamma>0$, representing D, M and H after a transformation of dependent variable, preserving the condition $q > 0$. In fact, only one such norm needs to be finite, in order to get A or B in $L(H)$.

In the next few sections we mainly shall focus on first order expresssions D with finite $\|D\|_H$, so that $A=D\Lambda$ is bounded in H. Our aim will be the investigation of algebras generated by classes of operators $D\Lambda$ together with classes of <u>multiplication operators</u> $Au(x) = a(M)u(x) = a(x)u(x)$. Of crucial interest will be the commutators of two generating operators $[A,B]$ with A and B of either form $D\Lambda$, or $a(M)$. In that respect we look at commutators of the forms $[a,D]$, $[D,F]$, $[H,a]$, and $[H,D]$, for $a = a(M)$, and first order expressions

(2.12) $$D = b^j\partial_{x^j} + p \quad , \quad F = c^j\partial_{x^j} + r \quad .$$

One calculates that

(2.13) $$[D,F] = (b^j c^k{}_{|x^j} - c^j b^k{}_{|x^j})\partial_{x^k} + (b^j r_{|x^j} - c^j p_{|x^j}) \ ,$$

which is another first order expression.

For the commutator $[H,D]$ it is useful to introduce covariant derivatives with respect to the Riemannian connection of the metric tensor h_{jk}. With the Christoffel symbols (cf.app.B)

(2.14) $$\Gamma^i_{jk} = \tfrac{1}{2}h^{il}(h_{jl|x^k} + h_{kl|x^j} - h_{jk|x^l})$$

write

(2.15) $$b^i{}_{|j} = b^i{}_{|x^j} - b^k\Gamma^i_{kj} \ , \ b^{j|k} = b^j{}_{|l}h^{lk} \ .$$

<u>Proposition 2.1.</u> We have

(2.16) $[H,D] = -\frac{1}{\kappa}\partial_{x^j}\kappa(b^{j|k}+b^{k|j})\partial_{x^k} - D_{\overline{p}^*} + D_{\overline{p}}^* - b^j q_{|x^j}$,

where we define

(2.17) $D_a = h^{jk} a_{|x^j}\partial_{x^k}$, $a \in C^1(\Omega)$,

and where p^* is defined by

(2.18) $D^* = -\frac{1}{\kappa}\partial_{x^j}\kappa\overline{b}^j + \overline{p} = -\overline{b}^j\partial_{x^j} + p^*$, $p^* = \overline{p} - \frac{1}{\kappa}(\kappa\overline{b}^j)_{|x^j}$.

The proof is a somewhat lengthy calculation. For $D_0 = b^j\partial_{x^j}$, $\Delta_\mu = \kappa^{-1}\partial_{x^j}\kappa h^{jk}\partial_{x^k}$ write $[H,D] = -[\Delta_\mu, D_0] - [\Delta_\mu, p] + [q, D_0]$, where $[q,D_0] = -b^j q_{|x^j}$. For $u,v \in C_0^\infty(\Omega)$ we get

$$-(u,[\Delta_\mu,p]v) = (\nabla u,\nabla(pv)) - (\nabla(\overline{p}u),\nabla v) = (\nabla u,(\nabla p)v)$$

$$-(u,(\nabla p).\nabla v) = (D_{\overline{p}}u,v) - (u,D_p v) \quad ,$$

with D_a as in (2.17). Accordingly,

(2.19) $-[\Delta_\mu, p] = D_{\overline{p}}^* - D_p$.

Similarly ,

$$-(u,[\Delta_\mu,D_0]v) = (\nabla u,\nabla D_0 v) - (\nabla(D_0^* u),\nabla v) \quad ,$$

using that $D_0^* = -\overline{D}_0 - \overline{c}$, $c = \kappa^{-1}(\kappa b^j)_{|x^j}$ .Therefore we get

$$-(u,[\Delta_\mu,D_0]v) = (\nabla u,\nabla D_0 v) + (\nabla\overline{D}_0 u,\nabla v) + (\nabla(\overline{c}u),\nabla v)$$

$$= \int (h^{ik}b^j(\overline{u}_{|x^i}v_{|x^k})_{|x^j} + h^{ik}\overline{u}_{|x^i}b^j_{|x^k}v_{|x^j} + h^{ik}b^j_{|x^i}\overline{u}_{|x^j}v_{|x^k})d\mu$$

$$+ (\nabla u, c\nabla v) + (u, D_c v)$$

$$= \int (h^{ij}b^k_{|x^j} + h^{jk}b^i_{|x^j} - h^{ik}_{|x^j}b^j)\overline{u}_{|x^i}v_{|x^k}d\mu + (u,D_c v) \quad .$$

Now we use (2.15) to express the derivatives $b^j_{|x^k}$ by covariant derivatives. In particular we may differentiate $h^{ij}h_{kj} = \delta^i_k$ for x_l and solve for $h^{im}_{|x^l}$ to obtain

(2.20) $h^{im}_{|x^l} = -h^{km}\Gamma^i_{kl} - h^{ij}\Gamma^m_{jl}$, using (2.14) . With (2.20) we get

$$(2.21)\qquad \begin{aligned} & h^{ij}b^{k}{}_{|x^j} + h^{jk}b^{i}{}_{|x^j} - h^{ik}{}_{|x^j}b^{j} \\ & = h^{ij}b^{k}{}_{|j} + h^{jk}b^{i}{}_{|j} = b^{i|k} + b^{k|i} \quad . \end{aligned}$$

Accordingly,

$$(2.22)\qquad -[\Delta_\mu, D_0] = -\kappa^{-1}\partial_{x^i}\kappa(b^{i|k} + b^{k|i})\partial_{x^k} + D_c \quad .$$

and formula (2.16) follows. Q.e.d.

Since we introduced covariant derivatives for the commutator $[H,D]$ we may as well use them in (2.13). This gives the formula

$$(2.23)\qquad [D,F] = (b^j c^k{}_{|j} - c^j b^k{}_{|j})\partial_{x^k} + (b^j r_{|j} - c^j p_{|j}) \quad ,$$

noting that first covariant derivatives of scalars are ordinary derivatives. The symmetry $\Gamma^i_{jk} = \Gamma^i_{kj}$ was also used for (2.23).

A simple calculation also gives

$$(2.24)\qquad [D,a] = b^j a_{|j} \;, \quad [H,a] = D_{\bar{a}}^{*} - D_a \quad ,$$

with D_a of (2.17).

Let us note, finally, that (2.23) is valid with any convenient affine connection, while (2.16) only holds for the Riemannian connection of the metric tensor h^{ik} .

It is easy now to set up conditions for boundedness in H of the operators

$$(2.25)\qquad [D,a]\;,\; [H,a]\Lambda\;,\; [D,F]\Lambda\;, \text{ and } \Lambda[H,D]\Lambda\;,$$

by just arranging for finiteness of the corresponding norms (1.6) or (2.10).

Proposition 2.2. Let the tensors b^j and c^j be bounded , together with their first (Riemannian) co-variant derivatives. Assume that

p and r are $O(\sqrt{q})$, and that ∇p , ∇r , ∇p^* are $O(\sqrt{q})$. Assume $q \geq 1$ on Ω , and that $\nabla q = O(q)$. Also let a , ∇a , and $\Delta_\mu a$ be bounded . Then all operators (2.25) are bounded in H .

The proof is evident, after the above calculations.

The result will have to be refined, before it becomes useful. This we shall attempt in the sections, below.

3. Compactness criteria for commutators.

Throughout this section we shall assume a triple $\{\Omega,d\mu,H\}$ given with $H = q - \Delta_\mu$ of (1.1) satisfying (1.2), and such that condition (w) holds: the minimal operator H_0 is essentially self-adjoint (which implies that im $H_0 = H(C_0^\infty(\Omega))$ is dense in $\mathcal{H}$). For most results we also will refer to a sub-extending $\{\Omega^\wedge,d\mu^\wedge,H^\wedge\}$ (c> $\{\Omega,d\mu,H\}$ (which may be $\{\Omega,d\mu,H\}$ itself), and assume here in addition to III,(3.24) that the tensors h^{jk} and $h^{\wedge jk}$ coincide on Ω, and that $q^\wedge \geqslant 1$ outside a compact set $K \subset \Omega^\wedge$ (we may achieve $K=\emptyset$ by (1.2) and III,prop.6.2, changing variables). In particular the norm of a tensor is unambiguously defined, since the two metric tensors h^{jk} and $h^{\wedge jk}$ coincide. Generally (w) will be required for $\{\Omega,d\mu,H\}$, but not for $\{\Omega^\wedge,d\mu^\wedge,H^\wedge\}$. In view of IV,thm.1.5 it is natural to expect q to tend to ∞ near $\partial\Omega \subset \Omega^\wedge$ whenever (w) holds. Accordingly thm.3.1, lemma 3.2 and lemma 3.3 often will have the best use for $\Omega = \Omega^\wedge$, although they are formulated for general $\Omega \subset \Omega^\wedge$. (Indeed the condition $\nabla a=o_\Omega(q^{\varepsilon/2})$ is weaker than $\nabla a^\wedge = o_{\Omega^\wedge}(q^{\wedge\varepsilon/2})$ near a point $x_0 \in \partial\Omega$, whenever $q \to \infty$ as $x \to x_0$, since $a^\wedge$ and $q^\wedge$ must stay continuous near such a point.) On the other hand, thm.3.4 and the lemmata following it strongly depend on condition (3.13), below, which cannot be satisfied for q near a point $x_0 \in \partial\Omega \subset \Omega^\wedge$ unless q stays bounded there (which is impossible if (w) is to hold). (This reflection assumes Ω to look like a manifold with smooth boundary near x_0.) Note that near such point x_0 the conditions (3.16) still allow some degeneration of the coefficients b^j, p of a differential expression D, but weaker than near infinity (of $\Omega^\wedge$).

The sequence of results, below, discusses conditions on the generating sets $\mathcal{A}^\#$ and $\mathcal{D}^\#$ implying compactness of commutators in the comparison algebra $C(\mathcal{A}^\#,\mathcal{D}^\#)$. Thm.3.1 deals with commutators of the form $[a,D\Lambda]$, while thm.3.4 looks at $[D\Lambda,F\Lambda]$, $a \in \mathcal{A}^\#$, $D,F \in \mathcal{D}^\#$. We use the notation $o_\Omega(.)$, etc., of app.A.

Theorem 3.1. Let $a\in C^\infty(\Omega)$ satisfy the condition $\nabla a=o_{\Omega^\wedge}(q^{\wedge\varepsilon/2})$ (in Ω) for some $0\leq\varepsilon<1$, and let the first order expression D (over Ω), of the form (2.2) satisfy $b = O(1)$, $p = O(\sqrt{q})$. Then we have

$$(3.1)\qquad [a,D\Lambda] \in \mathcal{K}(\mathcal{H}) .$$

More precisely, the unbounded operator $[a,D\Lambda] = aD\Lambda - D\Lambda a$ is densely defined, and extends continuously to a compact operator.

Proof. For $u \in \text{im } H_0 \subset C_0^\infty(\Omega)$ write $[a,D\Lambda]u = -b^j a_{|x^j}\Lambda u + D[a,\Lambda]u$.

Here the first term is of the form Cu , with $C \in K(H)$, due to our conditions on a and b (we get $b^j a_{|x^j} = o_{\Omega^\wedge}(q^{\wedge\varepsilon/2}) = o_{\Omega^\wedge}(q^{\wedge 1/2})$ (in Ω)), by III,thm.3.7. For the second term we will need the two lemmata, below.

Lemma 3.2. Let $0 < \varepsilon \leq 1$ be given, and let the function $a \in C^\infty(\Omega)$ satisfy

$$\nabla a = o_{\Omega^\wedge}((q^\wedge)^{\varepsilon/2}) \quad . \tag{3.2}$$

Then the commutator $[a,\Lambda^2] = a\Lambda^2 - \Lambda^2 a$, defined on $\text{im } H_0 = HC_0^\infty(\Omega)$, extends to an operator in $K(H)$. Moreover, there exists compact operators $C_j \in K(H)$, $j=1,2$, such that

$$[a,\Lambda^2] = \Lambda^{2-\varepsilon}C_1\Lambda + \Lambda C_2\Lambda^{2-\varepsilon} \quad . \tag{3.3}$$

Proof. Use App.A, L.A.1 to construct $0 < \gamma \in C^\infty(\Omega^\wedge)$, $\gamma = o_{\Omega^\wedge}(q^{\wedge\varepsilon/2})$, $\nabla a = O(\gamma)$, and define $\beta = \gamma^{1/\varepsilon}$, so that $\nabla a = O(\beta^\varepsilon)$, $\beta = o_{\Omega^\wedge}(\sqrt{q^\wedge})$, i.e., $\beta^\varepsilon\Lambda^\varepsilon \in K(H)$, by III,cor.3.8. For $u \in \text{im } H_0$ we have $u, \Lambda^2 u \in C_0^\infty$. Therefore, using (2.26), we get

$$a\Lambda^2 u - \Lambda^2 au = \Lambda^2 Ha\Lambda^2 u - \Lambda^2 aH\Lambda^2 u = \Lambda^2[H,a]\Lambda^2 u = \Lambda^2 D_{\bar{a}}^{*}\Lambda^2 u - \Lambda^2 D_a\Lambda^2 u$$

$$= \Lambda(\beta^{-\varepsilon}D_{\bar{a}}\Lambda)^*(\beta^\varepsilon\Lambda^\varepsilon)\Lambda^{2-\varepsilon}u - \Lambda^{2-\varepsilon}(\Lambda^\varepsilon\beta^\varepsilon)(\beta^{-\varepsilon}D_a\Lambda)\Lambda u \ .$$

Since $\beta^\varepsilon\Lambda^\varepsilon \in K(H)$, and $\beta^{-\varepsilon}D_a$ has bounded coefficients, so that (2.6) implies $\beta^{-\varepsilon}D_a\Lambda \in L(H)$, we conclude that (3.3) holds. It is trivial then, that also $[a,\Lambda^2] \in K(H)$, q.e.d.

Lemma 3.3. Let $0 < s < 2$, and let the constants $\varepsilon,r,\rho,t,\tau$ satisfy

$$0 \leq \varepsilon,r,t \leq 1 \ , \ 0 \leq \rho,\tau < 2-\varepsilon \ , \ r+\rho \ , \ t+\tau < 1+s-\varepsilon \ . \tag{3.4}$$

Then, if we have (3.2) with ε of (3.4) for some $a \in C^\infty(\Omega)$, we get

$$[a,\Lambda^s] = a\Lambda^s - \Lambda^s a = \Lambda^\rho C_1\Lambda^r + \Lambda^t C_2\Lambda^\tau \ , \tag{3.5}$$

where C_1, C_2 are operators in $K(H)$.

Proof. The self-adjoint operator Λ^2 has its spectrum on $[0,1]$, and its resolvent $R(\lambda) = (\Lambda^2 - \lambda)^{-1}$ is a holomorphic function from

$\mathbb{C} \setminus [0,1]$ to $L(H)$. Accordingly ,

$$\Lambda^s = i/2\pi \int_\Gamma R(\lambda)\, \lambda^{s/2} d\lambda \quad , \quad s > 0 , \tag{3.6}$$

with $\lambda^{s/2} = e^{s/2(\log|\lambda| + i \arg \lambda)}$, $|\arg \lambda| < \pi$,and the positively oriented circle $\Gamma = \{|\lambda - 1| = 1\}$.

The integral in (3.6) exists as an improper Riemann integral in norm convergence of $L(H)$. Its integrand is continuous, except at $\lambda = 0$, where we have $R(\lambda)\, \lambda^{s/2} = O(|\lambda|^{-1+s/2})$ as $\lambda \in \Gamma$. We get

$$[a,R(\lambda)] = R(\lambda)(\Lambda^2-\lambda)aR(\lambda) - R(\lambda)a(\Lambda^2-\lambda)R(\lambda) = R(\lambda)[\Lambda^2,a]R(\lambda),$$

as follows for $u \in \text{im}(H_0-\lambda) = (H-\lambda)C_0^\infty$, which is also dense in H, for all λ in the resolvent set, since H_0 is essentially self-adjoint (cf.I,thm.3.1). Then the same follows for $u \in H$, by continuous extension.Next we apply (3.3) with ε as in (3.2), and $\varepsilon,r,t,\rho,\tau$ satisfying (3.4) . It follows that

$$\lambda^{s/2}[a,R(\lambda)] = \Lambda^\rho C_3(\lambda)\Lambda^r + \Lambda^t C_4(\lambda)\Lambda^\tau \quad , \tag{3.7}$$

with

$$\begin{aligned} C_3(\lambda) &= -\lambda^{s/2}\Lambda^{2-\varepsilon-\rho}R(\lambda)C_1\Lambda^{1-r}R(\lambda) \\ C_4(\lambda) &= -\lambda^{s/2}\Lambda^{1-t}R(\lambda)C_2\Lambda^{2-\varepsilon-\tau}R(\lambda) \quad . \end{aligned} \tag{3.8}$$

Clearly C_3 and C_4 are compact for all $\lambda \in \Gamma$. Also we have

$$\Lambda^s R(\lambda) = O(|\lambda|^{-1+s/2}) \ , \ 0 \le s \le 2 \ , \ \lambda \in \Gamma \ , \tag{3.9}$$

from which we get

$$\|C_3\| = O(|\lambda|^{-1+(1+s-\varepsilon-r-\rho)/2}) \ , \ \|C_4\| = O(|\lambda|^{-1+(1+s-\varepsilon-t-\tau)/2}). \tag{3.10}$$

It follows that the integrals (3.11) exists as improper Riemann integrals in $L(H)$, and that

$$\int C_j(\lambda) d\lambda = C_j \in K(H) \ , \ j = 3,4 \ , \tag{3.11}$$

with the ideal $K(H)$ of compact operators of H. Thus (3.5) follows if we substitute (3.6) into the commutator, and use the above. This completes the proof of lemma 3.3. Also we note that lemma 3.3 ,for $s=1$, $\rho=t=1$, $r=\tau=0$, and any $\varepsilon<1=1+s-r-\rho=1+s-t-\tau$ gives

$$D[a,\Lambda]u = D\Lambda(C_1+C_2)u = Cu \ , \ C \in K(H) \ , \tag{3.12}$$

which completes the proof of thm.3.1 as well.

Let us emphasize that the function a in thm.3.1 does not have to be bounded. Only the derivative is restricted by (3.2). This will be useful in thm.3.4 below for commutators of the form $[q^{\varepsilon/2},\Lambda]$, using the restriction (3.13) . On the other hand in sec.4 we will discuss a class $A^{\#}_{c}$ of function generators giving compact commutators, using thm.3.1.

While we note that no restrictions at all on the comparison operator H were needed for the commutator of thm.3.1, if one wants compactness of commutators of two generators $D\Lambda$ of the second kind then it will be practical to require the condition

$$\text{(3.13)} \qquad \nabla q^{\wedge} = o_{\Omega^{\wedge}}(q^{\wedge}) \quad .$$

on the 'potential' $q^{\wedge}$. No further condition on $q^{\wedge}$ will be imposed, and none on q as well, if $\{\Omega,d\mu,H\} = \{\Omega^{\wedge},d\mu^{\wedge},H^{\wedge}\}$. On the other hand, if Ω is a proper subdomain of $\Omega^{\wedge}$, then at the boundary $\partial\Omega$ of Ω the function q must be singular, in order to get (w) , while $q^{\wedge}$ is smooth there. (Otherwise the Dirichlet and Neumann operators are different self-adjoint extensions of H_0 ; hence H_0 cannot be essentially self-adjoint.) For that case we will need a condition on the expression $b^j q_{|x^j}$, included in (3.16).

Note that (3.13) holds trivially for the Laplace operator.

The comparison operator $1-\Delta$ not only has $q=1$, but also $\zeta=1$, with the function $\zeta = \kappa/\sqrt{h}$, $h = \det((h^{jk}))^{-1}$. Note that $\zeta \in C^{\infty}(\Omega)$, due to the invariance properties of κ and $\sqrt{h}$. Similarly we define $\zeta^{\wedge}$ for the triple $\{\Omega^{\wedge},d\mu^{\wedge},H^{\wedge}\}$. For an expression D of the form (2.2) the adjoint D^* is given by (2.20). If we want D and D^* to satisfy the same type of estimates we must look at the term

$$\text{(3.14)} \qquad \kappa^{-1}(\kappa b^j)_{|x^j} = \zeta^{-1}(b^j\zeta)_{|j} \quad ,$$

with Riemannian covariant derivatives. We will require that the coefficients b^j , p of D satisfy the conditions

$$\text{(3.15)} \quad (b^j) = O(1) \ , \ p = O(\sqrt{q}) \ , \ \zeta^{-1}(b^j\zeta)_{|j} = O(\sqrt{q}) \ (\text{in } \Omega) \ ,$$

for D of (2.2) , just to insure boundedness of D and D^* , i.e., finiteness of c_1 in (2.6) for D and D^* . For $H=1-\Delta$ the term (3.14) equals the 'divergence' $b^j{}_{|j}$. Note that a change of dependent variable can make either q=const or $\zeta=1$, as was seen in III,6.

For a result on compactness of commutators between two generators $D\Lambda$, $D \in \mathcal{D}^{\#}$ we also introduce the following conditions:

$$\begin{aligned} &b^{j|k} + b^{k|j} = o_{\Omega^\wedge}(q^{\wedge\varepsilon/2}) \ , \ \nabla(\zeta^{-1}(b^j\zeta)_{|j}) = o_{\Omega^\wedge}(q^{1/2}q^{\wedge\varepsilon/2}), \\ &\nabla p = o_{\Omega^\wedge}(q^{1/2}q^{\wedge\varepsilon/2}) \ , \ b^j q_{|x^j} = o_{\Omega^\wedge}(q) \ , \ \text{for some } 0 \le \varepsilon < 1, \end{aligned} \tag{3.16}$$

<u>Theorem 3.4.</u> Let the comparison operator $H^\wedge$ satisfy (3.13), and let the expression D of (2.2) satisfy (3.15) and (3.16). Then the operators $D\Lambda$, $D^*\Lambda \in L(H)$ (by (2.6),(3.15)) satisfy

$$(D\Lambda)^* - D^*\Lambda \in K(H) \ , \ D\Lambda - (\Lambda D)^{**} \in K(H) \ , \tag{3.17}$$

where in particular ΛD, as an unbounded operator with domain H_1 , admits a continuous extension $(\Lambda D)^{**}$. Moreover, for D and F of (2.12) both satisfying (3.15) and (3.16), we get

$$[D\Lambda, F\Lambda] \in K(H) \ . \tag{3.18}$$

The proof again requires some preparations.

<u>Lemma 3.5.</u> Let D satisfy (3.16) for some $0 \le \varepsilon < 1$. Then the commutator $D\Lambda^2 - \Lambda^2 D$ (defined in HC_0^∞) extends to a bounded operator in H , which is compact. Moreover, we get

$$[D,\Lambda^2] = \Lambda^{1-\varepsilon}C_1\Lambda = \Lambda C_2\Lambda^{1-\varepsilon} \ , \ C_j = C_j(\varepsilon) \in K(H) \ , \ j = 1,2. \tag{3.19}$$

<u>Proof.</u> Again we get

$$D\Lambda^2 u - \Lambda^2 D u = \Lambda^2[H,D]\Lambda^2 u \quad u \in \operatorname{im} H_0 = HC_0^\infty \ , \tag{3.20}$$

where we may express $[H,D]$ by (2.16). Accordingly, we must discuss the operators $\Lambda^2 X \Lambda^2$ with X being one of the four terms of (2.16).

First consider the term $\Lambda^2 M_0 \Lambda^2$, with

$$M_0 = -\kappa^{-1}\partial_{x^j}\kappa m^{jk}\partial_{x^k}, \ m^{jk} = b^{j|k} + b^{k|j} = \beta m_1^{jk}, \ ((m_1^{jk})) = o_{\Omega^\wedge}(1) \ , \tag{3.21}$$

where $\beta > 0$ denotes a $C^\infty(\Omega^\wedge)$-function with $\beta = q^{\wedge\varepsilon/2}$, $x \in \Omega^\wedge \setminus K$. Write

$$M_0 = \beta M_1 - \beta m_1^{jk}\theta_{|j}\partial_{x^k} \ , \ M_1 = -\kappa^{-1}\partial_{x^j}\kappa m_1^{jk}\partial_{x^k} \ , \ \theta = \log\beta \ , \tag{3.22}$$

and observe that (3.13) implies

$$c^k = m_1^{jk}\theta_{|j} = \varepsilon/2 \ m_1^{jk} q^\wedge_{|j}/q^\wedge \ (\text{for } x \in \Omega^\wedge \setminus K) = o_{\Omega^\wedge}(1) \ (\text{on } \Omega). \tag{3.23}$$

For the second term in (3.22) we get $\Lambda^2\beta c^j\partial_{x^j}\Lambda^2 = \Lambda^{2-\varepsilon}(\Lambda^\varepsilon\beta)c^j\partial_{x^j}\Lambda\Lambda = \Lambda^{2-\varepsilon}C\Lambda$, of the proper form for (3.19). For the first term apply lemma 3.3 with $s = 1$, $\rho = t = 0$, $r = \tau = 1$, for $a = \beta$, noting that $\nabla\beta = \beta\nabla\theta = o_{\Omega^\wedge}(q^{\wedge\varepsilon/2})$, i.e., (3.2) holds. We get $[\Lambda,\beta]=C\Lambda$, with $C\in K(H)$, hence $\Lambda^2M_1\Lambda^2 = \Lambda^{1-\varepsilon}(\Lambda^\varepsilon\beta)(\Lambda M_1\Lambda)\Lambda + \Lambda C(\Lambda M_1\Lambda)\Lambda$, where the second term is of the form $\Lambda C\Lambda$, by (2.10) . The first term is of the form $\Lambda^{1-\varepsilon}C\Lambda$. Indeed, we get $\Lambda^\varepsilon\beta \in L(H)$, by (3.2) and I,lemma 5.1, and $\Lambda M_1\Lambda \in K(H)$ can be proven as follows: For $\delta > 0$ write $m_1^{jl} = \omega m_1^{jl} + \chi m_1^{jl}$, where $\omega \in C_0^\infty(\Omega^\wedge)$ is chosen such that the tensor χm_1^{jl}, with $\chi=1-\omega$, has norm $\leq\delta$. Correspondingly write $M_1 = M_2+M_3$, and note that $\|\Lambda M_3\Lambda\|\leq\delta$. On the other hand, $M_2 = \omega M_1 + D_2$, where the differential expression D_2 is of the form γD_4, with $D_4\Lambda$ bounded and $\gamma \in C_0^\infty(\Omega^\wedge)$. It follows that $\Lambda M_2\Lambda = C\Lambda M_1\Lambda + \omega(\Lambda M_1\Lambda)+\Lambda D_2\Lambda = C+\omega\Lambda M_1\Lambda$. Accordingly, $(\Lambda^\varepsilon\beta)(\Lambda M_1\Lambda)=E_\delta+\Lambda^\varepsilon\beta\omega(\Lambda M_1\Lambda)+C$,

where $E_\delta = (\Lambda^\varepsilon\beta)(\Lambda M_3\Lambda)$ has norm $= O(\delta)$, while the other two terms are compact. Since δ may be chosen arbitrarily, we get $(\Lambda^\varepsilon\beta)(\Lambda M_1\Lambda) \in K(H)$, using that $K(H)$ is norm closed. Thus we have shown that $\Lambda^2M_0\Lambda^2 = \Lambda^{1-\varepsilon}C\Lambda$ is in the proper form for (3.19).

Next let us focus on the term $\Lambda^2 b^j q_{|j}\Lambda^2 = J$. Using (3.16) we may write $J = \Lambda(\Lambda\beta)\gamma(\beta\Lambda)\Lambda$, where $0< \beta \in C^\infty(\Omega^\wedge)$, $\beta = o_{\Omega^\wedge}(q^{\wedge 1/2})$ is constructed according to lemma A.1 such that $\gamma=O(1)$. This shows that $J=\Lambda C\Lambda$, with compact C , using III, thm.3.7 .

Finally, for the two terms $\Lambda^2 D_{\bar p^*}\Lambda^2$ and $\Lambda^2 D_{\bar p}^*\Lambda^2$, it is sufficient to look at $J' = \Lambda^2 D_p\Lambda^2$ only, for symmetry reasons. With β as

in the discussion of $\Lambda^2M_0\Lambda^2$ and, similarly, $\gamma=q^{1/2}$ in $\Omega\setminus K$, $0<\gamma \in C^\infty(\Omega)$ we write $D_p = \beta\gamma D_3$, where D_3 has coefficients $o_{\Omega^\wedge}(1)$, by the third formula (3.16). Then, $J' = \Lambda^2\beta\gamma D_3\Lambda^2 = \Lambda\beta\Lambda\gamma D_3\Lambda^2 + \Lambda C\Lambda\gamma D_3\Lambda^2$, $=\Lambda^{1-\varepsilon}(\Lambda^\varepsilon\beta)(\Lambda\gamma D_3\Lambda)\Lambda + \Lambda^{1-\varepsilon}(\Lambda^\varepsilon\beta)C(\Lambda\gamma D_3\Lambda)\Lambda = \Lambda^{1-\varepsilon}C\Lambda$, using the same techniques over again. In particular we used lemma 3.3 on β again. Note finally that we also may push the factor β to the right : $\beta\gamma D_3 = \gamma D_3\beta - \varepsilon/2\ \gamma(D_3\theta)\beta$, with the function $\theta = \log q^\wedge$ again. Thus we also get $J' = \Lambda C\Lambda^{1-\varepsilon}$. This completes the proof of lemma 3.5.

<u>Lemma 3.6.</u> Under the assumptions of lemma 3.5 we have

(3.24) $[D\Lambda,\Lambda^s] = \Lambda^r C\Lambda^\rho$, $0 \leq s \leq 2$, $C = C_{r,\rho,s} \in K(H)$.

for every pair r,ρ of reals satisfying

(3.25) $r+\rho+\varepsilon<1+s$, and either $0\le r+\varepsilon\le 1$, $0\le\rho\le 2$, or $0\le r\le 1$, $0\le\rho+\varepsilon\le 2$.

Proof. We substitute (3.6) into the commutator of (3.24) for

$$[D\Lambda,\Lambda^s] = i/2\pi \int_\Gamma \lambda^{s/2}R(\lambda)[\Lambda^2,D]\Lambda R(\lambda)\, d\lambda \quad .$$

Now (3.19) may be used in the integrand $I(\lambda)$, for

$$I(\lambda) = \lambda^{s/2}\,\Lambda^{1-\varepsilon}R(\lambda)C_1\,\Lambda^2 R(\lambda) = \Lambda^r C_3(\lambda)\Lambda^\rho \text{ , with}$$

$$C_3(\lambda) = \lambda^{s/2}\Lambda^{1-r-\varepsilon}R(\lambda)C_1R(\lambda)\Lambda^{2-\rho} = \lambda^{s/2}\Lambda^{1-r}R(\lambda)C_2R(\lambda)\Lambda^{2-\rho-\varepsilon} \ .$$

Using (3.25) and the appropriate (3.19) we get $C_3(\lambda)\in K(H)$, $I(\lambda) = O(|\lambda|^{1+(1+s-r-\rho-\varepsilon)/2})$ on Γ . Thus (3.24) follows as earlier (3.5) . The proof is complete.

Lemma 3.7. For some $0\le\varepsilon\le 1$, let the two differential expressions D,F , as in (2.12), satisfy the conditions

(3.26) $b^jc^k{}_{|j} - c^jb^k{}_{|j} = o_{\Omega^\wedge}(q^{\varepsilon/2})$, $b^jr_{|j} - c^jp_{|j} = o_{\Omega^\wedge}(q^{(1+\varepsilon)/2})$,

Then the operator $\Lambda[D,F]\Lambda$ is bounded in H, and may be written as

(3.27) $$\Lambda[D,F]\Lambda = \Lambda^{1-\varepsilon}C_1 \ ,$$

with a compact operator C_1 .

Proof. Use lemma A.1 to construct a function $0\le\beta\in C^\infty(\Omega^\wedge)$ with $\beta = o_{\Omega^\wedge}(q^{\varepsilon/2})$, such that $\beta^{-1}[D,F]\Lambda$ is bounded, using (2.23), (2.6) and (3.26). Therefore $\Lambda[D,F]\Lambda = \Lambda^{1-\varepsilon}(\Lambda^\varepsilon\beta)(\beta^{-1}[D,F]\Lambda) = \Lambda^{1-\varepsilon}C_1$, with $C_1\in K(H)$, using III, cor.3.8.

Proof of theorem 3.4. For D satisfying (3.15),(3.16) we apply lemma 3.6 with $s=\rho=1$, $r=0$, for $[D\Lambda,\Lambda] = D\Lambda^2 - \Lambda D\Lambda = C\Lambda$. This implies $D\Lambda u = \Lambda Du + Cu$, $u\in H_1 = \text{im}\,\Lambda$. In particular ΛD ,with domain H_1 , must have a continuous extension $(\Lambda D)^{**}\in L(H)$, and $D\Lambda = (\Lambda D)^{**} + C$ follows. Taking adjoints we get $(D\Lambda)^* = D^*\Lambda + C^*$, so that (3.17) follows. For (3.18) use (3.27) for $\varepsilon=1$ and the second (3.17) for $D\Lambda F\Lambda u = (\Lambda D)^{**}F\Lambda u + Cu = Cu + \Lambda DF\Lambda u = Cu + \Lambda FD\Lambda u + C'u = C''u + (\Lambda F)^{**}D\Lambda u = F\Lambda D\Lambda u + C'''u$, for all $u\in HC_0^\infty$, with C, C', C'', $C'''\in K(H)$, completing the proof of the theorem.

4. Comparison algebras with compact commutators.

In this section we will discuss some concrete cases of comparison algebras with compact commutators. Since all the discussion in sec.3 is based on the validity of condition (w) we first will engage one of the results of ch.4.

First assume that the assumptions of IV, thm.1.5 hold for m=1. In details, let $\Omega^\wedge$ be a smooth complete Riemannian manifold without boundary, with a positive C^∞-measure $dm^\wedge=\kappa^\wedge dx$, and let the C^∞-function $q^\wedge$ be ≥ 1 outside some compact set K, and satisfy condition (3.13), i.e. (with the limit in $\Omega^\wedge$),

(4.1) $\lim_{x\to\infty} h^{\wedge jk} q^\wedge_{|x^j} q^\wedge_{|x^k}/q^{\wedge 2} = \lim_{x\to\infty} h^{\wedge jk} (\log q^\wedge)_{|x^j} (\log q^\wedge)_{|x^k} = 0.$

We assume that $\Omega^\wedge$ has a countable locally finite atlas.

Our manifold Ω is a subdomain of $\Omega^\wedge$, and its boundary $\partial\Omega$ is assumed to be a smooth compact n-1-dimensional submanifold of $\Omega^\wedge$. On Ω we have the restricted measure $d\mu = d\mu^\wedge|\Omega = \kappa dx$, and the restriction to Ω of the (contravariant) metric tensor is called h^{jk}. A comparison operator H is introduced by

$$(4.2) \qquad H = -\kappa^{-1}\partial_{x^j}\kappa h^{jk}\partial_{x^k} + q ,$$

with a real-valued C^∞-function $q \geq q^\wedge$. Then condition (w) for the triple $\{\Omega, d\mu, H\}$ results if we require condition IV, (1.4), for q :

$$(4.3) \qquad q(x) \geq \gamma_1/\mathrm{dist}(x,\partial\Omega)^2 \quad , \quad \text{for all } x \in \Omega\setminus K' ,$$

with a suitable compact set $K'\subset\Omega$, where γ_1 is a certain positive constant defined in IV,thm.1.5.

Now the results of sec.3 suggest introduction of the following generating sets $A^\#$ and $\mathcal{D}^\#$.

$A^\#_c$ is defined as the class of all bounded functions $a \in C^\infty(\Omega)$, such that $\nabla a = o_\Omega(q^{\varepsilon/2})$, with some $0 \leq \varepsilon < 1$. (That is, the $C^\infty(\Omega)$-function $q^{-\varepsilon/2} h^{jk} \bar{a}_{|x^j} a_{|x^k}$ is bounded over $\Omega\setminus K$, and has limit zero, as $x\to\infty$ (in Ω).)

$\mathcal{D}^\#_c$ is defined as the set of all first order partial differential expressions $D=b^j\partial_{x^j}+p$ with C^∞-coefficients, defined over Ω, satisfying (3.15),(3.16) with some $0\leq\varepsilon<1$. That is, in detail, the 3 complex-valued $C^\infty(\Omega)$-functions

$$h_{jk}\bar{b}^j b^k\ ,\ p/\sqrt{q}\ ,\ \kappa^{-1}(\kappa b^j)_{|x^j}/\sqrt{q} \tag{4.4}$$

are bounded over Ω , and also, the 4 $C^\infty(\Omega)$-functions (with $\zeta=\kappa/\sqrt{h}$)

$$q^{\wedge-\varepsilon}(h_{jj'}h_{kk'}(\bar{b}^{j|k}+\bar{b}^{k|j})(b^{j'|k'}+b^{k'|j'})\ ,\ b^j(\log q)_{|x^j}\ , \tag{4.5}$$

$$q^{-1}q^{\wedge-\varepsilon}h^{jk}\bar{p}_{|x^j}p_{|x^k},\ q^{-1}q^{\wedge-\varepsilon}h^{jk}(\zeta^{-1}(\bar{b}^l\zeta)_{|l})_{|j}(\zeta^{-1}(b^l\zeta)_{|l})_{|k}$$

also are bounded over Ω (or at least $\Omega\setminus K$), and have limit zero as $x\to\infty$ (in $\Omega^\wedge$) .

It is trivial that $A^\#_c$ and $\mathcal{D}^\#_c$ satisfy the general assumptions (a_0), (a_1), (d_0) of sec.1. In addition $\mathcal{D}^\#_c$ contains its formal Hilbert space adjoints.

Now thm.4.1, below, is an immediate consequence of thm.3.1, thm.3.4, and IV,thm.1.5. (We use thm.3.1 for $\Omega=\Omega^\wedge$, but thm.4.4 with $\Omega^\wedge$ as above.)

Theorem 4.1. Under the above assumptions the comparison algebra $\mathcal{C} = \mathcal{C}(A^\#_c,\mathcal{D}^\#_c)$ of section 1, formed with the above classes $A^\#_c$, $\mathcal{D}^\#_c$, contains the class $K(H)$, and has compact commutators. Moreover, $\mathcal{C}$ is generated by $a \in A^\#_c$, $D\Lambda$, $D \in \mathcal{D}^\#_c$ alone, i.e., the generators $(D\Lambda)^*$ may be omitted.

Let us shortly focus on some special cases onto which thm. 4.1 applies. In particular, for technical reasons, one often looks at the algebras of smaller generating sets. The above sets $A^\#_c$ $\mathcal{D}^\#_c$ may be regarded as the largest practical such sets, allowing for compact commutators.

Example (A): $\Omega=\mathbb{R}^n$, $d\mu=dx$, $H=1-\Delta$, $\Delta=\sum_{j=1}^n\partial_{x^j}{}^2$, (i.e., the Euclidean Laplace comparison triple of $\mathbb{R}^n$) . This case has been studied extensively in $[C_1]$, ch.III,IV . We have condition (s), since $\mathbb{R}^n$ is a complete space (IV, cor.1.7). Since $q(x)\equiv 1$ is a constant, we also get $q=q^\wedge=q^\varepsilon=q^{\wedge\varepsilon}\equiv 1$, so that the number ε becomes redundant. $A^\#_c$ consists of all bounded C^∞-functions with gradient vanishing at infinity. The largest comparison algebra of $[C_1]$,IV,1 , called $\mathfrak{A}_0$ there, uses as $\mathcal{D}^\#$ the set of all folpde's with coefficients in $A^\#_c$. A calculation confirms that the class $\mathcal{D}^\#_c$ properly contains $\mathcal{D}^\#$.

In $[C_1]$ we obtain precise knowledge about the 'symbol space' (i.e.,the maximal ideal space of $\mathfrak{A}_0/K(H)$).

Problem 4.2. Use the techniques of $[C_1]$ to characterize the symbol space of $C(A^\#_c, \mathcal{D}^\#_c)$.

Example (B): The Laplace comparison algebra of the Euclidean $\mathbb{R}^n$ proves particularly simple because this space $\Omega=\mathbb{R}^n$ is a locally compact abelian group (under translation). In $[C_1]$ we essentially use the Fourier transform for most of the proofs. Of course, the Fourier transform appears as the Plancherel transform of the translation group, which explains the simplicity of the case.

Other cases of manifolds which are translation groups, are just as simple. For example, let $\Omega = \mathbb{C}^{jk} = \mathbb{R}^j \times T^k$, $j+k=n$, with the circle $T = \mathbb{R}/(2\pi\mathbb{Z})$. For the simplest case of $j=k=1$, $n=2$ we obtain the 2-dimensional straight circular cylinder.

One may introduce global coordinates $x=(x^1,\dots,x^n)$, appointing that functions on $\mathbb{C}^{jk}$ must be 2π-periodic in the last k variables. then a comparison triple again is given by the same $d\mu=dx$, $H=1-\Delta$, as under (A) .

In [CS],V,2 we investigate the comparison algebra $C(A^\#_c, \mathcal{D}^\#)$, where $\mathcal{D}^\#$ again is a proper subset of $\mathcal{D}^\#_c$ (defined as under (A)). Clearly our present triple satisfies (s), as the Laplace comparison triple of a complete Riemannian manifold, by IV,cor.1.7 . There is an important difference between the symbol spaces for $\mathbb{R}^2$ (as under (A)) and the cylinder $\mathbb{C}^{1,1}$, which will have to be discussed later on (cf.VII , 2, VIII,2.) .

Example (C): $\Omega=\mathbb{R}^n$, $d\mu=dx$, with the expression (where $\alpha,\beta \in \mathbb{R}$, $\alpha \geq 0$)

$$(4.6) \qquad H = \langle x\rangle^{2\alpha} - \operatorname{div}\langle x\rangle^{2\beta}\operatorname{grad} = \langle x\rangle^{2\alpha} - \textstyle\sum_{j=1}^n \partial_{x^j}\langle x\rangle^{2\beta}\partial_{x^j} ,$$

$\langle x\rangle=(1+|x|^2)^{1/2}$. We have $h^{jk} = \langle x\rangle^{2\beta}\delta^{jk}$, hence $h_{jk}= \langle x\rangle^{-2\beta}\delta_{jk}$.

Proposition 4.3. The manifold $\Omega=\mathbb{R}^n$, with metric tensor $h_{jk} = \langle x\rangle^{-2\beta}\delta_{jk}$ is complete if and only if $\beta \leq 1$.

Proof. For any smooth curve $\{x(t):t\in J\}$ we obtain as arc length

$$(4.7) \qquad l = \int_J (\langle x(t)\rangle^{-2\beta}|x^\bullet(t)|^2)^{1/2}dt = \int_J \langle x\rangle^{-\beta}|x^\bullet(t)|dt .$$

For a moment define the real-valued function $\xi(t)=|x(t)|$ over the interval J . We have

$$(4.8) \qquad |\xi^\bullet(t)| = |(x(t)\cdot x^\bullet(t))/|\xi(t)|| \leq |x^\bullet(t)| .$$

Accordingly, if $J=(\rho,\rho')$, we get

$$(4.9) \qquad \left|\int_{\xi(\rho)}^{\xi(\rho')}(1+\xi^2)^{-\beta/2}d\xi\right| \leq \int_J \langle\xi(t)\rangle^{-\beta}|\xi^\bullet(t)|dt \leq l .$$

This proves that the geodesic distance between two point x , $x^\sim$ is bounded below by the integral $\int_\rho^{\rho^\sim} \langle\xi\rangle^{-\beta} d\xi$, $\rho=|x|$, $\rho^\sim=|x^\sim|$. This distance tends to ∞ ,as $x^\sim\to\infty$ if and only if $\int_0^\infty \langle\xi\rangle^{-\beta} d\xi = \infty$.

This happens precisely for $\beta\le 1$. Also (4.8) implies that $|\xi^\bullet(t)| = |x^\bullet(t)|$ whenever the curve runs in a radial direction. Accordingly, if $\beta>1$, then indeed all curves $\{tx_0 : 0<t<\infty\}$ have finite length, so that then Ω cannot be complete. Q.E.D.

Considering the expression (4.6) we thus have cdn.(s) whenever $\beta\le 1$, by IV, cor.1.7. On the other hand, for $\beta>1$ we will change the dependent variable by setting $u=\langle x\rangle^{-\lambda}v$, with $\lambda>0$ to be chosen. This transformation yields a new expression $H^\sim$ with potential given by III,(6.6). That is, with $\beta^\sim=\beta-1$,

$$
\begin{aligned}
q^\sim &= q - \langle x\rangle^{\lambda}\mathrm{div}(\langle x\rangle^{2\beta}\,\mathrm{grad}\,\langle x\rangle^{-\lambda}) \\
&= \langle x\rangle^{2\alpha} + \lambda(n+2\beta^\sim-\lambda)\langle x\rangle^{2\beta^\sim} - \lambda(2\beta^\sim-\lambda)\langle x\rangle^{2\beta^\sim-2} .
\end{aligned}
\tag{4.10}
$$

Let us first consider the case $\beta>\alpha+1$. Then the second term at right is leading, as $|x|$ is large. Its coefficient becomes largest if we choose $\lambda=\beta^\sim+n/2 = \beta-1+n/2$. Now, although we no longer have a complete Riemannian manifold we still may apply IV, thm.1.1.

The proper choice of the function σ will be $\sigma(x)=\tau\log\langle x\rangle$, with suitable $\tau>0$. We then get

$$
h^{jk}\sigma_{|x^j}\sigma_{|x^k} = \tau^2|x|^2\langle x\rangle^{2\beta^\sim-2} \le q^\sim, \text{ as } \tau=(\beta^\sim-\varepsilon+n/2), \text{ any } \varepsilon>0, \tag{4.11}
$$

(at least outside a large compact set, which is sufficient). We trivially have $\sigma(0)=0$, and (ii) of thm.1.1 holds, since $\tau>0$. Checking on $(iv)_\eta$, we find that

$$
e^{-\eta\sigma}|\nabla\sigma| = O(\langle x\rangle^{\beta^\sim-\eta(\beta^\sim-\varepsilon+n/2)}) = O(1) \tag{4.12}
$$

whenever $\beta^\sim<\eta(\beta^\sim+n/2)$, for sufficiently small $\varepsilon>0$. In other words, we must require that

$$
\eta > \beta^\sim/(\beta^\sim+n/2) \quad . \tag{4.13}
$$

Notice that the right hand side of (4.13) is <1 , for all $\beta^\sim>0$, while that expression tends to zero, as $\beta^\sim\to 0$,and $\beta^\sim>0$. Thus we always can satisfy $(iv)_\eta$ for $\eta<1$, enough for (w), for every choice of β whatsoever.

So far we have looked at the case $\alpha<\beta^{\sim}$ only. If $\alpha=\beta^{\sim}$, then in (4.11) we may set $\tau=((\beta^{\sim}-\varepsilon+n/2)^2+1)^{1/2}$. Similarly, for $\alpha>\beta^{\sim}$ we can allow any choice of τ . Correspondingly (4.13) improves to

(4.14) $\quad \eta>\beta^{\sim}/(1+(\beta^{\sim}+n/2)^2)^{1/2}$, as $\alpha=\beta^{\sim}$, $\eta>0$, as $\alpha>\beta^{\sim}$.

We have proven the result,below.

Theorem 4.4. For a comparison operator of the form (4.6) we always have cdn.(w) satisfied, regardless of the choice of α and β. Moreover, we have cdn.(s) for all $\beta\le 1$, regardless of the choice of α . If $0\le\alpha\le\beta-1$, we at least have cdn.(w). As β decreases from ∞ to +1 , we will have cdn.'s(s_k) for larger and larger k , such that for every integer $k_0>0$ there exists $\beta_{k_0}>1$ such that (s_{k_0}) holds for every $\beta\le\beta_{k_0}$.

For a concrete example we now set $q^{\wedge}=1$, and then, with $A^{\#}_c$, $\mathcal{D}^{\#}_c$ as above, introduce the classes $A^{\#}=A^{\#}_c$, and

(4.15) $\quad \mathcal{D}^{\#}=\{D\in\mathcal{D}^{\#}_c\colon \langle x\rangle^{-\beta}b^j\in A^{\#}_c,\ p=O(\langle x\rangle^{\alpha}),\ p_{|x^j}=o(\langle x\rangle^{-\beta+\alpha})\}$.

The symbol spaces for some such algebras have indeed been worked out by Sohrab [S_3] . We shall discuss this in VII,thm.4.8.

Example (D): This example, and example (E), below, only point to special cases of thm.4.1 where the various conditions can be translated into a simpler (or different) form. None of the algebras have been investigated in generality, regarding their symbol or symbol space.

Let Ω be an open subdomain of $\mathbb{R}^n$ with compact connected smooth boundary ∂Ω , where ∂Ω is an n-1-dimensional smooth submanifold of $\mathbb{R}^n$. Let $H = q - \Delta$, with the Euclidean Laplace operator Δ , above. Here the real-valued C^∞-potential $q\ge 1$ must satisfy (4.3), above, with the Euclidean distance. Set $\Omega^{\wedge} = \mathbb{R}^n$, $d\mu = d\mu^{\wedge} = dx$, $H^{\wedge} = 1-\Delta$, to get all conditions satisfied, including (4.1).

In other words, we have Ω either the interior or the exterior of a compact subdomain of $\mathbb{R}^n$.

Now the class $A^{\#}_c$ consists of all bounded $C^\infty(\Omega)$-functions a such that, for suitable ε , $0\le\varepsilon<1$, and all j=1,...,n, we have $a_{|x^j} = o(q^{\varepsilon/2})$ in Ω . That is, we have $|a_{|x^j}/q^{\varepsilon/2}| <\delta$ outside a suitable compact subset K_δ of Ω , for every $\delta>0$.

Also the class $\mathcal{D}_c^\#$ consists of all (globally defined) $D=b^j\partial_{x^j}$ $+ p$, with b^j , $j=1,..,n$, $p \in C^\infty(\Omega)$ and with the functions

$$b^j,\ p/\sqrt{q},\ b_{j\,|x^j}/\sqrt{q} \text{ bounded over } \Omega\ ,\ j=1,\dots,n\ , \text{ and with}$$

$$b^j_{\ |x^k}+b^k_{\ |x^j}\ ,\ b^j q_{|x^j}/q\ ,\ p_{|x^j}/\sqrt{q}\ ,\ (b^l_{\ |x^l})_{|x^j}/\sqrt{q}\ , \text{ all, } = o(1),$$

for $j,k=1,\dots,n$, in the sense described for $A_c^\#$.

As a consequence of thm.4.1 we then get a comparison algebra with compact commutators, containing $K(H)$. The symbol space of such an algebra seems to be unexplored.

Example (E): Let Ω be the interior of a complete Riemannian manifold $\Omega\cup\partial\Omega$ with compact boundary $\partial\Omega$. It is known that then $\Omega\cup\partial\Omega$ may be imbedded, in the sense required for thm.5.1, into a manifold $\Omega^\wedge$ which is a doubling of Ω (i.e., in two copies of $\Omega\cup\partial\Omega$ one identifies corresponding points of $\partial\Omega$, using as chart of a boundary point the union of the chart of $\Omega\cup\partial\Omega$ in the half space $\{x\in\mathbb{R}^n: x^n>0\}$ together with its reflection into $\{x^n\leq 0\}$).

Assuming the metric tensor $h^{jk} \in C^\infty(\Omega\cup\partial\Omega)$, we also may double h^{jk} , i.e., reflect it at the boundary. However, the resulting extension in general is no longer continuous at $\partial\Omega$. On the other hand, it is easy to arrange for a smooth transition, accross $\partial\Omega$, by modifying the reflected h^{jk} in a strip along $\partial\Omega$, "below $\partial\Omega$ " (cf. also the proof of III,thm.2.2) .

Similarly for a measure $d\mu$ on Ω . Furthermore, if a potential q is given on Ω (normally singular on $\partial\Omega$), it also may be reflected and suitably modified in a (two-sided) neighbourhood of $\partial\Omega$ in the doubled manifold, called $\Omega^\wedge$, to obtain a $q^\wedge \in C^\infty(\Omega^\wedge)$.

In this way, if a triple $\{\Omega,d\mu,H\}$ is given, on the above Ω , it is possible to satisfy all assumptions of thm.4.1, regarding the imbedding into $\Omega^\wedge$, with a sub-extending triple $\{\Omega^\wedge,d\mu^\wedge,H^\wedge\}$. We must require that q satisfies (4.3), and that $\nabla q/q = o(1)$ in $\Omega\setminus K$, with a compact neighbourhood of $\partial\Omega$ in $\Omega\cup\partial\Omega$, in order to insure that (4.1) can be satisfied.

This shows that the case of a manifold with boundary is naturally included in our discussion.

Example (F): Finally we point to the case of $\Omega = \mathbb{R}^n$, with $d\mu=dx$, and a (globally defined) comparison operator H, under the assumptions of IV, thm.1.9 .

In this case we will assume $\Omega^\wedge = \Omega = \mathbb{R}^n$, and $q^\wedge=1$. Then we again have all assumptions of thm.4.1 satisfied, for the classes $A_c^\#$, $\mathcal{D}_c^\#$. Again, the structure of the corresponding symbol space seems to be unexplored.

5. A discussion of one-dimensional problems.

In this section we get restricted to the case of a 1-dimensional manifold Ω. That is, our comparison triple assumes the form

$$(5.1) \qquad \{\Omega,d\mu,H\} = \{I,\kappa dx,\kappa^{-1}(-d/dx\ p\ d/dx + q)\} \qquad ,$$

with a finite or infinite open interval $I = \{\alpha < x < \beta\} \neq \emptyset$, and globally defined C^∞-functions $\kappa(x)$, $p(x)$, $q(x)$, $x \in I$, $\kappa(x)$, $p(x) > 0$, as $x \in I$.

In this case the two normal forms of III, prop.'s 6.1 and 6.2 may be further simplified by also introducing a special independent variable . A simple calculation, left to the reader, gives the following result.

Proposition 5.1. a)If new dependent and independent variables (v,y) instead of (u,x) are introduced by setting

$$(5.2)\ u=\gamma v,\ \gamma=(p\kappa)^{-1/4},\ y(x)=\int_{x_0}^{x}\sqrt{\kappa/p}\,dt,\ \alpha^\sim=\lim_{x\to\alpha}y(x),\ \beta^\sim=\lim_{x\to\beta}y(x)$$

with any given $x_0 \in I$,then the triple (5.1) assumes the form

$$(5.3) \qquad \{\Omega^\sim,d\mu^\sim,H^\sim\} = \{I^\sim,dy,-d^2/dy^2 + q^\sim\}$$

with $I^\sim = (\alpha^\sim,\beta^\sim)$, and $q^\sim$ of the form, with $\gamma^- = \gamma^{-1} = (p\kappa)^{1/4}(x(y))$,

$$(5.4) \qquad q^\sim(y) = (q/\kappa)(x(y)) - (d^2\gamma^-/dy^2/\gamma^-)(y) = H\gamma/\gamma(x(y)) \ .$$

b) Let $\gamma \in C^\infty(I)$ be a positive solution of the differential equation $H\gamma=\gamma$ (which exists according to III, prop.6.2, since I is noncompact, and the minimal operator H_0 has the lower bound 1) . Then, if new dependent and independent variables are introduced by setting

$$(5.5) \qquad u = \gamma v \ , \ y(x) = \int_{x_0}^{x}\kappa(t)\gamma^2(t)dt \ ,$$

then the triple (5.1) assumes the form

$$(5.6) \qquad \{I^\sim,dy,H^\sim\} = \{(\alpha^\sim,\beta^\sim),dy,-d/dy\,p^\sim d/dy+1\} \ ,$$

with

$$(5.7)\quad p^{\sim}(y) = (\kappa p\gamma^4)(x(y)) \ , \ \alpha^{\sim} = \lim_{x\to\alpha} y(x) \ , \ \beta^{\sim} = \lim_{x\to\beta} y(x).$$

Generally a transformation of dependent and independent variable of the form

$$(5.8)\qquad u = \gamma v \ , \quad y = \phi(x) \ , \ \phi'(x) > 0 \ ,$$

with a positive C^∞-function γ and an invertible C^∞-map $\phi : I \to I^{\sim}$, may be applied to the triple (5.1) as well as to any first order differential operator

$$(5.9)\qquad F = bD + c \quad , \ b,c \in C^\infty(I) \ , \ D = d/dx \ ,$$

and the Hilbert spaces H and H_1.

In view of proposition 5.1 it is no loss of generality to depart from a triple in <u>Sturm-Liouville normal form</u> :

$$(5.10)\quad I = (\alpha,\beta) \ , \ d\mu = dx \ , \ H = - D_x^2 + q \quad , \ D_x = d/dx \ .$$

The proposed change of variables will carry (5.10) into

$$I^{\sim} = (\alpha^{\sim},\beta^{\sim}) \ , \ \alpha^{\sim} = \phi(\alpha+0) \ , \ \beta^{\sim} = \phi(\beta-0) \quad ,$$

$$d\mu^{\sim} = \kappa^{\sim} dy \quad , \quad \kappa^{\sim}(y) = \gamma^2(\phi(y))\phi'(y)$$

$$(5.11)\quad H^{\sim} = -\kappa^{\sim -1} D_y \kappa^{\sim} p^{\sim} D_y + q^{\sim} \ , \ F^{\sim} = b^{\sim} D_y + c^{\sim} \ , \ D_y = d/dy \ ,$$

$$p^{\sim}(y) = (\phi'(y))^{-2} \ , \ q^{\sim}(y) = (H\gamma/\gamma)(\phi(y)) \ ,$$

$$b^{\sim}(y) = b(\phi(y))/\phi'(y) \ , \ c^{\sim}(y) = (c + b\gamma'/\gamma)(\phi(y)) \ .$$

All this assumes that the transformation amounts to an isometry $H \leftrightarrow H^{\sim} = L^2(I^{\sim},d\mu^{\sim})$, carrying the equations $Hu=f$,$Fv=g$ into $H^{\sim}u^{\sim}=f^{\sim}$, $F^{\sim}v^{\sim}=g^{\sim}$, with the transformed functions $u^{\sim},v^{\sim},f^{\sim},g^{\sim}$ of u,v,f,g, respectively.

While we will work with the triple in the form (5.10) we also note that our conditions on the coefficients of H and F will not remain invariant under a transformation (5.8). Hence we shall tend to repeat the construction of a C^*-algebra with compact commutator of the preceeding sections,while trying to optimize these conditions, or else trying to bring them onto a simpler form.

First the requirement of condition (w) is invariant under (5.8) as well as that of (s) or (s_k), because (5.8) amounts to a unitary transformation U: $H \to H^\sim$, taking the operator H into $H^\sim = U^{-1}HU$, and $UC_0^\infty(I) = C_0^\infty(I^\sim)$. An application of IV, thm.1.8 to (5.10) yields the following

Proposition 5.2. a) If $\alpha = -\infty$, $\beta = \infty$, then the assumed semi-boundedness of H alone implies that (s) = (s_∞) holds.

b) If α or β (or both) are finite then there exists a nondecreasing sequence η_k, $k = 1,2,\dots$, such that (s_k) holds whenever

(5.12) $q(x) \geq \eta_k(x - \alpha)^{-2}$ near α, or $q(x) \geq \eta_k(x - \beta)^{-2}$, near β,

respectively. (Both for 2 finite ends). Specifically (w)=(s_1)

holds for q with (5.12), for $\eta_1 = \sqrt{3}/2$, and (s) = (s_∞) holds for

(5.13) $(x-\alpha)^2/q(x)=o(1)$, as $x\to\alpha$, or, $(x-\beta)^2/q(x)=o(1)$, as $x\to\beta$,

respectively. (Both conditions, if α and β both are finite, but if only one end is finite then the infinite end never requires a special condition,except the semi-boundedness of the operator).

From (5.11) and (1.7) we know that the operator $F\Lambda$, with $\Lambda = H^{-1/2}$, is bounded whenever, for some $\gamma>0$, $\gamma \in C^\infty(I)$,

(5.14) $$b = O(1) \quad , \quad c + b\gamma'/\gamma = O(\sqrt{q-\gamma''/\gamma}) \quad .$$

This condition may be simplified by introducing the function

(5.15) $$\delta = \gamma'/\gamma \ , \ \delta' + \delta^2 = \gamma''/\gamma \ , \ \gamma = \gamma(x_0)\exp\Big(\int_{x_0}^{x}\delta(t)dt\Big) \ .$$

Proposition 5.3. The operator $A = (F\Lambda)^{**}$ is in $L(H)$ whenever there exists a real-valued function $\delta \in C_0^\infty(I)$ such that

(5.16) $$b = O(1) \ , \quad c + b\delta = O(\sqrt{q-\delta'-\delta^2}) \quad .$$

The proof is immediate. Note that change of coordinates does not give an improvement of conditions for b. To evaluate the second condition (5.16) we assume that the operator bound, hence the O(.) -constant in (5.16) is 1, as always can be achieved by multiplication with a positive constant. Then $(5.16)_2$ is equivalent to the Riccati type differential inequality

(5.17) $$\delta' + 2\delta\,\mathrm{Re}(\bar{c}b) + (1+|b|^2)\delta^2 \leq q - |c|^2 \quad .$$

For example, consider the case of $b \equiv 0$, where (5.17) yields

$$0 \leq |c|^2 \leq q - \delta' - \delta^2 . \tag{5.18}$$

For $\delta \equiv 0$ (5.18) gives the trivial condition $|c| \leq \sqrt{q}$.
A more general δ will improve this only if

$$\delta' + \delta^2 < 0 \tag{5.19}$$

over the interval I , or at least near one of the endpoints. (or in a sequence of intervals clustering at an endpoint), in view of the fact that a $C_0^\infty(I)$ -coefficient b or c always gives a bounded operator $F\Lambda$. In fact, $q_0 = -(\delta+\delta^2)$ also is the improvement of the potential q under the change of variable (with $\delta = \gamma'/\gamma$).

Notice that a real-valued δ satisfying (5.19) in some open interval J must be a decreasing function of x which can vanish at most once.At its only possible zero we have $\delta' < 0$. On the other hand for all $x \in J$ with $\delta \neq 0$, (5.19) amounts to

$$y' = (1/\delta)' = - \delta'/\delta^2 > 1 , \text{ for } y(x) = 1/\delta(x) . \tag{5.20}$$

Consider a right endpoint $\beta = \infty$, for example. If $\delta(x_0) < 0$, for some x_0 , then $1/\delta$, having slope > 1 , by (5.20), must get zero after a finite x-increase, leading to a contradiction. Hence $\delta > 0$ for all $x \geq x_0$, x_0 sufficiently large. Then (5.20) yields $1/\delta - 1/\delta_0 > x - x_0$, for all $x \geq x_0$, with $\delta_0 = \delta(x_0)$. Hence,

$$0 < \delta(x) \leq \delta_0/(1+\delta_0(x-x_0)) \quad , \text{ for all } x \geq x_0 , \tag{5.21}$$

so that $\delta \to 0$, as $x \to \infty$. One then concludes that also

$$\left| \int_x^{x+h} (\delta'+\delta^2)dt \right| \leq \delta_0/(1+\delta_0(x-x_0)) + h\delta_0^2/(1+\delta_0(x-x_0))^2 , \ h>0. \tag{5.22}$$

Accordingly, while the possible improvement of q may be large at some specific point, it will become small in the average, over a finite interval $(x,x+h)$, as $x \to \infty$, and can be significant only if $q \to 0$, as $x \to \infty$. Since we get $q \equiv 1$ in suitable coordinates, it follows that q, in the average cannot be too different from 1 . Accordingly the improvement of the boundedness estimates by coordinate changes may be of little use, if the interval is $(-\infty,+\infty)$.

On the other hand, if the right end β of the interval $I=\Omega$ is finite, then one may think of possibly significant changes of q by change of coordinates.

Example 5.4. The choice $\delta(x) = (\beta-x)^{\lambda}$, $0<\lambda<1$, near β -i.e.,

(5.23) $\gamma(x) = \exp(-(\beta-x)^{\lambda+1}/(\lambda+1))$, near β , $x < \beta$,

gives the additive improvement

(5.24) $q_0(x) = \lambda(\beta-x)^{\lambda-1} - (\beta-x)^{2\lambda} \approx \lambda(\beta-x)^{\lambda-1} \to \infty$, $x\to\beta$.

Similarly $\delta(x) = -(\beta-x)^{\lambda}$, $-1<\lambda<0$, near β , resulting in the even faster increasing

(5.25) $q_0(x) = |\lambda|(\beta-x)^{\lambda-1} + (\beta-x)^{2\lambda} \approx |\lambda|(\beta-x)^{\lambda}$, $x<\beta$, near β.

However, cdn.(w), if installed by IV,thm.1.5, requires $q\geq\gamma(\beta-x)^{-2}$, which again makes the possible gain by either (5.24) or (5.25) appear insignificant.

Coming to some more explicit examples in 1 dimension, let us examine examples (A) and (C) of sec. 4 , in this connection. Both give some typical 1-dimensional examples. While (A) already is in Sturm-Liouville normal form, (C) is not. Its Sturm-Liouville transform is very similar to that used in the proof of thm.4.4, in the case of $\beta>1$. We will take up the discussion of the corresponding algebras $\mathcal{C}$ in VII,4. Also there we will discuss certain more general examples of Sohrab, around the quantum mechanical harmonic and anharmonic oscillator equation in one dimension.

Finally, in this section, consider the following:
Example (G): Define

(5.26) $\Omega=I=\mathbb{R}=(-\infty,+\infty)$, $d\mu=dx$, $H=1-D_x^{\,2}$, $D_x=d/dx$.

Let $A^{\#}$ be the class of all $C^{\infty}(\mathbb{R})$-functions

(5.27) $a=b+c$, $b,c\in C^{\infty}(\mathbb{R}^n)$, where $b\in CS(\mathbb{R})$, and c is 2π-periodic .

Here $CS(\mathbb{R})$ denotes the class of continuous functions over $\mathbb{R}$ with limits at $+\infty$, and at $-\infty$, where the limits need not to be equal. Also, $\mathcal{D}^{\#}$ is the class of all folpde's with coefficients in $A^{\#}$.

Here we want to show that the corresponding algebra $\mathcal{C}(A^{\#},\mathcal{D}^{\#})$ does not have compact commutators. Indeed, consider the commutator

(5.28) $[e^{ix},D\Lambda] = [e^{ix},s(D)]$, $s(x) = x/\langle x\rangle$.

If we conjugate with the inverse Fourier transform F^{-1} we get

(5.29) $Fs(D)F^{-1} = s(x)$, $Fe^{ix}F^{-1} = e^{-iD} = T_{-1}$, with the translation operator $T_{-1}u(x) = u(x-1)$. Therefore,

$$[e^{ix},D\Lambda] = (\nabla_{-1}s)(D)e^{ix} \ , \ \nabla_{-1}s(x)=s(x-1)-s(x) \ . \tag{5.30}$$

This is not a compact operator, since multiplication by e^{ix} is a unitary map of $H=L^2(\mathbb{R})$, while the function $\nabla_{-1}s(x)$ does not vanish identically, so that $\nabla_{-1}s(D)$ cannot be compact.

Similarly we get ,(for $j=0,\pm1,\pm2,\dots$, with $\lambda(x)=\langle x\rangle^{-1}$)

$$[e^{ijx},D\Lambda] = \nabla_{-j}s(D)e^{ijx} \ , \ [e^{ijx},\Lambda] = \nabla_{-j}\lambda(D)e^{ijx} \ . \tag{5.31}$$

where $\nabla_j b(x)=b(x+j)-b(x)$. On the other hand, for $a\in CS(\mathbb{R})$ we get

$$[a(x),D\Lambda] \in K(H) \ , \ [a(x),\Lambda] \in K(H) \ , \tag{5.32}$$

(cf.[C_1] , ch.III , or also by application of sec.3, above.)

Proposition 5.5. The commutator ideal E of the present algebra C is the norm closure of the algebra of all operators of the form

$$\chi^-(x)\sum_{j=-N}^{N}a_j^+(D)e^{ijx} + \chi^-(x)\sum_{j=-N}^{N}a_j^-(D)e^{ijx} + K, \text{ where} \tag{5.33}$$

$$a_j^\pm \in CO(\mathbb{R}),\ K \in K(H),\ \chi^\pm=1 \text{ for } \pm x\geq\gamma>>1,\ =0 \text{ for } \pm x\leq\gamma-1.$$

Proof. It is clear that all commutators of generators of C are of the form (5.33), by the above. Any product of an operator (5.33) with a generator of the algebra again is of that form. Especially, a product $a(D)b(x)$ or $b(x)a(D)$, for $a,b \in CO(\mathbb{R})$, is compact, by [C_1],III,L.8.1, or III, thm.3.7. By the Stone-Weierstrass theorem one concludes that every operator $a(D)e^{ijx}$ is in E , for arbitrary $a\in CO$. Similarly E contains the entire compact ideal, by lemma 1.1. This completes the proof.

In VIII,1 we will look at this algebra again, trying to determine its symbol space and ideal chain.

6. An expansion for expressions within reach of an algebra C.

In the present section we shall prepare relating a differential expression L to a comparison algebra in a manner which will allow us to connect an invertibility or Fredholm statement of a certain realization of L to a corresponding statement involving a bounded operator in the algebra. The general technique appears to be common in theory of singular integral operators (cf. Calderon and Zygmund [CZ_2], for example).

Given a comparison triple $\{\Omega, d\mu, H\}$ on a manifold Ω, and any subalgebra $\mathcal{A}$ of $\mathcal{L}(\mathcal{H})$, $\mathcal{H}=L^2(\Omega,d\mu)$. Let us assume cdn.(s) for all of the present section.

<u>Definition 6.1</u>. A differential expression L of order $\leq$ N (for some integer N=0,1,...,) is said to be <u>within comparison reach</u> (or shortly 'within reach') of the algebra $\mathcal{A}$ if the unbounded operator $L_0\Lambda^N$, with the minimal operator of L and $\Lambda=H^{-1/2}$, has a continuous extension A to $\mathcal{H}$, where $A \in \mathcal{A}$.

More precisely we should speak of an 'N-reach', since the condition obviously may depend on N. We will omit this distinction, however, since the choice of N always will be clear.

In particular we note that cdn.(s) implies that the space $\operatorname{dom} L_0\Lambda^N = \Lambda^{-N}C_0^\infty(\Omega)$ is dense in $\mathcal{H}$. Indeed, for even N=2k we get $\Lambda^{-N}C_0^\infty = H^kC_0^\infty = \operatorname{im} H_0^k$ dense in $\mathcal{H}$, since H_0^k is ≥ 1 and essentially self-adjoint, using I,thm.3.1. For odd N=2k-1, if $f\in\mathcal{H}$, $(f,\Lambda^{-N}u)=0$ for all $u\in C_0^\infty$, then we set $f=\Lambda^{-1}g$, using that Λ^{-1} is self-adjoint and ≥ 1, hence $\operatorname{im}\Lambda^{-1}=\mathcal{H}$. It follows that $(\Lambda^{-1}g,\Lambda^{-N}u)=(g,\Lambda^{-N-1}u)=$ $=(g,H_0^ku)=0$ for all $u\in \operatorname{dom} H_0^k$, hence again g=0 and f=0, so that $\Lambda^{-N}C_0^\infty$ again is dense.

Since all spaces $\Lambda^{-N}C_0^\infty$ are dense it follows that L of order N is within reach of $\mathcal{L}(\mathcal{H})$ whenever we get, with a constant c,

$$(6.1) \qquad \|Lu\| \leq c\|\Lambda^{-N}u\| \text{ for all } u \in C_0^\infty(\Omega).$$

It is trivial that all 0-order expressions (i.e. multiplications) $a\in \mathcal{A}^\#$ and all folpdes $D\in\mathcal{D}^\#$ are within reach of the comparison algebra $\mathcal{C}= \mathcal{C}(\mathcal{A}^\#,\mathcal{D}^\#)$, by definition of $\mathcal{C}$. In VI,3 we shall see that every differential expression of compact support is within reach of the minimal comparison algebra. The general question for expressions within reach of a comparison algebra $\mathcal{C}$ is vital for our applications and will be studied in IX,3 and X,2.

For a differential expression L of order N within reach of a comparison algebra $\mathcal{C}$ there is a natural realization Z defined:

<u>Definition 6.2</u>. Let $A\in\mathcal{C}$ be the continuous extension of $L_0\Lambda^N$ to $\mathcal{H}$ requested in def.6.1. Then we set

$$(6.2) \qquad \operatorname{dom} Z = \operatorname{im}\Lambda^N = \operatorname{dom}\Lambda^{-N}, \; Zu = A\Lambda^{-N}u, \; u \in \operatorname{dom} Z,$$

i.e., $Z=A\Lambda^{-N}$, in the sense of I,(1.3).

Remark 6.3. The space im Λ^N = dom $\Lambda^{-N} = H_N$ is commonly referred to as the (L^2-) Sobolev space of order N of the triple $\{\Omega, d\mu, H\}$ (assuming cdn.(s), as we presently do). Here H_N is a Hilbert space under the graph norm I,(2.9) of the (closed) self-adjoint operator Λ^{-N} . Since $\Lambda^{-N} \geq 1$ it follows that the graph norm is equivalent to

$$(6.3) \qquad \|u\|_N = \|\Lambda^{-N} u\| \ , \ u \in H_N \ ,$$

called the Sobolev norm of order N . With this norm the operator $\Lambda^{-N}: H_N \to H$ becomes an isometry between the Hilbert spaces H and H_N. Then the unbounded operator $Z \in P(H)$ may just as well be regarded as a bounded operator of $L(H_N, H)$. There exists an inverse $Z^{-1} \in L(H, H_N)$ if and only if A^{-1} exists in $L(H)$. Similarly the unbounded operator Z admits a bounded closed inverse if and only if $A \in L(H)$ is invertible. We also set $H_\infty = \cap H_N$ and note that all L and Λ^N take $H_\infty \to H_\infty$. For more details cf.VI,3 and X,2.

In order to facilitate a later discussion of expressions within reach and their realization Z we next will discuss a certain Taylor type expansion with integral remainder, valid for certain expressions within reach.

Definition 6.4. A (C^∞-)differential expression L of order $\leq$N is said to be H-compatible (or said to satisfy cdn.(p)) if for all j=0,1,2,... the iterated commutators

$$(6.4) \quad (\mathrm{ad}\ H)^0 L = L, \ (\mathrm{ad}\ H)^j L = [H,[H,\ldots[H,L]..], \text{ with j commutators},$$

are within reach of $L(H)$ (as expressions of order N+j).

Proposition 6.5. Let again $R(\lambda) = (\Lambda^2 - \lambda)^{-1} \in L(H)$, and let L be H-compatible. Then the operator $A = L\Lambda^N \in L(H)$ satisfies the identity

$$(6.5) \quad \begin{aligned} R(\lambda)A - AR(\lambda) &= \textstyle\sum_{j=1}^N \lambda^{j-1}((\mathrm{ad}\ H)^j L)\Lambda^{2j+N+2} R^{j+1}(\lambda) \\ &+ \lambda^M \Lambda^2 R(\lambda)((\mathrm{ad}\ H)^{M+1} L)\Lambda^{2M+N+2} R^{N+1}(\lambda) \ , \ \lambda \in \Gamma \ , \end{aligned}$$

for all M=0,1,2,.... .

Proof. We get

$$(6.6) \quad R(\lambda)Lu - LR(\lambda)u = \Lambda^2 R(\lambda)[H,L]\Lambda^2 R(\lambda)u \ , \ u \in \mathrm{im}(1-\lambda H_0) \ ,$$

as used before (cf. sec.3) , and may extend this to $u \in H_\infty$. Note that (6.6) amounts to (6.5) for N=0. Suppose (6.5) holds for $N\geq 0$ then write $\Lambda^2R(\lambda)=1+\lambda R(\lambda)$, and $L_{N+1} = (\text{ad } H)^{N+1}L$, for a moment. For $u \in \text{im}(1-\lambda H_0)$ we get,

$$
(6.7)\qquad \begin{aligned} \Lambda^2R(\lambda)u &= (1+\lambda R(\lambda))L_{N+1}u = L_{N+1}(1+R(\lambda))u \\ &+ \lambda(R(\lambda)L_{N+1}u-L_{N+1}R(\lambda)u) = L_{N+1}\Lambda^2R(\lambda)u+\lambda\Lambda^2R(\lambda)L_{N+2}\Lambda^2R(\lambda)u \ , \end{aligned}
$$

and again we may extend this to H_∞ . Substituting (6.7) into (6.5) yields (6.5) for N+1 , q.e.d.

Proposition 6.6. In the notation of lemma 3.3 we have

$$
(6.8)\qquad \int_\Gamma \Lambda^{2j}R^{j+1}(\lambda)\lambda^{s/2}d\lambda = 2\pi i(-1)^{j+1}\binom{s/2}{j}\Lambda^s \ , \ s>0, \ j=0,1,\ldots,
$$

with binomial coefficients $\binom{s/2}{j}$. Here the integral converges in norm convergence of $L(H)$, as an improper Riemann integral.

Proof. Integrate by parts in formula (3.6), using that $d/d\lambda R^k(\lambda) = k\ R^{k+1}(\lambda)$.

Proposition 6.7. With the notations of prop.6.5 we have

$$
(6.9)\qquad \Lambda^sA = \sum_{j=0}^M(-1)^j\binom{s/2+j-1}{j}((\text{ad } H)^jL)\Lambda^{s+2j+N} + R_M \ ,
$$

where

$$
(6.10)\qquad R_M = i/2\pi \int_\Gamma \Lambda^2R(\lambda)((\text{ad } H)^{M+1}L)\Lambda^{2M+2+N}R^{M+1}(\lambda)\lambda^{s/2+M}d\lambda \ .
$$

Proof. Multiply (6.5) by $i/(2\pi)\lambda^{s/2}$, and integrate over Γ . Note that the integral may be interchanged with the operators $(\text{ad } H)^jL = L_j$, which are preclosed. Then use (6.8) to evaluate some of the integrals.

Theorem 6.8. For all (real or complex) s with $-2M-2<\text{Re } s < M+1$, $M=1,2,\ldots$, and $A \supset L\Lambda^N$, with an H-compatible expression L of order $\leq N$ we have the expansion

$$
(6.11)\qquad \Lambda^sA\Lambda^{-s} - A = \sum_{j=1}^M(-1)^j\binom{s/2+j-1}{j}(L_j\Lambda^{N+2j}) + S_{M,s} \ ,
$$

where $L_j = (\text{ad } H)^jL \in L_{N+j}$, and the 'remainder' $S_{M,s}$ is given by

$$
(6.12)\qquad \begin{aligned} -2\pi iS_{M,s} &= \int_\Gamma \lambda^{\varepsilon-1}d\lambda S^1(\lambda)(\Lambda L_{M+1}\Lambda^{N+M+1})S^2_{M,s}(\lambda) \ , \\ S^1(\lambda)&=\Lambda R(\lambda)\lambda^{1/2} \ , \ S^2_{M,s}(\lambda) = \lambda^{M+1/2+s/2-\varepsilon}\Lambda^{M+1-s}R^{N+1}(\lambda) \ , \ 0<\varepsilon<1. \end{aligned}
$$

Here the two $L(H)$-valued functions $S^j(\lambda)$ of λ , Λ , (and s) are bounded and regular analytic for $\lambda \in \Gamma\setminus\{0\}$, $-2M-2+2\varepsilon < s < M+1$, as ε is kept fixed. The integral in (6.12) converges in $L(H)$, as an improper Riemann integral of a continuous integrand, even if the constant operator $\Lambda\Lambda_{M+1}\Lambda^{N+M+1}$ is replaced by an arbitrary operator $V \in L(H)$. Moreover, we even have the functions

$$(6.13) \qquad S^2_{M,s}(\lambda)\Lambda^{s-M-1} \text{ , as } s>0 \text{ , and } S^2_{M,s}(\lambda)\Lambda^{2\varepsilon-M-1} \text{ , as } s<2\varepsilon \text{ ,}$$

bounded and analytic on the (λ,s)-set specified.

We should emphasize that, by writing (6.11) we intend to imply that the (product of unbounded) operator(s) $\Lambda^s A \Lambda^{-s}$ is bounded, and that its continuous extension to H should be taken in its place.

Remark 6.9. It is an immediate consequence of thm.6.7 that the remainder $S_{M,s}$ is a holomorphic function of s (with values in $L(H)$), defined in the strip $-2M-2 < \mathrm{Re}\, s < M+1$, and that even the operators $S_{M,\varepsilon}\Lambda^{s-M-1}$, for $0 \le s < M+1$, and $S_{M,\varepsilon}\Lambda^{\varepsilon-M-1}$, for $-2M-2+\varepsilon \le s \le \varepsilon$, are in $L(H)$. Moreover, it should be observed that the two factors $S^1(\lambda)$ and $S^2_{M,s}(\lambda)$ both are functions of Λ , hence commute with every function of Λ .

The proof of thm.6.8 is a matter of analytic continuation of the identities (6.9), (6.10) from the positive real s-axis onto the s-strip specified. First we get (6.11) for $0 < s \le M+1$, by multiplying (6.9) from the right by Λ^{-s} . Next we note that all terms at right are entire functions of s except the remainder. On the other hand, the remainder may formally be written, as specified in (6.12), where the function S^j are analytic. Also, boundedness of the functions S^j , near their singularity $\lambda=0$ is a consequence of estimate (3.9). Thus one concludes that indeed the remainder integral admits an analytic extension into the strip $-2M-2 < \mathrm{Re}\, s < M+1$. Finally, we at least have the complex-valued function $((\Lambda^s A \Lambda^{-s} - A)u,u) = \phi(s)$ analytic in the strip, for every fixed $u \in H_\infty$. One thus concludes that (6.11),(6.12) holds in the same inner product sense. This, in turn, implies H-boundedness of the pre-closed operator $\Lambda^s A \Lambda^{-s}$ for s in the strip. The remaining estimates follow as part of the above proof, or are trivial consequences.

Remark 6.3'. Note that the above definition of Sobolev space is meaningful not only for integers $N \ge 0$ but also for general $s \ge 0$.

Similarly, for $s<0$ one defines the Sobolev space H_s as the completion of H under the norm $\|u\|_s = \|\Lambda^{-s}u\|$ (note that now $\Lambda^{-s} \in L(H)$).

For a more detailed investigation of the spaces H_s cf. IX,1. Here we only note that the spaces $\{H_s : s \in \mathbb{R}\}$ define an HS-chain in the sense of I,6. Then thm.6.8 above has the following consequence

Corollary 6.10. The operator $L_0\Lambda^N$, for an H-compatible differential expression of order $\leq N$, not only admits an extension A to $L(H)$ but the restriction $A|H_\infty$ extends to an operator $A_s \in L(H_s)$. In other words, A (or $A|H_\infty$) is of order 0 in the sense of I,6 .

Also, for every s,N there exists a constant c and an integer $M \geq 0$ such that the operator norm in $L(H_s)$ of A_s is bounded by

$$(6.14) \qquad c \sum_{0 \leq k \leq M} \|L_k \Lambda^{N+2k}\| \ , \ L_k = (\mathrm{ad}\ H)^k L \ ,$$

for all H-compatible expressions L of order $\leq N$.

The proof is an immediate consequence of thm.6.8: We must show that $\|L_0\Lambda^N u\|_s = \|\Lambda^{-s}L_0\Lambda^N u\| \leq c\|u\|_s = c\|\Lambda^{-s}u\|$, or, with

$v=\Lambda^{-s}u$, $\|\Lambda^{-s}(L_0\Lambda^N)\Lambda^s v\| \leq c\|v\|$, for $v \in \Lambda^{-s-N}C_0^\infty$. That, exactly, follows from our expansion (6.11) , if only we choose M large enough to include the real number s in the strip of analyticity, defined in the thm. In particular we find that $L_j\Lambda^{N+j} \in L(H)$, and the remainder $S_{M,s}$ as well, in view of the properties stated.

CHAPTER 6. MINIMAL COMPARISON ALGEBRA AND WAVE FRONT SPACE; SOBOLEV SPACES ON COMPACT MANIFOLDS.

Starting a general investigation of the symbol space of a comparison algebra, (i.e., the maximal ideal space of the commutative algebra C/E) , we first will focus on the minimal comparison algebra J_0 , introduced in V,1 (It is generated by the classes of a and D with compact support). We know that it always has the commutator ideal $K(H)$. Its symbol space will be called the <u>wave front space</u> , and be denoted by $\mathbb{W}$. We first will show that the wave front space is independent of the choice of H and dμ (sec.1). In fact, it proves to be homeomorphic to the bundle of spheres of radius infinity compactifying the cotangent space $T^*\Omega_x$ at a point $x \in \Omega$ (sec.2). For simplicity one commonly identifies these spheres with the corresponding unit spheres

(0.1) $$\{ (x,\xi) : \quad h^{jk}(x)\xi_j\xi_k = 1 \} \subset T^*\Omega_x \ .$$

Thus $\mathbb{W}$ then is called the 'co-sphere bundle' of Ω (cf.[AS],[Se_2]).

The wave front space is found naturally imbedded in the symbol space of every comparison algebra C meeting our general assumptions (cf. VII,1). The result described above is due to Seeley [Se_2] in the case of a compact manifold, and to Gohberg [Gb_1], in the case of $\mathbb{R}^n$ (cf. also [CI], Herman [H_1]).

For a compact manifold the minimal comparison algebra is the only comparison algebra, evidently. For noncompact manifolds one will ask the question about points of the symbol space $\mathbb{M}$, not contained in $\mathbb{W}$. This will be the focus of our attention in ch.VII . In fact, it will be found that such points are linked to the essential spectrum of the comparison operator H .

In sec.3 we show that all differential expressions of compact support are within reach of the minimal comparison algebra J_0 . This leads to a result about the Fredholm property of the realization of V,6 (thm.3.4) on compact manifolds. In fact, we even get a <u>Green inverse</u>, i.e., an inverse modulo $O(-\infty)$ (thm.3.7).

This fact is exploited in sec.4 to obtain the familiar Sobolev estimates on compact manifolds.

1. The local invariance of the minimal comparison algebra.

In this section we focus on the minimal comparison algebra $\mathcal{J}_0$ introduced in V,1 . Let $A_m^{\#} = C_0^{\infty}(\Omega)$ and $\mathcal{D}_m^{\#} = \{D = b^j \partial_{x^j} + p :\ b^j$ and p are C^{∞} , and of compact support $\}$ be the corresponding sets of functions and folpde's. Essential self-adjointness of H_0 is not required in this section.

Theorem 1.1. The algebra $\mathcal{J}_0$ remains unchanged if the measure $d\mu$, and the comparison operator H are modified over a compact subset $\mathcal{U} \subset \Omega$. That is, if $\{\Omega, d\mu, H\}$ and $\{\Omega, d\nu, K\}$ are two triples and if there exists a compact set $\mathcal{U} \subset \Omega$ such that H=K and $d\mu = d\nu$ outside $\mathcal{U}$, then both triples have the same minimal comparison algebra $\mathcal{J}_0$.

Remark. Notice that the two Hilbert spaces $L^2(\Omega, d\mu)$, and $L^2(\Omega, d\nu)$ coincide as sets, and have equivalent norms, but in general have different inner products. Accordingly there also are two different adjoints. Our proposition states that the two corresponding algebras $\mathcal{J}_0$ of bounded operators coincide. They necessarily are C^*-algebras with respect to either adjoint.

The proof of thm.1.1 will require some preparations.

In order to use methods similar to those developed in V,3 it will be necessary to have condition (w), or even (s_k) , for larger k, available. Since we focus on a and D with compact support in this section, it will be convenient to relate the results to a modified comparison operator $H^{\blacktriangle} = H + q^{\blacktriangle}$, where the nonnegative function $q^{\blacktriangle}$ is chosen such that $\{\Omega, d\mu, H^{\blacktriangle}\}$ satisfies (s_k) , for suitable k .

First of all, by III, prop.6.2 we may change the dependent variable, and then obtain a constant nonnegative potential $q = \text{const} \geq 1$. In this representation define

$$(1.1) \qquad q^{\blacktriangle} = \chi(\eta_k \tau)^{-2} h^{jk} \tau_{|x^j} \tau_{|x^k} \quad , \quad \tau = \textstyle\sum_{j=1}^{\infty} 2^{-j} \varepsilon_j \omega_j \ ,$$

with our partition $\{\omega_j\}$ of app.A, where $\varepsilon_j \to 0$, as $j \to \infty$, and $0 < \varepsilon_j \leq (\mathrm{Max}\{1, \sup\{|\nabla \omega_j(x)| : x \in \Omega_j\}\})^{-1}$. Also, η_k denotes the positive real number specified in IV, thm.1.1, and χ denotes any $C^\infty(\Omega)$-function with $0\le\chi\le 1$, $\chi=1$ outside some compact set. In particular χ may be made to vanish in some given compact set, so that $H=H^\blacktriangle$ there. Then a calculation confirms that the expression $H^\blacktriangle$ satisfies the conditions of IV, thm.1.1, for m=k , so that we have (s_k) satisfied for $H^\blacktriangle$. Finally we may switch back to the old dependent variable, which does not disturb (s_k), and leaves the function $q^\blacktriangle$ unchanged as well. We have proven:

Proposition 1.2. For every comparison operator H and k=1,2,... there exists a nonnegative function $q^\blacktriangle \in C^\infty(\Omega)$ such that $H^\blacktriangle = H+q^\blacktriangle$ satisfies condition (s_k) of V,1 . Moreover, $q^\blacktriangle$ may be chosen to vanish in any given compact set of Ω .

Remark. It is easy to also obtain a $q^\blacktriangle$ such that $H^\blacktriangle$ satisfies (s), For example, one just might replace the constant η_k in (1.1) by a positive $C^\infty(\Omega)$-function $\eta(x)$ satisfying $\lim_{x\to\infty}\eta(x)=0$.

Proposition 1.3. Let $\Lambda=H^{-1/2}$ and $\Lambda^\blacktriangle = H^{\blacktriangle\,-1/2}$. For every function $a \in A^\#_m$ and folpde $D \in \mathcal{D}^\#_m$ we have

$$(1.2) \quad a\Lambda - a\Lambda^\blacktriangle = \Lambda C = C'\Lambda \ , \quad D\Lambda - D\Lambda^\blacktriangle = C'' \ , \ C,C',C'' \in K(H) \ .$$

Proof. For $u \in C_0^\infty(\Omega)$ write

$$(1.3) \qquad D(H^\blacktriangle+\lambda)u = (H+\lambda)Du + Dq^\blacktriangle u + [D,H]u = (H+\lambda)Du + Mu \ ,$$

with a second order expression M of compact support. For any $\lambda>0$ write $v = (H^\blacktriangle+\lambda)u$, and multiply (1.3) by $(H+\lambda)^{-1}$, to get

$$(1.4) \ R\ \lambda)Dv=DS(\lambda)v+R(\lambda)MS(\lambda)v=DS(\lambda)v+(\Lambda^{-1}R(\lambda))\Lambda M\Lambda(\Lambda^{-1}\Lambda^\blacktriangle)\Lambda^{\blacktriangle\,-1}S(\lambda)v$$

with $R(\lambda) = (H+\lambda)^{-1}$, $S(\lambda) = (H^\blacktriangle+\lambda)^{-1}$. By (w) the space $(H^\blacktriangle+\lambda)C_0^\infty$ is dense, so that (1.4) may be extended continuously to $v \in H$. In particular $\Lambda^{-1}R(\lambda)$,and $\Lambda^{\blacktriangle\,-1}S(\lambda)$ are bounded and $O((1+\lambda)^{-1/2})$, as easily seen, while $\|\Lambda^{-1}\Lambda^\blacktriangle\| \le 1$ follows from the inequality

$$(1.5) \qquad \|\Lambda^{-1}u\|^2 = (u,Hu) \le (u,H^\blacktriangle u) = \|\Lambda^{\blacktriangle\,-1}u\|^2 \ , \ u \in C_0^\infty(\Omega) \ .$$

Now a resolvent integral may be employed again. Let us use the resolvent of the operator H , not the resolvent of Λ^2 , for a change. We have the formula

$$(1.6) \quad H^{-s} = \pi^{-1}\sin \pi s \ / \binom{s-1}{m-1} \int_0^\infty (R(\lambda))^m \lambda^{m-s-1} d\lambda, \ s>0 \ , \ m=1,2,\ldots$$

with binomial coefficients $\binom{s-1}{m-1}$, where the denominator may have a simple zero, for certain integers s, but then the numerator $\sin \pi s$ vanishes also, and (1.6) holds with the limit of the quotient, as s approaches such points. The proper and improper Riemann integrals exist in norm convergence, as usual.

Formula (1.6) is a direct consequence of V,(6.8). One simply must introduce the new resolvent and make a substitution $\mu=-1/\lambda$ of the integration variable. Also the integration path must be deformed into the positive real axis, chosen as slit of the branch of λ^{m-s-1}, and then the integrand must be evaluated along the slit. Finally an analytic continuation in s must be made, using the spectral theorem for the self-adjoint H.

We use (1.6) on H and $H^{\blacktriangle}$ for s=1/2 to obtain

$$\Lambda = 1/\pi \int_0^\infty R(\lambda)\lambda^{-1/2}d\lambda \ , \ \Lambda^{\blacktriangle} = 1/\pi \int_0^\infty S(\lambda)\lambda^{-1/2}d\lambda \ ,$$

Integrating (1.4) it follows that

$$(1.7) \qquad (\Lambda D)^{**} = D\Lambda^{\blacktriangle} + 1/\pi \int_0^\infty C(\lambda)d\lambda \ ,$$

with

$$(1.8) \quad C(\lambda)=\lambda^{-1/2}(\Lambda^{-1}R(\lambda))(\Lambda M\Lambda)(\Lambda^{-1}\Lambda^{\blacktriangle})(\Lambda^{\blacktriangle -1}S(\lambda))=O(\lambda^{-1/2}(1+\lambda)^{-1}),$$

as is derived, using the fact that $\Lambda^{-1}R(\lambda) = O((1+\lambda)^{-1/2})$, and $\Lambda^{\blacktriangle -1}S(\lambda) = O((1+\lambda)^{-1/2})$. Also $C(\lambda)$ is compact for all $\lambda > 0$ and norm continuous in λ again. (One may write $(\Lambda^{-1}R(\lambda))(\Lambda M\Lambda) = (\Lambda^{-1-\varepsilon}R(\lambda))(\Lambda^{1+\varepsilon}M\Lambda)$, where the first term is bounded,the second is compact, due to the compact support of M.) It follows that the integral in (1.7) is a compact operator. Taking adjoints in (1.7) we get $D\Lambda = (\Lambda^{\blacktriangle}D)^{**} + C$, $C \in K(H)$ .This may be combined with the second formula V,(3.17), applied to $H^{\blacktriangle}$ and $\Lambda^{\blacktriangle}$, for the second relation (1.2). For the first relation we assume the expression D of order 0 (i.e., $b^j = 0$). Then M will be of first order only, and we may repeat the argument with $\Lambda M\Lambda$ replaced by ΛM or $M\Lambda$, while the remaining factor Λ (or $\Lambda^{\blacktriangle}$) may be taken out of the integral, either left or right. This completes the proof.

Proposition 1.4. The minimal comparison algebras J_0 of $\{\Omega,d\mu,H\}$, and $\{\Omega,d\mu,H^{\blacktriangle}\}$ are identical.

This proposition is an immediate consequence of V,lemma 1.1, and the second formula (1.2).

We will need (1.2) in a slightly more general setting,

expressed by the corollary, below.

Corollary 1.5. Formula (1.2) also holds if H and $H^{\blacktriangle}$ are two general comparison operators of the form V,(1.1) satisfying V,(1.2), perhaps with respect to different measures $d\mu$ and $d\mu^{\blacktriangle}$, as long as the two Hilbert spaces over Ω coincide (i.e., the function $\psi = d\mu/d\mu^{\blacktriangle}$ is bounded and bounded below), and as long as H and $H^{\blacktriangle}$ have the same principal part (i.e., $H-H^{\blacktriangle}$ is a first order expression). Also (w) must hold for $H^{\blacktriangle}$, not necessarily for H , and (1.5) must hold, at least with an additional multiplicative constant at right

Proof. The important point is that (1.3) is still valid, i.e., $D(H^{\blacktriangle}+\lambda) = (H+\lambda)D + M$, with a second order expression M of compact support. Also the resolvent integrals are still well-defined, and satisfy the estimates required, so that the proof just may be repeated, q.e.d.

It is clear now that, for the proof of thm.1.1 it may be assumed that both triples $\{\Omega, d\mu, H\}$ and $\{\Omega, d\nu, K\}$ satisfy (s_k), for an arbitrary k (or even $(s)=(s_\infty)$). Also we may assume that $d\mu = d\nu$. Indeed, let $\zeta = d\mu/d\nu$. Note that ζ is a positive $C^\infty(\Omega)$-function equal to 1 outside the compact set U. We get

$$K=-(\zeta\kappa)^{-1}\partial_{x^j}\zeta\kappa h^{\wedge jk}\partial_{x^k}+q=K^{\blacktriangle}+D^{\blacktriangle} \quad , \quad K^{\blacktriangle}=-\kappa^{-1}\partial_{x^j}\kappa h^{\wedge jk}\partial_{x^k}+q+\gamma \ ,$$

with $D^{\blacktriangle} = -\zeta^{-1}h^{\wedge jk}\zeta_{|x^j}\partial_{x^k} - \gamma$ having compactly supported principal part, while the constant γ may be chosen such that $K^{\blacktriangle}$ satisfies V, (1.2) . Clearly H and $K^{\blacktriangle}$ now are self-adjoint with respect to the same inner product, while K and $K^{\blacktriangle}$ generate the same minimal algebra J_0, in view of cor.1.5. If necessary, prop. 1.4 must be used again, to get (s) or (s_k) for $K^{\blacktriangle}$.

Now theorem 1.1, will be a consequence of the following result, first published in [CS], ch.V, lemma 4.10.

Theorem 1.6. Let condition (s) hold for H and K , and assume that $d\mu = d\nu$, so that we only have one adjoint. Then the operators $V_t=(H^tK^{-t})^{**}$ are in $L(H)$, for every $t \in \mathbb{R}$. If $U_t = V_t+K(H)$ denotes the coset mod $K(H)$ of V_t, then $\{U_t : t\in\mathbb{R}\}$ is a 1-parameter abelian subgroup of positive self-adjoint elements of the linear group $GL(L(H)/K(H))$ in the C^*-algebra $L(H)/K(H)$.

Let us first show that thm.1.6 implies thm.1.1. For a moment write $J_0(H)$ and $J_0(K)$ for the minimal algebras of H and K. We conclude that $V_1-1 = HK^{-1}-1 = (H-K)K^{-1}\in J_0(K)$, because G=H-K has compact support, and may be written as a finite sum of products DF, with folpde's $D,F \in \mathcal{D}^{\#}_m$. (Then $DFK^{-1} = DK^{-1/2}FK^{-1/2} + DK^{-1/2}CK^{-1/2} \in J_0$ follows from V, lemma 3.6.) Thus $U_1 - 1 \in J_0/K$, and for the coset $U_{-1/2}$ of $V_{1/2} = (H^{-1/2}K^{1/2})^{**}$ we also must have $U_{-1/2} - 1 \in J_0/K$. Indeed, $U_{-1/2}$, as positive hermitian element of L/K must be the unique inverse positive square root of U_1, since the group property of U_t gives $U_1 = (U_{-1/2})^{-2}$. This means that $U_{-1/2}$ may be obtained as a resolvent integral

$$(1.9) \qquad U_{-1/2} = i/2\pi \int_\Gamma (U_1-\lambda)^{-1}\lambda^{-1/2}d\lambda ,$$

with a curve Γ surrounding $Sp(U_1) \subset (0,\infty)$, and staying in the right half-plane $Re\ \lambda > 0$. In particular $\lambda^{1/2}$ denotes the branch of the root assuming the value 1 at $\lambda=1$. It is important that 0 is not in $Sp(U_1)$, since U_1 is invertible (its inverse is U_{-1}, by thm.1.6). Write $U_1 = 1+J$, $J \in J_0/K$. One may select Γ such that 1 is inside (it will belong to $Sp\ U_1$ anyway). Then

$$(1.10) \qquad U_{-1/2} - 1 = i/2\pi \int_\Gamma ((J+1-\lambda)^{-1} - (1-\lambda)^{-1})\lambda^{-1/2}d\lambda ,$$

by the residue theorem, where the integral converges in norm of L/H. Clearly,

$$(1.11) \quad (J+1-\lambda)^{-1} - (1-\lambda)^{-1} = -J(1-\lambda)^{-1}(J+1-\lambda)^{-1} \in J_0(K)/K,\ \lambda \in \Gamma .$$

Since J_0/K is closed it follows that $U_{-1/2} - 1 \in J_0(K)/K$. Thus also

$$(1.12) \qquad V_{-1/2} = (H^{-1/2}K^{1/2})^{**} = 1 + J^\wedge ,\ J^\wedge \in J_0(K) .$$

Now from the generators $a \in A^{\#}_m$, $DH^{-1/2}$, $D \in \mathcal{D}^{\#}_m$, of $J_0(H)$ the first type trivially is in $J_0(K)$ as well. On the other hand,

$$(1.13) \quad DH^{-1/2} = DK^{-1/2}(K^{-1/2}H^{1/2}) = DK^{-1/2}(1+J^{\wedge *}) \in J_0(K) ,$$

as well, by (1.12) so that $J_0(H) \subset J_0(K)$. For reason of symmetry we then get the opposite inclusion, hence $J_0(H) = J_0(K)$, q.e.d.

Notice that the full strength of thm.1.6 was not used in the proof of thm.1.1. In particular, only the classes $U_{\pm 1}$ and $U_{\pm 1/2}$ were ever used, together with the group property $(U_{-1/2})^{-2} = U_1$. Thus, while stating thm.1.6 in its elegant form, we only discuss the following weaker result (for full proof of thm.1.6 cf.IX,5).

<u>Lemma 1.7</u>. Assume only (s_2), for H and K. Then we have $V_t \in L(H)$ for all $|t| \leq 1$. The cosets U_t mod K are hermitian and positive in

L/K , and we have $(U_{s/2})^2 = U_s$, $U_sU_{-s}=U_0=1$ for all $|s|\leq 1$.

Proof. We already proved that $HK^{-1}\in K(H)$, so that $U_1 = HK^{-1}+K(H)$ is well defined. Similarly $KH^{-1}\in L(H)$, thus also

$$(1.14)\qquad V_{-1} = (H^{-1}K)^{**} = (KH^{-1})^{*} \in L(H) \quad ,$$

so that U_{-1} is well-defined. Moreover, using I,lemma 5.1, we now get H^tK^{-t} and $K^tH^{-t} \in L(H)$, $0\leq t\leq 1$, and hence get U_t well defined for all $|t|\leq 1$. Also we get

$$(1.15)\qquad HK^{-1} - (HK^{-1})^{*} = GK^{-1} - (GK^{-1})^{*} \; , \; G = H - K \; .$$

The compactly supported second order expression G is a sum of products of ≤ 2 folpde's. We get $DFK^{-1} = DK^{-1/2}FK^{-1/2} + C$, $C\in K$, as already used above. Accordingly $DFK^{-1} - (DFK^{-1})^{*} \in K$, by V,thm. 3.6, and $U_1^{*} = U_1$ follows . Similarly $U_{-1}^{*} = U_{-1}$.

Next we need a commutator relation.

Proposition 1.8. For $0\leq s\leq 1/2$ we have

$$(1.16)\qquad H^s[H^{-s},K^{-s}]H^s \in K(H) \;\; , \text{ and } H^{2s}[H^{-s},K^{-s}] \in K(H) \; .$$

Proof. With the resolvents $R(\lambda)=(H+\lambda)^{-1}$, $S(\lambda)=(K+\lambda)^{-1}$, and integrals as in (1.6) (and either $(\sigma,\tau)=(s,t)$, or $(\sigma,\tau)=(2s,0)$) write

$$(1.17)\; H^{\sigma}[H^{-s},K^{-s}]H^{\tau} = (\sin^2\pi s)/\pi^2 \int_0^\infty \int_0^\infty \lambda^{-s}d\lambda\mu^{-s}d\mu H^{\sigma}[R(\lambda),S(\mu)]H^{\tau},$$

(cf. (1.6) for more detail) where

$$(1.18)\; [R(\lambda),S(\mu)]v = R(\lambda)S(\mu)[K,H]S(\mu)R(\lambda)v \;\; , \;\; v \in \mathrm{im}(H_0+\lambda)(K_0+\mu),$$

by a simple calculation.

We now need condition (s_2) , at least, to insure that $\mathrm{im}(H_0+\lambda)(K_0+\mu)$ is dense for all $\lambda,\mu > 0$. (cf. prop.1.9, below). The expression $J = [K,H]$ is of order 3 , and has compact support. Accordingly we may write J as a finite sum of products of at most 3 folpde's with compact support. Accordingly, for evaluation of the term at right of (1.17) we will substitute $[K,H]$ in (1.18) by a product DEF of 3 such folpde's , to indicate the procedure. We then will have to commute $S(\mu)$ with D :

$$(1.19)\qquad V_D = [S(\mu),D] = (K^{1/2}S(\mu))(K^{-1/2}[D,K]K^{-1/2})(K^{1/2}S(\mu)) \; ,$$

(using (w)) , where the middle bracket is bounded (and compact, if multiplied by $K^{-\varepsilon}$, with arbitrarily small $\varepsilon>0$) , while the two outer brackets decay with μ, so that $[S(\mu),D]$ is compact, and

$$V_D = [S(\mu),D] = O((1+\mu)^{-1}) . \tag{1.20}$$

Moreover, the commutator V_D not only is bounded, but even GV_D, with an arbitrary folpde G of compact support, is bounded. Therefore we now can write

$$\begin{aligned} R(\lambda)S(\mu)DEFS(\mu)R(\lambda)v &= R(\lambda)DS(\mu)ES(\mu)FR(\lambda)v \\ &+ R(\lambda)V_D ES(\mu)FR(\lambda)v - R(\lambda)DS(\mu)ES_F R(\lambda)v - R(\lambda)V_D EV_F R(\lambda)v , \end{aligned} \tag{1.21}$$

where the right-hand side contains only bounded operators, so that a continuous extension to $v \in H$ is possible.

The remainder of the proof consists of just balancing all the various effects occurring in the integrand

$$\lambda^{-s}\mu^{-s}H^{\sigma}R(\lambda)S(\mu)DEFS(\mu)R(\lambda)H^{\tau} , \tag{1.22}$$

where we focus on the first term at right of (1.21). The integral needs a power $(1+\lambda)^{s-1-\delta}(1+\mu)^{s-1-\delta}$, with $\delta>0$ to converge. That is, in view of $H^{\nu}R(\lambda)=O((1+\lambda)^{\nu-1})$, $K^{\nu}S(\mu)=O((1+\mu)^{\nu-1})$, we can absorb a total power (left and right) $H^{s-\delta+1}$, using the two factors $R(\lambda)$. For $0\le s<1/2$ this means that both $\sigma=\tau=s$ and $\sigma=2s$, $\tau=0$ give a converging integral of compact operators, as long as two of the folpde's D,E,F can be absorbed by the factors $S(\mu)$. That can be done, because it still leaves a factor $(1+\mu)^{-1}$, hence a total power $\mu^{-s}(1+\mu)^{-1}$ to generate a convergent integral. For $s=1/2$, on the other hand, one may split $K^{-1/4}$ from the second $S(\mu)$ and $H^{-1/4}$ from the last $R(\lambda)$, for a term $K^{-1/4}FH^{-1/4}=(K^{-1/4}H^{1/4})(FH^{-1/2}+[H^{-1/4},F]H^{-1/4}) \in L(H)$, by V, lemma 3.6, while we still have the integrand $O(\lambda^{-1/2}(1+\lambda)^{-3/4}\mu^{-1/2}(1+\mu)^{-3/4})$. The other three terms in (1.21) are similar, although easier to treat.

This completes the proof of prop.1.8. Then relation $(1.16)_1$ gives $(K^{-s}H^{s})^{**} - H^{s}K^{-s} \in K$, so that U_s is hermitian, $0\le s\le 1/2$. From the second (1.16) we get $(H^{s}K^{-s})^2 = H^{2s}K^{-2s} + H^{2s}[H^{-s},K^{-s}]H^{s}K^{-s} = H^{2s}K^{-2s}+C$, $C\in K$, so that $U_s^{\,2}=U_{2s}$. This also gives $U_t=(U_{t/2})^2$ positive hermitian, for all $0\le t\le 1$. Interchanging H and K then yields the same relations for $-1\le t\le 0$. Finally it is trivial that $U_tU_{-t} = U_{-t}U_t = U_0 = 1$. This completes the proof of lemma 1.7.

<u>Proposition 1.9</u>. If condition (s_2) holds for K then it also holds

for H , and we have $\mathrm{im}(H_0+\lambda)(K_0+\mu)$ dense in H , for all $0\le\lambda,\mu<\infty$.
Proof. Let $f\in H$, and $(f,(H+\lambda)(K+\mu)u)=0$, for all $u\in C_0^\infty(\Omega)$. Write this as $(f,(K+\mu)^2u) = (f,((K-H)+(\mu-\lambda))(K+\mu)u)$, $u\in C_0^\infty(\Omega)$. Observe that $(H+\lambda)(K+\mu)$ is an elliptic, hence hypo-elliptic expression. Thus the distribution solution f of $(K+\lambda)(H+\lambda)f=0$ is $C^\infty(\Omega)$. Hence we can write $(f,(K+\mu)^2u)=(((K-H)+(\mu-\lambda))f,(K+\mu)u)$. Due to self-adjointness we may write $g=(K-H)f+(\mu-\lambda)f=(K+\mu)h$, with $h\in\mathrm{dom}(K+\mu)$. Conclusion: $(f-h,\mathrm{im}(K+\mu)^2)=0$, i.e., $f=h$, $(K+\mu)f-(H+\lambda)f=(K+\mu)f$, or $(K+\mu)f=0$, $f=0$, so that indeed $\mathrm{im}(H_0+\lambda)(K_0+\mu)$ is dense, q.e.d.

Remark: A proof of prop.1.9 avoiding hypo-ellipticity is possible, using the same arguments as in the proof of IX, prop.5.1.

2. The wave front space.

In this section we will discuss the structure of the symbol space $\mathbb{W}$ of the minimal comparison algebra J_0 . The space $\mathbb{W}$ is defined as the maximal ideal space of the commutative C^*-algebra $J_0/K(H)$ (note prop.2.1, below). This space will be seen to be the bundle of unit spheres in the cotangent space of Ω , while the continuous function $\sigma_A:\mathbb{W}\to\mathbb{C}$ assigned to a generator A=a (or A=DΛ) turns out to be the formal principal symbol of the (zero-or first-order) expression D or a .

Especially, for compact manifolds Ω, this result is well known (cf. Seeley [Se_2], also Gohberg [Gb], Cordes [CI], in case of the (non-compact) $\mathbb{R}^n$). (Note that for a compact Ω the algebra J_0, hence the space $\mathbb{W}$, is entirely independent of the operator H or measure dμ used, within the general restrictions made. This is an immediate consequence of thm.1.1.)

Proposition 2.1. The algebra J_0 contains the compact ideal $K(H)$, and has compact commutators.
Proof. It is sufficient to discuss compactness of commutators, since $J_0\supset K(H)$ was seen in V, lemma 1.1. Moreover, we may apply prop.1.2 again, to conclude that, for the commutator of two generators with compactly supported a or D , one may assume H replaced by $H^\blacktriangle$ satisfying (w). Then,of course, V,thm.3 .1 and V,thm.3.4 give the desired compactness of commutators, q.e.d.

The theorem,below, gives complete control of the wave front space , as we shall call the maximal ideal space $\mathbb{W}$ of the commutative C^*-algebra $J_0/K(H)$, henceforth. We use the notation 'symbol' (or symbol function) for the continuous function over $\mathbb{W}$ associated to the coset $A+K(H)$ of an operator $A \in J_0$ by the Gel'fand map.

Theorem 2.2. The space $\mathbb{W}$ is homeomorphic to the co-sphere bundle $S^*\Omega$ of Ω. Here $S^*\Omega$ is the collection of all (x,ξ) in the cotangent bundle $T^*\Omega$ of Ω (i.e. $x \in \Omega$, $\xi \in \mathbb{R}^n$) with

$$(2.1) \qquad h^{jk}(x)\xi_j\xi_k = 1 .$$

Moreover, in this interpretation of $\mathbb{W}$ the symbol functions of the algebra generators $a \in A^\#$ and $A = D\Lambda$, for $D = b^j\partial_{x^j} + p \in \mathcal{D}^\#$ are explicitly given by the formulas

$$(2.2) \qquad \sigma_a(x,\xi) = a(x) \quad , \sigma_{D\Lambda}(x,\xi) = ib^j(x)\xi_j \ , \ (x,\xi) \in S^*\Omega = \mathbb{W} .$$

Proof. First we observe that J_0 contains the commutative C^*-subalgebra $CO(\Omega)$ of all continuous functions $a(x)$ with $\lim_{x\to\infty}a(x) = 0$, which is obtained as closure of the algebra $A^\#_m \subset J_0$. The maximal ideal space of $CO(\Omega)$ is equal to Ω itself . The imbedding $CO(\Omega) \to J_0$ generates a corresponding 'associate dual map' $\iota:\mathbb{W} \to \Omega$, which must be surjective (cf. $[C_1]$, App.AII,5) . (In particular we use that a multiplication operator $u \to au$ is never compact unless $a \equiv 0$, so that also an injection $CO(\Omega) \to J_0/K$ is induced.) Later on we will recognize ι as the bundle projection in $S^*\Omega$, as defined.

Let Ω_j , $j=1,2,\ldots$, be a locally finite atlas of Ω. Assume that every Ω_j has compact closure $\Omega_j^{clos} \subset \Omega^\wedge_j$, where $\{\Omega^\wedge_j\}$ is an atlas again. Define J_j as the closed 2-sided ideal of J_0 containing all operators with symbol = 0 outside of $\iota^{-1}\Omega_j$.

Proposition 2.3. The ideal J_j is generated by $K(H)$ and all a, $D\Lambda$ with $a \in A^\#_m$, $D \in \mathcal{D}^\#_m$ having compact support, contained in Ω_j .

Proof. Evidently all the above generators are contained in J_j , since for $a \in C_0^\infty(\Omega_j)$ we have $\sigma_a(m) = a(\iota(m))=0$, as $m \in \mathbb{W}\setminus\iota^{-1}\Omega_j$, using the inherent properties of the dual map ι . Similarly, for D with compact support, contained in Ω_j a function $\chi \in C_0^\infty(\Omega_j)$ may be found with $\chi = 1$ on the support of D , so that $D\Lambda = \chi D\Lambda$, $\sigma_{D\Lambda} = \sigma_\chi.\sigma_{D\Lambda}$. We have $\sigma_\chi(m) = \chi(\iota(m))$ again, so that $\sigma_{D\Lambda}=0$, for all m with $\iota(m) \in \Omega\setminus\Omega_j$, follows as well.

Vice versa, let $A \in J_j$. We choose functions $\chi_l \in C_0^\infty(\Omega_j)$, $l=1,2,\ldots$, with $0 \le \chi_l \le 1$, $\chi_l=1$ in K_l , where K_l denotes an increasing sequence of compact sets with $\cup K_l = \Omega_j$. Such a sequence may be constructed, since Ω_j may be regarded as an open subset of $\mathbb{R}^n$ with compact closure. Since $\sigma_A(m)$ is continuous and $=0$ outside $\iota^{-1}\Omega_j$, we conclude that

$$|\sigma_{(1-\chi_l)A}(m)|=(1-\chi_l(\iota(m)))|\sigma_A(m)|\le\sup\{|\sigma_A(m)|:\iota(m)\in\Omega_j\setminus K_l\}\to 0,\ l\to\infty,$$

in view of the compactness of $\partial\Omega_j$, and uniform continuity of σ_A.

It follows that the coset mod K of $A_l=(1-\chi_l)A$ converges to 0 as $l\to\infty$. Or, there exists a sequence $\{C_j\}$ of compact operators such that $\|A-\chi_l A-C_l\|\to 0$, $l\to\infty$. However, it is clear that $\chi_l A+C_l \in J_j$, since A_j can be approximated by a finite sum $C+\sum P$ with $C\in K$, and $P=D_1\Lambda D_2\Lambda\ldots D_N\Lambda$, $D_j\in \mathcal{D}^{\#}_m$, so that

$$(2.3)\qquad \chi_l P = \chi_l D_1\Lambda\chi_{l+1}D_2\Lambda\ldots\chi_{l+N-1}D_N\Lambda + C',\ C'\in K,$$

where $\chi_{l+k-1}D_k$ have support in Ω_j. For (2.3) we used prop.1.2, and $\chi_l = \chi_l\chi_{l+1}\ldots\chi_{l+N-1}$, of course. The proof is complete.

Note that a partition of unity $1=\sum_j\omega_j$, supp $\omega_j\subset\Omega_j$, may be used to verify that $J=\dotplus_j J_j$. That is, every $A\in J$ may be obtained as a limit (for $m\to\infty$) of finite sums $\sum_j A^m_j = A^m$, where $A^m_j\in J_j$.

Again it is clear that each J_j is a C^*-algebra without unit (except in case $\Omega_j=\Omega$ compact) containing K, and that J_j/K has maximal ideal space $\mathbb{M}_j=\iota^{-1}\Omega_j\subset\mathbb{W}$. With this identification the functions in $CO(\mathbb{M}_j)$ correspond to the cosets $A^\vee=\{A+K\}$ of J_j/K, for $A\in J_j$. Clearly we get $\mathbb{W}=\cup_j\mathbb{M}_j$, due to $\Omega=\cup_j\Omega_j$.

For the description of $\mathbb{W}$ it therefore suffices to discuss $\mathbb{M}^j$ and the homomorphisms $J_j\to J_j/K\to C(\mathbb{M}_j)$. Since $\Omega\hat{}_j\supset\Omega_j^{clos}$ is a chart, we have global coordinates $x^1,\ldots,x^n$, defined by a homeomorphism $\Omega\hat{}_j\to O\hat{}_j\subset\mathbb{R}^n$, mapping Ω_j onto $O_j\subset\hat{O}_j$. An explicit set of generators mod K of J_j then is given by

$$(2.4)\qquad \{\ \phi, -i\phi\partial_{x^k}\Lambda\ ,k=1,\ldots,n\ ,:\ \phi\in C_0^\infty(O_j)\ \}\ .$$

(Here we break with our usual convention to silently identify Ω_j with the open subset O_j of $\mathbb{R}^n$, for reason of clarity.)

Proposition 2.4. The space $\mathbb{M}_j$ is homeomorphic to the set

$$(2.5)\qquad \{(x,\xi)\in O_j\times\mathbb{R}^n\ ,\ h^{jk}(x)\xi_j x_k = 1\ \}\ ,$$

equipped with its natural topology as a subset of $\mathbb{R}^{2n}$. Moreover, under this homeomorphism, we have

(2.6) $\sigma_\phi = \phi(x)$, $\sigma_{A_j}(x,\xi) = \phi(x)\xi_j$, for $A_j = -i\phi\partial_{x^j}\Lambda$, $j=1,\dots,n$.

<u>Proof.</u> For an $m \in \mathbb{M}_j$ let $x = \iota(m)$, $\xi_k = \sigma_{A_k}(m)$, where $\phi \in C_0^\infty(\Omega_j)$ (real-valued) satisfies $\phi(x)=1$ near $x=\iota(m)$, $l=1,\dots,n$. We claim that this defines the desired homeomorphism. Indeed, first of all we note that (x,ξ) is well-defined : if ϕ and ψ both are = 1 near $x = \iota(m)$, and if $B_j = -i\psi\partial_{x^j}$, then we get $A_j-B_j = (1-\chi)(A_j-B_j)$, with a C_0^∞-function χ , equal to 1 near $\iota(m)= x$, hence $\sigma_{A_j}(m) = \sigma_{B_j}(m)$, showing that ξ_j is well defined. Also ξ_j is real, due to

$\overline{\xi}_j = \sigma_{A_j^*}(m)$, where $A_j^*=(-i\phi\partial_{x^j})^*\Lambda + C = -i\phi\partial_{x^j}\Lambda + C'$, $C,C' \in K$,

using (1.2) and V, thm.3.4 , so that $\overline{\xi}_l = \xi_l$. Furthermore, with ϕ as above we note that $h^{jk}\phi \in A_m^\#$. For $u,v \in \Lambda C_0^\infty$ we then get

$$(u,A_j^*(h^{jk}\phi)A_kv) = (u,\Lambda(-\tfrac{i}{\kappa}\partial_{x^j}\kappa\phi)h^{jk}\phi(-i\phi\partial_{x^k})\Lambda v) = (\Lambda u,\phi^3 H\Lambda v)$$

$$- (u,\Lambda(D_{\phi^3} + q\phi^3)\Lambda v) .$$

Since ΛC_0^∞ is dense we conclude that

(2.7) $A_j^*(h^{jk}\phi)A_k = \phi^3 + [\Lambda,\phi^3]H\Lambda - \Lambda(D_{\phi^3}+q\phi^3)\Lambda = \phi^3+C$, $C \in K$,

using (1.2), V, lemma 3 .3, V, lemma 3.6, and that $\Lambda H\Lambda u=u$, $u\in\Lambda C_0^\infty$. Taking symbols one obtains the relation

(2.8) $$h^{jk}(x)\xi_j\xi_k = 1 , \quad \text{as } x = \iota(m) .$$

With the above we have defined a map $\upsilon:\mathbb{M}_j \to M_j$, with

(2.9) $$M_j = \{(x,\xi) : x \in O_j , \xi \in \mathbb{R}^n , h^{jk}(x)\xi_j\xi_k = 1 \} .$$

We shall prove that υ defines a homeomorphism onto M_j. Assuming this to be true let us observe that the proper transformation law is valid,since the ξ_j will inherit the co-variant tensor property of the ∂_{x^j} .

To verify that υ is 1-1 we recall that the maximal ideal $m\in\mathbb{M}_j$ is characterized by its corresponding homomorphism $J_j \to \mathbb{C}$, given by $A \to \sigma_A(m)$, $A \in J_j$. Let m, m' be such that $\iota(m) = \iota(m') = x$, and as above. Note that this implies $\sigma_\phi(m)=\sigma_\phi(m')$, and $\sigma_{A_j}(m)=\sigma_{A_j}(m')$ for all $\phi \in C_0^\infty(\Omega_j)$, regardless whether $\phi(x) = 1$, or not. Since σ is a homomorphism, and since the ϕ and A_j listed form a set of generators of J_j, modulo K, it follows that $\sigma_A(m) = \sigma_A(m')$, for all $A \in J_j$. Thus the homomorphisms corresponding to m and m' are identical, which means that $m = m'$. Thus the map υ is injective.

Finally we focus on surjectivity of υ. It is clear that, above every $x \in \Omega_j$ there will be at least one point $(x,\xi) \in \upsilon(\mathbb{M}_j)$. For the map ι maps onto $\mathbb{W}$, so that some $m \in \mathbb{M}_j$ with $x=\iota(m)$ can be found. For this m we then get the ξ_j by our above construction. To show that every (x,η), for general $\eta \in \mathbb{R}^n$ is an image, we construct an automorphism of J_j, leaving K invariant, which carries (x,ξ) into (x,η), in form of a coordinate transform. Let $M=((m_j^{\ k}))$ be a constant real $n\times n$-matrix with $M\xi=\eta$, $M^* J^2 M=J^2$, where $J=((h^{jk}(x)))^{1/2}$ (i.e., JMJ^{-1} is orthogonal). Such a matrix exists for every $\xi,\eta \in \mathbb{R}^n$, and one even may choose $\det M = 1$, so that JMJ^{-1} is a rotation, and can be smoothly deformed into the identity matrix. Denoting $x=x_0$, for a moment we design a diffeomorphism $\theta:\Omega_j \to \Omega_j$ as follows: Let $R(\tau)$, $0 \le \tau < \infty$, be a C^∞-family of rotations such that $R(\tau)=JMJ^{-1}$ for $0 \le \tau < \varepsilon$, $R(\tau)=1$ for $\tau \ge 2\varepsilon$. Define $\theta(x)=x_0+JR^T(|J(x-x_0)|)J^{-1}(x-x_0)$, $x \in \Omega_j$. By construction θ maps each (Euclidean) ellipsoid $\tau^2=|J^{-1}(x-x_0)|^2=h_{jk}(x_0)(x^j-x_0^j)(x^k-x_0^k)$ onto itself, and is equal to the identity as $|J(x-x_0)| \ge 2\varepsilon$. Therefore, for $0<\varepsilon$ sufficiently small, θ defines a diffeomorphism of Ω_j onto itself, and θ is linear near x_0, and equal to the identity near $\partial\Omega_j$. In particular we may extend θ to a diffeomorphism $\Omega \to \Omega$ equal to the identity outside Ω_j.

Using θ we define the unitary operator $\Theta : H \to H$ by

$$\Theta u(x) = (\det(\partial\theta/\partial x))^{1/2} u(\theta(x)) \quad . \tag{2.10}$$

Then $A \to A^\blacktriangle = \Theta^{-1} A \Theta$ is a $*$-automorphism $J_j \to J_j$ leaving $K(H)$ invariant.

Indeed this automorphism trivially leaves K invariant, and $a^\blacktriangle = \Theta^{-1} a \Theta = a \circ \theta^{-1}$, where $a \circ \theta^{-1} \in C_0^\infty(\Omega_j)$ if $a \in C_0^\infty(\Omega_j)$. Similarly,

$$\Theta^{-1} u = (u/\nu)\circ\theta^{-1} \ , \ \nu = (\partial\theta/\partial x)^{1/2} \ , \tag{2.11}$$

and, for $b \in C_0^\infty(\Omega_j)$, $D_r = b\partial_{x^r}$, we get

$$D_r^\blacktriangle u = \Theta^{-1} b \partial_{x^r} \Theta u = (b\circ\theta^{-1})(((\nu_{|x^r}/\nu)\circ\theta^{-1})u + (\theta^l_{\ |x^r}\circ\theta^{-1})\partial_{x^l} u), \tag{2.12}$$

which again is an expression with compact support, contained in Ω_j. Accordingly, the automorphism Θ leaves the generating sets of J_j invariant.

In fact, near $x=x_0$, the expression $D_r=-i\phi\partial_{x^r}$ goes into $D_r^{\blacktriangle} = -i\phi m_r{}^j\partial_{x^j} = m_r{}^j D_j$, modulo a zero order term.

We now must find out how the operator Λ changes under conjuation by Θ. Clearly the differential expression H goes into

$$\text{(2.13)}\qquad \begin{gathered} H^{\blacktriangle} = \Theta^{-1}H\Theta = -\kappa^{\blacktriangle-1}\partial_{x^j}\kappa^{\blacktriangle}h^{\blacktriangle jk}\partial_{x^k} + q^{\blacktriangle} \ , \ d\mu^{\blacktriangle} = \kappa^{\blacktriangle}dx \ , \\ \kappa^{\blacktriangle} = \kappa\circ\theta^{-1} \ , \ h^{\blacktriangle jk} = (h^{lm}\theta^j{}_{|x^l}\theta^k{}_{|x^m})\circ\theta^{-1} \ , \ q^{\blacktriangle} = (H\nu/\nu)\circ\theta^{-1} \ . \end{gathered}$$

Indeed, since Θ is unitary (i.e., $\Theta^*=\Theta^{-1}$), we get for $u \in C_0^\infty(\Omega)$,

$$\text{(2.14)}\qquad (u,H^{\blacktriangle}u) = \int_\Omega (h^{jk}(\nu(u\circ\theta))_{|x^j}(\nu(u\circ\theta))_{|x^k} + |q\circ\theta|^2\nu^2)d\mu \ .$$

For the first term we use the formula

$$\text{(2.15)}\qquad (\nu(u\circ\theta))_{|x^j} = \nu_{|x^j}(u\circ\theta) + \nu(u_{|x^l}\circ\theta)\theta^l{}_{|x^j} \ ,$$

and a partial integration similar to that of III.(6.7), to get

$$\text{(2.16)}\qquad (u,H^{\blacktriangle}u) = \int (h^{lm}\theta^j{}_{|x^l}\theta^k{}_{|x^m}((\overline{u}_{|x^j}u_{|x^k})\circ\theta) + (H\nu/\nu)|u\circ\theta|^2)\nu^2 d\mu \ .$$

Using that $\nu^2=|\partial\theta/\partial x|$ we may transform the integral backward. Then the hermitian form may be polarized to to get an expression for $(u,H^{\blacktriangle}v)$. The second order expression $H^{\blacktriangle}$ is uniquely determined by the values of $(u,H^{\blacktriangle}v)$, for all $u,v \in C_0^\infty$, and (2.13) results.

Clearly $H^{\blacktriangle} = H$ and $d\mu = d\mu^{\blacktriangle}$ outside a small compact neighbourhood of the point x_0. Accordingly we may apply thm.1.1 and the triples $\{\Omega,d\mu,H\}$, and $\{\Omega,d\mu^{\blacktriangle},H^{\blacktriangle}\}$ generate the same J_0. In fact, by definition of Λ and $\Lambda^{\blacktriangle}$ as inverse positive square roots of H and $H^{\blacktriangle} = \Theta^{-1}H\Theta$ (where now H and $H^{\blacktriangle}$ mean the self-adjoint closures of the minimal operators) it is clear that $\Lambda^{\blacktriangle} = \Theta^{-1}\Lambda\Theta$. We thus get

$$\text{(2.17)}\qquad \begin{gathered} \Theta^{-1}a\Theta = a\circ\theta^{-1} = a^{\blacktriangle} \ , \ \Theta^{-1}D\Lambda\Theta = D^{\blacktriangle}\Lambda^{\blacktriangle} \ , \\ D^{\blacktriangle} = \Theta^{-1}D\Theta = ((b^l\theta^j{}_{|x^l})\circ\theta^{-1})\partial_{x^j} + (D\nu/\nu)\circ\theta^{-1} \ . \end{gathered}$$

Now consider the associate dual map $m \to m^{\blacktriangle} = \Theta^{-1}m\Theta$ of the automorphism $A \to A^{\blacktriangle}$. By construction we get $\iota(m) = \iota(m^{\blacktriangle})$, whenever $\iota(m)=x_0$, since $a^{\blacktriangle}(x_0) = a(x_0)$ for all a . Therefore for our (x_0,ξ) we have $m^{\blacktriangle} \in \iota^{-1}(x_0)$ as well, and we now will show that $\upsilon(m^{\blacktriangle}) = (x_0,\eta)$. Indeed, $\sigma_{A_r}(m^{\blacktriangle}) = \sigma_{A_r^{\blacktriangle}}(m)$, by the properties of the associate dual map $m \to m^{\blacktriangle}$. We have seen that

$$(2.18) \qquad A_r^{\blacktriangle} = D_r^{\blacktriangle}\Lambda^{\blacktriangle} = m_r^{\ 1}D_1\Lambda^{\blacktriangle} + C = m_r^{\ 1}A_1(\Lambda^{-1}\Lambda^{\blacktriangle}) + C \ .$$

Accordingly, $\sigma_{A_r^{\blacktriangle}}(m) = m_r^{\ 1}\xi_1\sigma_{(\Lambda^{-1}\Lambda^{\blacktriangle})}(m) = \eta_r\sigma_{(\Lambda^{-1}\Lambda^{\blacktriangle})}(m)$. Thus we are only left with showing that $\sigma_{(\Lambda^{-1}\Lambda^{\blacktriangle})}(m) = 1$. However, from lemma 1.8 we know that the coset $U_{-1/2}$ of $\Lambda^{-1}\Lambda^{\blacktriangle}$ is the inverse square root of $U_1 = H^{\blacktriangle}H^{-1} = \sum h^{rl}A_rA_1 + C$, near x_0. Thus, over x_0 we get $\sigma_{U_1} = h^{rl}\xi_r\xi_1 = 1$. This completes the proof.

3. Differential expressions within reach of the algebra J_0.

In chapter X we shall exploit the information on the symbol space of the general comparison algebra to its full extent. Presently let us discuss differential expressions within reach of the minimal comparison algebra. For the special case of a compact manifold Ω we then will discuss the implications of our theory, such as existence of a Fredholm and Green inverse, the Sobolev estimates etc.

First let Ω be general, not necessarily compact. Consider a given triple $\{\Omega,d\mu,H\}$ on Ω , with its Hilbert space $H=L^2(\Omega,d\mu)$. Let J_0 be the corresponding minimal comparison algebra. Using prop.1.2 (and the remark following it) and prop.1.4 we may modify the potential q of H such that cdn.(s) holds, without changing J_0. Moreover, using thm.1.1 we also may modify H and dμ on compact subsets of Ω (In fact, applying later results of algebra surgery (cf.VIII,thm.3.1) it will follow that J_0 remains unchanged as long as the measure dμ, hence the Hilbert space H is kept the same outside some compact set).In particular, in the following, it is no loss of generality to assume that cdn.(s) holds.

Specifically for a compact manifold there is only one Hilbert space H, and the choice of H and dμ will not affect J_0 at all. Then we have cdn.(s) for just any H, regardless of the potential q , by IV,thm.1.1.

Proposition 3.1. Let L be an N-th order differential expression with C^∞-coefficients and of compact support on Ω. Then the unbounded linear operator $L\Lambda^N$ with domain $\Lambda^{-N}C_0^\infty(\Omega)$ admits a continuous extension A to all of the Hilbert space H , and we have $A\in C=J_0$. In other words, every differential expression with compact support is within reach of the minimal comparison algebra J_0 .

Proof. For N=0 and N=1 the statement is trivial, since the operators $a\Lambda^0 = a$ and $D\Lambda$, for a multiplier a or a folpde D belong to the generators of $C=J_0$. For the general case it is essential that every L of order N and of compact support can be obtained as a finite sum of (not more than N) first order expressions. Indeed, refer to some finite open chart cover of the compact set supp L , a subordinate partition of unity, and explicit form of L in the local coordinates

$$\text{(3.1)} \quad \operatorname{supp} L \subset \cup\Omega_j,\ 1=\sum\omega_j \text{ on } \operatorname{supp} L,\ L=L^j= \sum_{|\alpha|\le N} a_\alpha^j(x)D_x^\alpha \text{ on } \Omega_j,$$

and write

$$\text{(3.2)} \quad L = \sum\omega_j L_j = \sum_j \sum_{|\alpha|\le N} (\omega_j a_\alpha^j)\Pi_{\nu=1}^n \Pi_{\rho=1}^{\alpha}(\omega_{j\nu\rho}D_{x^\nu}) ,$$

using the auxiliary functions $\omega_{j\nu\rho}\in C_0^\infty(\Omega_j)$, =1 near supp $\omega_j\subset\Omega_j$. Here the expressions $\omega_{j\nu\rho}D_{x^\nu}$ may be interpreted as folpdes in $\mathcal{D}_m^\#$, and, for $|\alpha|=N$, we may multiply the factor $(\omega_j a_\alpha^j)$ into the last such folpde. Accordingly each summand in (3.2) is a product of at most N folpdes. Trivially we may assume exactly N folpdes in each product. Accordingly is is sufficient to prove the statement for products

$$\text{(3.3)} \quad L = D_1\cdots\cdots D_N ,\ D_j \in \mathcal{D}_m^\# .$$

Notice also that $\Lambda^{-N}C_0^\infty(\Omega)$ is dense in H for every N, as implied by cdn.(s), as seen in V,6, for example.

Therefore it is sufficient to show that $L\Lambda^N u$, for $u\in\Lambda^{-N}C_0^\infty$, can be written as a sum of terms $A_1\ldots A_R u$, with generators A_j of J_0 .

For an induction proof consider $K = D_1D_2$, $M = D_3\ldots D_N$. Assuming that an operator $C\supset K\Lambda^2$, $C\in L(H)$ exists, we get

$$\text{(3.4)} \quad KM\Lambda^N u = CM\Lambda^{N-2}u + C[H,M]\Lambda^N u ,\ u \in \Lambda^{-N}C_0^\infty(\Omega) .$$

(Confirm in details that $CHM\Lambda^N u=KM\Lambda^N u$ and $CM\Lambda^{N-2}u=CMH\Lambda^N u$.)

Note that the commutator is an expression of order $\leq N-1$, so that $[H,M]\Lambda^N=[H,M]\Lambda^{N-1}\Lambda \in L(H)$, by induction hypothesis. Thus we only must prove the statement for N=2 to make the induction complete.

On the other hand, for a twofold product DF , with folpdes D,F, we write (with exactly the same reasoning)

$$(3.5) \qquad DF\Lambda^2 u = D\Lambda^2 Fu + D\Lambda^2[H,F]\Lambda^2 u \ , \ u \in \Lambda^{-2}C_0^\infty(\Omega) \ .$$

Again the first term $D\Lambda^2 F = (D\Lambda)(F^*\Lambda)^*$ is a product of generators, while [H,F] is of order ≤ 2 , hence a sum of products of at most two folpdes. It follows that the second terms operator is a finite sum of terms

$$(3.6) \qquad (D\Lambda)(\Lambda D')(F'\Lambda)\Lambda \quad \in C \ ,$$

with folpdes D', F', where F'=1 may occur as well. This completes the proof.

Next let us ask for the symbol of $A \supset L\Lambda^N$, as in prop.3.1. We find that $A \supset D_1 \ldots\ldots D_N\Lambda^N$ has symbol

$$(3.7) \qquad \sigma_A = \Pi_{j=1}^N \sigma_{A_j} \ , \quad A_j = D_j\Lambda \ .$$

Indeed the second terms in both (3.4) and (3.5) carry an extra power of Λ which may be written as $\chi\Lambda$, mod $K(H)$, with $\chi \in C_0^\infty$ since there always is a expression with compact support in the same term. Recall that $\chi\Lambda \in K(H)$, by III,thm.3.7. Also we get $D\Lambda - \Lambda D \in K(H)$ by V,thm.3.4, so that $\sigma_{D\Lambda} = \sigma_{\Lambda D}$.

Now we may combine (3.7) and the product representation (3.2) of an expression L and apply thm.2.2 to obtain the symbol of $A \supset L\Lambda^N$. The proof of prop.3.2 below is a formal verification left to the reader:

Proposition 3.2. Let L be as in prop.3.1. Assume that, in local coordinates of $\Omega' \subset \Omega$ L has the form

$$(3.8) \qquad L = \sum_{|\alpha| \leq N} a_\alpha(x) D_x^\alpha \ , \ x \in \Omega' \ ,$$

with coordinates x of the chart $\Omega' \subset \Omega$. Then the symbol of the operator $L\Lambda^N \subset A \in C$, on the set $\iota^{-1}(\Omega') \subset \mathbb{W} = \mathbb{M} = S^*\Omega$, is explicitly given by the formula

$$(3.9) \qquad \sigma_A(x,\xi) = \sum_{|\alpha|=N} a_\alpha(x)\xi^\alpha \ , \ (x,\xi) \in \iota^{-1}(\Omega') \ .$$

Notice that the expression at right of (3.9) is commonly

referred to as the principal part polynomial (or the principal symbol) of the expression L . Since we know that σ_A , for $A \in C$, is a continuous function over the subset $S^*\Omega$ of the cotangent space, it follows that also the right-hand side of (3.9) is independent of the coordinates used, as also easily confirmed directly. In particular it sometimes is more convenient to use tensor notation instead of multi-index notation: We may write

$$(3.10) \qquad \sum_{|\alpha|=N} a_\alpha(x)\xi^\alpha = a^{\sim j_1 \dots j_N} \xi_{j_1} \dots \xi_{j_N} ,$$

with summation convention and a unique symmetric contravariant tensor $a^\sim$ of N variable indices.

Remark 3.3. Observe that the expression L is elliptic over $\Omega' \subset \Omega$ if and only if the symbol of $A \supset L\Lambda^N$ (with respect to the algebra J_0 does never vanish on the part of the wave front space $\iota^{-1}(\Omega')$.

From now on consider the case of a compact manifold Ω only. Then the algebra J_0 contains the identity operator of H and its wave front space $\mathbb{W}$ is compact. As a consequence of prop.3.1 every differential operator on Ω with C^∞-coefficients is within reach of the unique (minimal) comparison algebra $C=J_0$. Then let $A \supset L_0\Lambda^N$ and $Z=A\Lambda^{-N}$ be the operator of V,def.6.1 and realization of V,def. 6.2, respectively.

Theorem 3.4. The operator $A \supset L\Lambda^N$ has an inverse modulo the class $F(H) \subset L(H)$ of all operators of finite rank if and only if L is elliptic on Ω . In details, there exists an operator $B \in L(H)$ with

$$(3.11) \qquad AB = 1 + F \quad , \ BA = 1 + G \ , \quad F, G \in F(H) \ ,$$

if and only if L is uniformly elliptic. Moreover, we then get

$$(3.12) \qquad \dim \ker A < \infty \ , \ \dim H/\mathrm{im}\, A < \infty \ ,$$

and B may be chosen as operator in J_0 such that -F and -G coincide with the orthogonal projections onto the finite dimensional spaces $\ker A^*$ and $\ker A$, respectively. On the other hand, we instead may choose $B \in J_0$ such that for an arbitrarly given $\varepsilon > 0$ we get

$$(3.13) \quad \|B\| \le \|1/\sigma_A\|_{L^\infty(\Omega)} + \varepsilon, \ \sigma_A(x,\xi) = \sum_{|\alpha|=N} a_\alpha(x)\xi^\alpha \text{ in local coord's.}$$

Proof. For a bounded operator A an inverse modulo $F(H)$ exists if and only if A has an inverse modulo $K(H)$. Indeed, since $F(H) \subset K(H)$, we only must show that (3.11) for a $B \in L(H)$ with $F, G \in K(H)$ instead of $F(H)$ implies (3.11) for some other $B=C$. By Rellich's criterion we get orthogonal projections P , Q with finite rank such that

$$(3.14) \qquad \|F(1-P)\| \le 1/2 \ , \ \|G(1-Q)\| \le 1/2 \ .$$

Then we may write

$$(3.15) \qquad AB = (1+F(1-P)) + FP \ , \ BA = (1+G(1-Q)) + GQ \ ,$$

where the first terms at right are invertible (an inverse is given by the norm convergent infinite series $\sum_{j=0}^{\infty}(F(P-1))^j$, in the first case, for example). Thus we have

$$(3.16) \quad A(B(1+F(1-P))^{-1}) = 1 + F' \ , \ F' = FP(1+F(1-P))^{-1} \in F(H),$$

showing that $B' = B(1+F(1-P))^{-1}$ is a right inverse mod F of A . Similarly we use the second (3.15) to obtain a left inverse. It is known that existence of both a left and a right inverse implies that both, left and right inverse are inverses, so that indeed A is invertible mod F .

This shows that we only must discuss invertibility of $A \supset L\Lambda^N$ mod $K(H)$. But $K(H)$ is a closed 2-sided $*$-ideal of the C^*-algebra $L(H)$. Moreover, $C = J_0$ is a C^*-subalgebra of $L(H)$ containing $K(H)$. It therefore follows that $L^\vee = L(H)/K(H)$ and $C^\vee = C/K(H)$ are C^*-algebras, and that $C^\vee \subset L^\vee$, using a well known result on C^*-algebras (cf. [R_1] or any general book on C^*-algebras). Moreover, using another standard result on C^*-algebras, $C^\vee$ is inverse closed in $L^\vee$. That is, $A^\vee = A+K(H) \in L^\vee$ is invertible in $C^\vee$ if and only if it is invertible in $L^\vee$ (cf.[C_1],AII,L.7.15).

It follows that $A \supset L\Lambda^N$ is invertible mod $F(H)$ if and only if its coset $A^\vee = A+K(H)$ has an inverse in $C^\vee$. Here we finally use the fact that the commutator ideal E of $C=J_0$ equals $K(H)$, and that $C^\vee$ is isometrically isomorphic to $C(\mathbb{W})$. Since the cosphere bundle $\mathbb{W}=S^*\Omega$ is compact, a function $\sigma \in C(\mathbb{W})$ is invertible if and only if it never vanishes. Hence $A \supset L\Lambda^N$ is invertible mod $F(H)$ if and only if L is elliptic, by remark 3.3 above. This completes the proof of the first statement.

Observe that $\ker A \subset \ker(1+G)$ and $\ker A^* \subset \ker(1+F^*)$, where both right hand sides are finite dimensional. Hence $\dim \ker A < \infty$, $\dim \ker A^* < \infty$. Notice that im A is closed: For $Au_k \to v$, $k \to \infty$ get $(1+G)u_k \to Bv$. If w_k is the component of u_k orthogonal to $\ker(1+G)$

then we must have $w_k \to w$ and $BAw = Bv$. Hence $v=Aw + z$, $z\in \ker B$. Then let $p_k=u_k-w_k \in \ker(1+G)$. We get $Ap_k \to z$. If q_k denotes the component of p_k othogonal to $\ker A \subset \ker(1+G)$ then we also have $Aq_k \to z$, hence $q_k \to q$, since A is invertible in the finite dimensional space $\ker(1+G) \cap (\ker A)^{\perp}$. Then $v=A(w+q)$, so that im A is closed hence $\mathrm{im}\, A = (\ker A^*)^{\perp}$. Thus $\dim(H/\mathrm{im}\, A) = \dim \ker A^* < \infty$, and we have (3.12) confirmed.

It then is clear that $A_0 = A|\mathrm{im}\, A^*$ constitutes an invertible map $\mathrm{im}\, A^* \to \mathrm{im}\, A$. Let its inverse be called B_0 and extend $B_0: \mathrm{im}\, A \to \mathrm{im}\, A^*$ to an operator $B': H \to H$ by setting $B'=0$ on $\ker A^* = (\mathrm{im}\, A)^{\perp}$. It follows that $B' \in L(H)$ and that AB' abd $B'A$ are the orthogonal projections onto im A and im A^* , respectively. Thus $1-AB'$ and $1-B'A$ project onto $\ker A^*$ and ker A, resp. Since B' is an inverse mod $K(H)$ of $A \in J_0$ we conclude that also $B' \in J_0$, using again that inverses in $L^{\vee}$ of elements in $C^{\vee}$ must be in $C^{\vee}$.

Finally it is found that '1/2' in (3.14) may be replaced by any arbitrary positive number δ . Then the conclusion leading to (3.16) gives an inverse mod $F(H)$, called B", for a moment such that $\|B''\| \leq \|B\| + \delta$ .On the other hand the isometry of the Gelfand Naimark isomorphism implies that

$$(3.17) \qquad \|1/\sigma_A\|_{L^\infty(\Omega)} = \inf\{\|B+K\| : K \in K(H)\} = \gamma$$

Since B+K is a K-inverse whenever B is, we can find a K-inverse B with $\|B\| \leq \gamma + \varepsilon/2$. Setting $\delta = \varepsilon/2$ we then get the desired F-inverse B" This completes the proof of thm.3.4.

An operator $A \in L(H)$ satisfying (3.12) is called a Fredholm operator. An inverse B of A mod F is called a Fredholm inverse of A . For more details cf.X,1 . The unique Fredholm inverse satisfying (3.11) with the orthogonal projections onto ker A and ker A^* will be called the distinguished Fredholm inverse.

Next it will be our intention to show that the distinguished Fredholm inverse has the properties of a "Green inverse", intimately related to the generalized Green's kernel of the realization Z .

Proposition 3.5. Every C_0^∞-differential expression of order $\leq N$ over Ω , whether compact or not, is H-compatible in the sense of V,def. 6.4.

The proof is evident, since indeed the successive commutators $L_0=L$, $L_1=[H,L]$, ..., $L_{j+1}=[H,L_j]$ all are compactly supported and of order $\leq N+j$.

For an H-compatible expression of order N we now may apply V,thm.6.8 and its corollaries.

Proposition 3.6. For $A\supset L_0\Lambda^N$ of prop.3.1 we have $\Lambda^s A\Lambda^{-s}\in L(H)$, and

$$\Lambda^s A\Lambda^{-s} - A \in K(H) \quad . \tag{3.18}$$

In other words, for every $\leq$ N-th order differential expression L with C^∞-coefficients on a compact manifold Ω the operator $A\supset L_0\Lambda^N$ is of order 0 and is K-invariant under H-conjugation.

The proof is immediate, in view of the above mentionned thm. since all terms at right of V,(6.11) are compact: Their differential expressions are compactly supported and there is an excess of Λ-powers.

Theorem 3.7. For an elliptic differential expression L on a compact manifold Ω the realization Z of V,(6.2) allows a 'distinguished Green inverse' G , defined as an operator $G\in O(-N)$ (in the sense of I,6) such that

$$ZG = 1-Q \ , \ Z^*G = 1-P \ , \tag{3.19}$$

with the H_0-orthogonal projections P and Q onto ker Z and ker Z^* , respectively. We have ker Z , ker $Z^* \subset H_\infty$, and P, Q are continuous operators $H_{-\infty}\to H_\infty$ of finite rank.

Proof. Using thm.3.4, if L is elliptic, then the distinguished Fredholm inverse B of $A\supset L_0\Lambda^N$ is well defined. This implies that the operator $A=A_0$ is Fredholm. Moreover, from prop.3.6 above we know that $A\in O(0)$ and that A is K-invariant under H-conjugation. Thus I,thm.6.6 may be applied and we get ker $A\subset H_\infty$, ker $A^*\subset H_\infty$. Accordingly the distinguished Fredholm inverse B gives a distinguished Green inverse of A . Let $G=\Lambda^N B$. then we get LG=AB=1-Q, as stated. Also $GL=\Lambda^N BA\Lambda^{-N} = 1-\Lambda^N P\Lambda^{-N}$, with P of thm.3.4. One then confirms that $P' = \Lambda^N P\Lambda^{-N}$ is the H_0-orthogonal projection onto ker $L^* = \ker(A^*\Lambda^N)$. In particular $(A^*\Lambda^N)_0$ is Fredholm and K-invariant under H-conjugation, so that ker $L^*\subset H_\infty$ again. Q.E.D.

4. The Sobolev estimates for elliptic expressions on compact Ω .

In this section we again consider a compact manifold under any H and $d\mu$ whatsoever. We have condition (s) and therefore the HS-chain of Sobolev spaces as in sec.3 above. First we note:

Lemma 4.1. For a compact manifold we have $H_\infty = C^\infty(\Omega)$, and we get Lu = Zu for all $u\in H_\infty$ and every expression L and the realization

Z of V, def.6.1.

This follows at once from the well known Sobolev imbedding lemma: For $s \geq n/2+k$ the space H_s is (compactly) imbedded in $C^k(\Omega)$ (cf. IX, prop.1.5). Here we shall discuss a somewhat weaker result also implying lemma 4.1. (For a Sobolev imbedding of L^p into L^q cf. IV, prop.4.1.)

Lemma 4.2. For $s \geq [n/2]+k+1$ the space H_s has a natural imbedding into the space $C^k(\Omega)$, defined as follows: The class $u \in H_s$ of squared integrable functions (with any two coinciding outside a set of measure 0) contains precisely one function in $C^k(\Omega)$. Moreover, there exists a constant $c_{s,k}$ independent of u such that

$$(4.1) \qquad \|u\|_{C^k(\Omega)} \leq c_{s,k} \|u\|_s \, , \text{ for all } u \in H_s \, , \; s \geq [n/2]+k+1 \, .$$

Proof. We use Taylor's formula with integral remainder: A function $\phi(t)$ of one variable t, infinite differentiable on $\mathbb{R}$, and with derivatives of all orders (including 0) vanishing at 0 satisfies

$$(4.2) \qquad \phi(t) = \int_0^t (t-\tau)^{N-1}/(N-1)! \; \phi^{(N)}(\tau)d\tau \, ,$$

with the N-th derivative $\phi^{(N)}$, since all terms of the Taylor series at 0 vanish and only the remainder in integral form remains. We apply (4.2) to the function $\phi(t) = u(x_0+(1-t)z_0)\chi((1-t)z_0)$, where $u \in C^\infty(\mathbb{R}^n)$ is arbitrary and $\chi \in C^\infty(\mathbb{R}^n)$, $\chi \geq 0$, denotes a cut-off function at 0; we have $\chi=1$ near 0, $\chi=0$ for $|x| \geq 1/2$. Also, $x_0, z_0 \in \mathbb{R}^n$ are arbitrary, but $|z_0| = \sum z_0^{j^2} = 1$. $N>0$ denotes an integer to be determined. With a change of integration variable $1-\tau \to t$ we get

$$(4.3) \qquad u(x_0) = \int_0^1 dt \; t^{N-1}/(N-1)! d^N/dt^N(\chi(tz_0)u(x_0+tz_0)) \, .$$

Now we estimate, writing

$$(4.4) \qquad |u(x_0)| \leq c \int_0^1 dt t^{N-1} \psi(tz_0) \sum_{|\theta| \leq N} |u^{(\theta)}(x_0+tz_0)| \, ,$$

where ψ denotes another cut-off functions with the same properties as χ and $|\chi^{(\theta)}| \leq \psi$ for all x and $|\theta| \leq N$, and with a constant c. Then we integrate (4.4) over the unit sphere of $\mathbb{R}^n$, with respect to the variable z_0, and the surface measure dS. Using that $dx = r^{n-1}drdS$, with $r=|x-x_0|$, we get

$$(4.5) \qquad |u(x_0)| \leq c \int dx |x-x_0|^{N-n} \psi(x-x_0) \sum_{|\theta| \leq N} |u^{(\theta)}(x)| \, .$$

Now we set $N=[n/2]+1$ and use Schwarz' inequality, noting that $\int_{|x-x_0|\le 1} |x-x_0|^{2(N-n)}\,dx = c_0<\infty$, with this choice of N. Thus,

$$(4.6)\qquad |u(x_0)| \le c \sum_{|\theta|\le N} \{\int |\psi(x-x_0)u^{(\theta)}(x)|^2 dx\}^{1/2},\ x_0\in\mathbb{R}^n ,$$

where c is independent of x_0 .

We now may implant (4.6) onto a compact manifold Ω. Choosing a finite atlas such that each chart Ω' is contained in a chart Ω'' in such a way that $\mathrm{dist}(\Omega',\partial\Omega'')\ge 2$ we find that

$$(4.7)\qquad |u(x_0)| \le c \sum \|L_{j,x_0}u\| \ ,\ x_0\in\Omega\ ,\ u\in C^\infty(\Omega)\ .$$

where the sum is finite and the differential operators L_{j,x_0} of order $\le N$ with C^∞-coefficients satisfy

$$(4.8)\qquad \|L_{j,x_0}\Lambda^N\| \le c\ ,\ c \text{ independent of } x_0 \text{ and } j\ .$$

(Also their number is independent of x_0). The verification of the latter fact requires some calculation involving decomposition of L_{jx_0} into sums of products of folpdes, similar as in (3.2), left to the reader.

Combining (4.7) and (4.8) we get

$$(4.9)\qquad \|u\|_{L^\infty(\Omega)} \le c\ \|u\|_N\ ,\ \text{for all } u\in C^\infty(\Omega)\ .$$

For an arbitrary $u\in H_N$, $N=[n/2]+1$, there exists a sequence $u_k\in C^\infty(\Omega)$ such that $\|u-u_k\|\to 0$, $k\to\infty$. We then conclude uniform convergence of u_k (i.e., convergence in $C(\Omega)$), applying (4.9) to the difference $u_j-u_k\in C^\infty$. Hence there exists a limit of u_k in $C(\Omega)$. It is clear that the limits in $C(\Omega)$ and in H must coincide whenever they both exists. Thus indeed $u\in C(\Omega)$, i.e., $H_N\subset C(\Omega)$.

Similarly, for $N=[n/2]+1+k$ we may apply (4.7) to $\omega u^{(\alpha)}$, $|\alpha|\le N$, where $u^{(\alpha)}$ is a local derivative of u and ω cuts off in a neighbourhood of the complement of the local chart used. Clearly the expressions $L_{j,x_0}\omega\partial_x^\alpha$ are of order $\le N+|\alpha| \le N+k$. Hence (4.8) may be replaced by

$$(4.10)\qquad \|L_{j,x_0}\omega\partial_x^\alpha\Lambda^{N+k}\| \le c\ ,\ c \text{ independent of } x_0 \text{ and } j\ .$$

This implies the result for general k. Q.E.D.

<u>Corollary 4.3</u>. If Ω is not necessarily compact, but the triple

$\{\Omega, d\mu, H\}$ satisfies cdn.(s), then we still have the imbedding $H_s \to C^k(\Omega)$, as $s \geq [n/2]+k+1$, although the estimate (4.1) holds only locally, with $\|u\|_{C^k(\Omega)}$ and $c_{s,k}$ replaced by $\|u\|_{C^k(\Omega')}$, and $c_{s,k,\Omega'}$, respectively, for each $\Omega' \subset \Omega$ with compact closure.

All above proofs may be carried over to prove cor.4.3.

A simple consequence of thm.3.7 is the result below.

Theorem 4.3 (The Sobolev estimate). Let the expression L on Ω be elliptic. For every $s \in \mathbb{R}$ there exists a constant c_s such that

(4.11) $$\|u\|_{N+s} \leq c_s(\|Lu\|_s + \|u\|_s) \text{ , for all } u \in H_\infty \quad ,$$

Proof. Using thm.3.7 we get

(4.12) $$\|(1-P)v\|_s = \|BAv\|_s \leq \|B\|_s \|Av\|_s, \ v \in H_\infty.$$

Here we may introduce $v = \Lambda^{-N}u$, $u \in C_0^\infty$. It follows that

(4.13) $$\|u\|_{N+s} = \|\Lambda^{-N}u\|_s \leq \|B\|_s \|Zu\|_s + \|P\Lambda^{-N}u\|_s.$$

Finally we use the fact that $P \in O(-\infty)$, so that $\|P\Lambda^{-N}\|_s < \infty$. Therefore (4.11) follows from lemma 4.1, q.e.d.

Remark 4.5. Note that the term $c_s\|u\|_s$ at right of (4.11) may be replaced by $c_{s,t}\|u\|_t$, for just any $t \in \mathbb{R}$. Indeed, the expression $P\Lambda^{-N}$ of (4.13) still belongs to $O(-\infty)$, so that $\|P\Lambda^{-N}\|_{s,t} < \infty$ for any pair $s,t \in \mathbb{R}$.

Remark 4.6. It is possible to prove the following stronger result For every $\varepsilon > 0$ there exists a constant $c(\varepsilon)$ such that for an elliptic expression L of order N we have

(4.14) $$\|u\|_{s+N} \leq \|\sigma_A^{-1}\|_{L^\infty(\Omega)} + \varepsilon)\|Zu\|_s + c(\varepsilon)\|u\|_s \ , \ u \in H_\infty \ .$$

To prove this we apply thm.3.7 to the self-adjoint elliptic expression L^*L . It is found that the bounded self-adjoint opera- $A_0^*A_0$ (in H_0) is Fredholm and admits compact conjugation. Similarly all the operators $A_0^*A_0 - \lambda = \Lambda^N(L^*L - \lambda H^N)_0\Lambda^N$, as long as $\lambda < m_L^2 = (\inf\{|\sigma_A| : (x,\xi) \in \mathbb{W}\})^2$. We conclude from there that $A_0^*A_0$ has only discrete spectrum below m_L^2 . Moreover, all eigen functions to eigenvalues less than m_L^2 are in H_∞ . From this fact one concludes the existence of a bounded self-adjoint B_0 with norm $\|B\|_0 \leq m_L^{-1} + \varepsilon$ such that

(4.15) $$A_0^* A B = B A_0^* A_0 = 1-R \ ,$$

with a finite dimensional orthogonal projection R onto a subspace of H_∞ , i.e., the span of all eigen spaces to eigenvalues $< m_L/(1+\varepsilon m_L)$. In particular B is positive self-adjoint. Now relation (4.15) may be used instead of the earlier BA=1-P to derive the improved estimate for s=0 . Similarly for general s, using the operator $A_s^{\langle s \rangle} A_s$, with the Hilbert space adjoint $A^{\langle s \rangle}$ of A in the space H_s .

CHAPTER 7. THE SECONDARY SYMBOL SPACE.

In the present chapter we turn our attention to the set $\mathbb{M}\setminus\mathbb{W}$ of the symbol space $\mathbb{M}$ of a general comparison algebra, consisting of all points not contained in the wave front space. As pointed out before, the wave front space $\mathbb{W}$ (i.e. the symbol space of the minimal comparison algebra) is found naturally imbedded as an open subset of the symbol space $\mathbb{M}$, for every comparison algebra C meeting our general assumptions. This now will be discussed in detail in sec.1, below. If the manifold Ω is compact, so that the minimal comparison algebra is the only comparison algebra, then we have $\mathbb{M}=\mathbb{W}$. In that case every differential expression on Ω is within comparison reach of the algebra J_0 , as was shown in VI,3. For an elliptic expression L then the realization Z of V,def.6.2 has the property that $Z-\lambda=(A-\lambda\Lambda^N)\Lambda^{-N}$ is Fredholm for every $\lambda\in\mathbb{C}$. In other words, there is no essential spectrum, in the sense of [CHe_1]). This fact is in direct relation to the fact that the set $\mathbb{M}\setminus\mathbb{W}$ is void. Essentially it follows that the latter set, or, rather its interior $\mathbb{M}_s = \mathbb{M}\setminus(\mathbb{W}^{clos})$, is the origin of the essential spectrum of elliptic operators within reach of a comparison algebra C, (assuming that $E = K(H)$) .

Accordingly we now engage in a detailed discussion of the set $\mathbb{M}_s$, which is called the secondary symbol space of the algebra C . In this respect it may first be noticed that so far, as long as only the minimal comparison algebra was under discussion, we did not require <u>any</u> geometrical assumptions on the space Ω under its Riemannian metric induced by H whatsoever, except for the C^∞-Riemannian structure. On the other hand, for the investigation of the secondary symbol we will have to impose assumptions at infinity. We choose to work with a set of analytically formulated conditions (cf. sec's 2,3,4 below) but will make little effort to translate these conditions into geometrical properties.

Notice that an extensive study of the space $\mathbb{M}_s$ for a variety of comparison algebras over $\mathbb{R}^n$ was done in [CHe$_1$] and [C$_1$],III,IV. It appears natural from these results to seek the space $\mathbb{M}_s$ as a subset of the boundary of a suitable compactification of the cotangent space $T^*\Omega$ of Ω .

To be more specific, we first note that the multiplications a(x), for $a \in A^\#$, alone generate a C^*-subalgebra $C_{A^\#}$ of C , isometrically isomorphic to the function subalgebra of $CB(\Omega)$ generated by $A^\#$. The maximal ideal space $M_{A^\#}$ of $C_{A^\#}$ will be a compactification of the manifold Ω . Actually, this is the smallest compactification of Ω onto which every function of $A^\#$ allows a continuous extension, as well known (cf.[Ke$_1$]).

If $a \in E \cap C_{A^\#}$ then $\chi a \in E \cap J_0 = K(H)$, for $\chi \in C_0^\infty(\Omega)$, hence $a=0$. Thus we also find $C_{A^\#}$ as a C^*-subalgebra of the commutative C^*-algebra C/E . Then the associate dual map of the injection $C_{A^\#} \to C/E$ defines a continuous surjective map $\iota: \mathbb{M} \to C_{A^\#}$, already used in ch.6 for the minimal comparison algebra J_0 .

Now one may assign a <u>formal algebra symbol</u> τ_A to each of the generators of C , by setting $\tau_a(x,\xi) = a(x)$, for $a \in A^\#$, $\tau_{D\Lambda}(x,\xi) = (b^j(x)\xi_j + p(x))/(h^{jk}(x)\xi_j\xi_k + q(x))^{1/2}$, for $D=-ib^j\partial_{x^j}+p \in \mathcal{D}^\#$. Clearly the formal symbols are bounded complex-valued C^∞-functions, defined over the cotangent bundle $T^*\Omega$ of Ω . Thus again, a smallest compactification $\mathbb{P}^*\Omega$ of $T^*\Omega$ exists onto which all formal algebra symbols can be continuously extended. Again $\mathbb{P}^*\Omega$ is the maximal ideal space of the C^*-subalgebra of $CB(T^*\Omega)$ generated by the formal algebra symbols. Again one obtains a continuous surjective map $\pi:\mathbb{P}^*\Omega \to M_{A^\#}$, as associate dual map of the injection $C_{A^\#} \to C(\mathbb{P}^*\Omega)$, since $C_{A^\#}$ also is a subalgebra of the formal symbol algebra.

We are interested in the boundary $\partial\mathbb{P}^*\Omega = \mathbb{P}^*\Omega \setminus T^*\Omega$ of the compactification $\mathbb{P}^*\Omega$. Notice that $\pi^{-1}(\Omega)$ is the bundle of cospheres of infinite radius, recognized as homeomorphic to the wave front space $\mathbb{W}$ in VI,2 and VII,1 .

In case of the algebras over $\mathbb{R}^n$ mentioned it proved that the above homeomorphism extends to a homeomorphism between all of $\mathbb{M}$ and all of $\partial\mathbb{P}^*\Omega$. On the other hand, in case of certain algebras over the cylinder $S^1\times\mathbb{R}$ one obtains no secondary symbol space at all, although the space $\partial \mathbb{P}^* \Omega \backslash \mathbb{W}$ is nonvoid. In each of these two cases, on the other hand, $\mathbb{M}$ may be found (naturally identified with) a compact subset of $\partial \mathbb{P}^* \Omega$. For a third example, however, - i.e., the half space algebra of $[C_1], V$ - this no longer is true: certain one-dimensional "filaments" no longer are subsets of $\partial \mathbb{P}^* \Omega$.

In [CS],VI a result, identifying the entire $\mathbb{M}$ with a compact subset of $\partial \mathbb{P}^* \Omega$, was discussed under a number of special assumptions, and for a compact commutator algebra. The proof was simplified in $[CM_2]$. McOwen $[M_0]$, $[M_1]$ extended this result to a larger class of Laplace comparison triples and algebras.

Here we adopt some principles of McOwen's discussion by imposing a set of conditions (m_j) , naturally satisfied in McOwen's case.

In essence, we request that the pointwise 'inner product'

$$(0.1) \qquad \{D,F\} = h_{jk}\bar{b}^j c^k + \bar{p}r/q \ , \ x \in \Omega,$$

defined for folpde's $D = b^j \partial_{x^j} + p$, $F = c^j \partial_{x^j} + r \in \mathcal{D}^{\#}$, also is well-defined over $\partial M_{A^{\#}} = M_{A^{\#}} \backslash \Omega$. In other words, the function $\{D,F\}(x)$ of (0.1) must be in $C_{A^{\#}}$. In addition, certain separation conditions are required. One finds that such a condition has the effect that, over each point $x_0 \in M_{A^{\#}}$, the set $\pi^{-1}(x^0)$ retains the vector space structure of the cotangent space, or that of a lower dimensional space. This, in essence, makes the indicated result possible: $\mathbb{M}$ is found as a compact subset of the space $\partial \mathbb{P}^* \Omega$.

In sec.2 we specify cdn's (m_j) , and first show a local result, identifying $\iota^{-1}(x_0)$, for $x_0 \in M_{A^{\#}}$, with a compact subset of a $\mathbb{C}^k$. In sec.3 we add more conditions and then find $\iota^{-1}(x_0)$ as a compact subset of a hemisphere. This also allows a proof of $\mathbb{M} \to \partial \mathbb{P}^* \Omega$ (sec.3) . On the other hand, in sec.4 it is found that the compactification $\partial \mathbb{P}^* \Omega$ is not unique over $\partial M_{A^{\#}}$, insofar, as the expression (0.1) may be replaced by the corresponding one for any subextending triple $\{\Omega, d\mu\hat{}, H\hat{}\}$, and then the formal symbol by the corresponding expression involving the coefficients of $H\hat{}$. Then one obtains an imbedding $\mathbb{M} \to (\partial \mathbb{P}^* \Omega)\hat{}$ if the conditions are right.

All of this is discussed for a variety of examples, especially those of V,4 and V,5 (cf. sec.2 and sec.4) . In particular some examples of Sohrab [S_1] , [S_2] , [S_3] , are discussed and partly generalized.

In general we do not assume that the algebra C has compact commutators.

1. The symbol space of a general comparison algebra.

In this section we will discuss a general comparison algebra $\mathcal{C} = C(A^{\#},\mathcal{D}^{\#})$ of a triple $\{\Omega,d\mu,H\}$ with generating classes $A^{\#}$, $\mathcal{D}^{\#}$ satisfying cdn.'s (a_0),(a_1), and (d_0) of V,1. Again no essential self-adjointness of H_0 will be required. In general we cannot expect the commutators of C to be compact. For example let $\Omega = \mathbb{R}^n$, $d\mu=dx$, $H=1-\Delta$, with the Euclidean Laplace operator Δ . Let us choose $A^{\#}$ as the class of $a=b+c \in C^{\infty}(\mathbb{R}^n)$ with $c \in CO^{\infty}(\mathbb{R}^n) = \{a \in C^{\infty}(\mathbb{R}^n): a^{(\alpha)}(x)=o(1) \text{ for all } \alpha\}$, and with b an n-fold periodic function with periods $e^j=(\delta_{jl})_{l=1,\dots,n}$. Let $\mathcal{D}^{\#}$ be the class of folpde's $b^j\partial_{x^j}+p$, with coefficients b^j, $p \in A^{\#}$. Then it is easily seen (along the proof of V,prop.5.5, discussing the 1-dimensional case) that not all commutators of operators in C are compact.

Generally we denote by E the smallest norm closed 2-sided ideal of C containing all commutators of elements in C .

Lemma 1.1. We have $[A,B] \in K(H)$ whenever $A \in C$, $B \in J_0$.

Proof. It is sufficient to give the proof for the case that A and B are generators of their corresponding algebras. For $a \in A^{\#}_m$, $D \in \mathcal{D}^{\#}$ we get $[a,D\Lambda] \in K$ by V,thm.3.1. For $a \in A^{\#}$, $D \in \mathcal{D}^{\#}_m$ let $\chi \in C_0^{\infty}$ be such that $\chi D = D$. Then $aD\Lambda = a\chi D\Lambda \equiv D\Lambda a\chi \equiv \chi D\Lambda a \pmod K$, so that again $[a,D\Lambda] \in K$ (we used V, thm.3.1 twice here). For $D\in\mathcal{D}^{\#}_m$, $F \in \mathcal{D}^{\#}$ we get $D\Lambda F\Lambda v = \Lambda DF\Lambda v + Cv = \Lambda FD\Lambda v + C'v = \Lambda F\chi D\Lambda v + C'v = F\chi\Lambda D\Lambda v + C''v = F\Lambda\chi D\Lambda v + C'''v = \Phi\Lambda D\Lambda v + C'''v$, for all $v\in\Lambda C_0^{\infty}$, where we used V,lemma 3.6, V,lemma 3.7, V,lemma 3.6, and V,thm.3.1, in that order. Thus we also get $[D\Lambda,F\Lambda]$ compact. Note that not even condition (w) is required, for each commutation, in view of VI,(1.2), since we always were commuting expressions or functions with compact support. This completes the proof.

Lemma 1.2. Let $\chi_j\in C_0^{\infty}(\Omega)$, $0\le\chi_j\le1$, and $\chi_j=1$ in $\Omega_1\cup\dots\cup\Omega_j$, for j =1,

2,..., referring to a locally finite atlas $\{\Omega_j\}$ with properties as in VI,2 again. Then an operator $A \in C$ is in the minimal algebra J_0 if and only if there exists a sequence of operators $C_l \in K$ with

$$(1.1) \qquad A = \lim_{l\to\infty}(\chi_l A \chi_l + C_l)$$

in operator norm convergence.

Proof. If $A \in J_0$ then its symbol σ_A (with respect to J_0/K) is a function in $CO(\mathbb{W})$, so that we must have $\lim_{l\to\infty}(1-\chi_l)\sigma_A = 0$ in uniform convergence over $\mathbb{W}$. This directly implies (1.1).

Vice versa, for any $A \in C$ we have $\chi_l A \chi_l \in J_0$. Indeed, it is sufficient to show this for the generators of C , in view of the relation $\chi_l = \chi_l \chi_{l+1} \cdots \chi_{l+j}$, and the commutator property of lemma 1.1. But for $a \in A^\#$ we trivially get $a\chi_l^2 \in J_0$. Also for $A = D\Lambda$, $D \in \mathcal{D}^\#$ we get $\chi_l D\Lambda \chi_l \in J_0$ as well, q.e.d.

Theorem 1.3. We have $E \cap J_0 = K(H)$.

Proof. Note that $T = E \cap J_0$ is a closed 2-sided ideal of the algebra J_0 , and we observed before that T contains nontrivial compact operators, while J_0 was seen to be irreducible in V,1. It follows that $T \supset K$, by V,lemma 1.1, so that we only must show that $T \subset K$. Let $A \in T$. There exists a sequence $A_j \to A$ such that A_j is a finite sum of terms $X=B[C,D]E$, where $B,C,D,E \in C$. Also, for a sequence of functions χ_l with the properties of lemma 1.2. we must have $\lim_{l\to\infty}(\chi_l A \chi_l + C_l) = A$ in norm convergence, by lemma 1.2.

Combining these two facts, we get

$$(1.2) \quad \|A-C_l-\textstyle\sum_{j=1}^{N^l}\chi_r X_{lj}\chi_r\| \le \|A-C_l-\chi_r A\chi_r\|+\|\chi_r(A-\sum_{j=1}^{N^l}X_{lj})\chi_r\| \le \varepsilon ,$$

as r and j are sufficiently large. This means that we only must show that terms of the form $\chi X \chi = \chi B[C,D]E\chi$ are compact, where $B,C,D,E \in C$, $\chi \in C_0^\infty$, due to closedness of K in norm convergence.

Applying lemma 1.1 we get $\chi X \chi \equiv B\chi[C,D]\chi E \pmod{K}$. Hence we only must show that $\chi[P,Q]\chi \in K$ for every $P,Q \in C$. In turn then this may be reduced to the same statement where now P and Q are generators of C , again using lemma 1.2. Let again $\omega \in C_0^\infty$, $\omega\chi=\chi$. For $a \in A^\#$, $D,F \in \mathcal{D}^\#$ we have $\chi[a,D\Lambda]\chi = \chi[\omega a,D\Lambda]\chi \in K$, again by lemma 1.2. Finally, $\chi[D\Lambda,F\Lambda]\chi = \chi[\omega D\Lambda,F\Lambda\omega]\chi$, where $F\Lambda\omega \equiv F\omega\Lambda + C$, by V,(3.12) again. Accordingly, $\chi[D\Lambda,F\Lambda]\chi \equiv \chi[\omega D\Lambda,F\omega\Lambda]\chi \in K$, which completes the proof of thm.1.3.

The significance of thm.1.3 is realized when we now start discussing the symbol space $\mathbb{M}$ of the algebra C . We noted before that $C/E = C^{\vee}$, as a commutative C^*-algebra with unit, is isometrically isomorphic to an algebra $C(\mathbb{M})$ of continuous complex-valued functions over a compact Hausdorff space $\mathbb{M}$, by the Gel'fand Naimark theorem. The space $\mathbb{M}$ is explicitly given as the collection of all maximal ideals of $C^{\vee}$. Its topology is the weak topology of the function class induced by the elements of $C^{\vee}$. The space $\mathbb{M}$ was called the symbol space of the comparison algebra C, and the continuous function corresponding to $A \in C$ the symbol σ_A of A. We will recognize the wave front space $\mathbb{W}$ of the corresponding minimal algebra as an open subset of $\mathbb{M}$.

Consider the injection $J_0 \to C$. The commutator ideals of J_0 and C are $K(H) = K$ and E, respectively, by V, lemma 1.1. We have $K \subset E$, by thm.1.3, hence a map $J_0^{\vee} \to C^{\vee}$ is induced by $A^{\vee} = A+K \to A+E$, (i.e. we have the inclusion $A+K \subset A+E$, so that the map is independent of $A \in A^{\vee}$, hence well defined). Now by thm.1.3 this map is an injection. For if $A, B \in J_0$ and $A+K$, $B+K$ both are subsets of $A+E$, then $B-A \in J_0 \cap E = K$, so that $A^{\vee} = B^{\vee}$.

This means that inclusion of cosets defines a natural injection $J_0^{\vee} \to C^{\vee}$, which is a *-isomorphism, and will be considered as an identification of $J_0^{\vee}$ with a subalgebra of $C^{\vee}$.

Proposition 1.4. J_0 is a closed two-sided ideal of C. Correspondingly $J_0^{\vee}$, with above identification, is a closed two-sided ideal of $C^{\vee}$.

Proof. We have $A \in C$ in J_0 if and only if $A = \lim_{l\to\infty}(\chi_l A \chi_l + C_l)$, in norm convergence, with some $C_l \in K$, and with the sequence χ_l of lemma 1.2. For $B \in C$ we get $B(\chi_l A \chi_l + C_l) = \chi_l BA \chi_l + C_l'$, $C_l' \in K$, by lemma 1.1. Therefore $BA = \lim_{l\to\infty}(\chi_l BA\chi_l + C_l') \in J_0$, q.e.d.

For a closed ideal I of a commutative C^*-algebra A one defines the hull $h(I)$ as the collection of all maximal ideals containing I. The hull is a subset of the maximal ideal space M of A, and a closed subset, since we clearly have

$$h(I) = \cap\{\ \{m \in M : \phi_x(m) = 0\} : x \in I\ \},$$

with the symbol homomorphism $\phi: A \to C(M)$, so that $h(I)$ appears as an intersection of closed subsets of M.

The complement $R = M \setminus h(I)$ is an open (locally compact) subset of M, and it follows that the restriction $\phi|I$ of the symbol homomorphism ϕ to I constitutes an isometric isomorphism $I \to C(M)$, separating the points of R, but vanishing outside of R. One thus may identify the functions ϕ_x with their restrictions to R, arriving at an isometric isomorphism from I onto $CO(R)$ which must be the symbol homomorphism of the commutative C^*-algebra I.

In other words, the space R may be regarded as the maximal ideal space of I, and the symbol homomorphism of I appears as the restriction of the symbol homomorphism of A to R and I.

In our case of $I = J_0^{\vee} \subset C^{\vee}$ the maximal ideal space of $J_0^{\vee}$ was the wave front space $\mathbb{W}$, so that we have a natural identification of $\mathbb{W}$ with an open subset of the symbol space $\mathbb{M}$ of C.

On the other hand we again may consider the C^*-function algebra $C_{A^\#}$ obtained as norm closure of the algebra generated by $A^\#$ (either in L^∞-norm of functions over Ω or in operator norm of $L(H)$, which give the same, (cf.[C_1], III,(8.14)). In the case of J_0 (i.e., $A^\# = A^\#_m$) we had $C_{A^\#} = CO(\Omega)$. For general $A^\#$ the class $C_{A^\#}$ will be a subalgebra of $CB(\Omega)$, the algebra of all bounded continuous functions over Ω. The maximal ideal space $M_{A^\#}$ of $C_{A^\#}$ will be a certain compactification of Ω. (For more details about such compactifications cf.[C_1],IV,lemma 1.5). As in VI,1 we get an injection $C_{A^\#} \to C$. Using that $C_{A^\#} \cap K = \{0\}$, we also get a corresponding injection $C_{A^\#} \to C/E = C(\mathbb{M})$. That injection again defines a corresponding associate dual map $\iota : \mathbb{M} \to M_{A^\#}$, and ι again is surjective, similarly as in VI,1 in the special case of J_0.

Now we have the following result.

Theorem 1.5. The symbol space $\mathbb{M}$ of a comparison algebra C contains the wave front space $\mathbb{W}$ of the corresponding minimal algebra J_0 as an open subset. Moreover with the map $\iota : \mathbb{M} \to M_{A^\#}$ defined above we get

$$\mathbb{W} = \iota^{-1}(\Omega) . \tag{1.3}$$

Proof. The only thing left to prove is (1.3). Note that we have $m \in \mathbb{W}$ if and only if there exists an operator $A \in J_0$ with $\sigma_A(m) \neq 0$. Using lemma 1.2 once more we get $A = \lim_{l\to\infty}(\chi_l A \chi_l + C_l)$, hence

$$\sigma_A(m) = \lim_{l\to\infty}(\chi_l(\iota(m))^2 \sigma_A(m)) = \sigma_A(m) \lim_{l\to\infty}(\chi_l(\iota(m)))^2 . \tag{1.4}$$

But then we must have $\iota(m) \in \Omega$, since the compactly supported functions χ_1 vanish at each point of $\partial M_{A^\#}$. Hence we get $\mathbb{W}\subset\iota^{-1}(\Omega)$.

Vice versa, for $m \in \iota^{-1}(\Omega)$ let $A \in C$ be an operator with $\sigma_A(m)\neq 0$, and let the function $\chi \in C_0^\infty(\Omega)$ be 1 at $x_0 = \iota(m) \in \Omega$. Then lemma 1.2 implies that $\chi A\chi \in J_0$ while we get $\sigma_{\chi A\chi}(m)=\sigma_A(m)\chi(\iota(m)) = \sigma_A(m)\neq 0$, hence $m \in \mathbb{W}$. This completes the proof.

Note that the wave front space was completely characterized as the cosphere bundle $S^*\Omega$ of the manifold Ω (cf. VI, thm.2.2) . We now denote the closure of $\mathbb{W}$ in $\mathbb{M}$ by $\mathbb{M}_p$, and call $\mathbb{M}_p$ the <u>principal symbol space</u> of the comparison algebra C . The complement $\mathbb{M}_s = \mathbb{M}\setminus\mathbb{M}_p$ will be called the <u>secondary symbol space</u> of C .

Our next effort will be aimed at the structure of the secondary symbol space. Sec. 2, below, will give some answers, in that respect.

2. The space $\mathbb{M}\setminus\mathbb{W}$, and some examples.

In this section we will look at the secondary symbol space $\mathbb{M}_s$. We will obtain some information about $\mathbb{M}_s$, but under certain restrictions on Ω and the triple at infinity, and not as complete as for the wave front space. Again no assumption on essential self-adjointness of H_0 will be made.

In addition to cdns.$(a_0),(a_1),(d_0)$ of V,1 the classes $A^\#$ and $\mathcal{D}^\#$ now will be assumed to satisfy the following.

<u>Condition</u> (m_1): For each pair $D=b^j\partial_{x^j}+p$, $F=c_j\partial_{x^j}+r$ of folpde's in $\mathcal{D}^\#$ the bounded $C^\infty(\Omega)$-function $\{D,F\}$ defined by

$$(2.1)\qquad \{D,F\} = h_{jk}\bar{b}^j c^k + \bar{p}r/q \ , \ x \in \Omega \ ,$$

is a function in $C_{A^\#}$, the norm closed subalgebra generated by $A^\#$.

<u>Condition</u> (m_2): $A^\#$ and $\mathcal{D}^\#$ are complex vector spaces, and $\mathcal{D}^\#$ is a left module of $A^\#$.(i.e., the pointwise product aD belongs to $\mathcal{D}^\#$, for all $a \in A^\#$ and $D \in \mathcal{D}^\#$) .

A condition like (m_1) was first used by McOwen [M_1] in his thesis written with the author. Also the approach, below, to use (2.1) as a local inner product is due to McOwen.

For any given $D,F\in\mathcal{D}^\#$ the function $\{D,F\}:\Omega\to\mathbb{C}$ admits a continuous extension to the space $M_{A^\#}$ which is a certain compactification of Ω. Then let us focus on a given fixed point $x_0 \in M_{A^\#}$, finite or infinite. The value $Q(D,F)=\{D,F\}(x_0)$ at x_0 of the function $\{D,F\}$, as D,F range over $\mathcal{D}^\#$, defines a positive semi-definite sesquilinear form on $\mathcal{D}^\#$, given explicitly by (2.1) at finite points, but only by continuity at points $x_0 \in M_{A^\#}\setminus\Omega$. The set

$$(2.2) \qquad \mathcal{D}^\#_{x_0} = \{D \in \mathcal{D}^\# : Q(D,D)=0\}$$

defines a linear subspace of $\mathcal{D}^\#$. Indeed, one obtains Schwarz' inequality $|Q(D,F)|^2 \leq Q(D,D)Q(F,F)$ for the semi-definite form Q, which implies that $Q(D,F)=0$ whenever either $Q(D,D)=0$ or $Q(F,F)=0$. Then we get $Q(cD+c'F,cD+c'F) = 0$ whenever $Q(D,D)=Q(F,F)=0$.

It then is easily seen that Q induces an inner product in the quotient space $\mathcal{D}^\#/\mathcal{D}^\#_{x_0} = S_{x_0}$. Clearly, for finite x, the space S_x is just the local $\mathbb{C}^{n+1}$ spanned by $(b^1(x),\ldots,b^n(x),p(x))$, with inner product given by (2.1).

<u>Proposition 2.1</u>. The space S_x is finite dimensional for $x \in M_{A^\#}$, and we have $\delta_x=\dim S_x \leq n+1$ (n= dimension of Ω). Moreover, δ_x is a lower semi-continuous function of x, defined on $\partial M_{A^\#}=M_{A^\#}\setminus\Omega$. (That is, $\liminf_{x\to x_0}\delta_x \geq \delta_{x_0}$.)

<u>Proof</u>. If $D_1,\ldots,D_\delta \in \mathcal{D}^\#$ is (a set of representatives for) an orthonormal base of S_{x_0} then $\{D_j,D_l\}(x_0) = \delta_{jl}$, $j,l=1,\ldots,\delta=\delta_{x_0}$. But the $\delta\times\delta$-matrix $V=((\{D_j,D_l\}(x)))$ has continuous entries over the space $M_{A^\#}$, hence will stay invertible in some neighbourhood of x_0. In fact, a nonsingular $\delta\times\delta$-matrix $W(x)=((w_{jl}))$ may be found such that $W^*VW = ((\delta_{jl}))$. Then $\nabla_j=\sum w_{jl}D_l$ defines an orthonormal set of S_x, so that $\dim S_x=\delta_x \geq \dim S_{x_0} = \delta_{x_0}$. Thus we get $\liminf_{x\to x_0}\delta_x \geq \delta_{x_0}$, or δ_x is lower semi-continuous. In particular, due to the basic properties of $A^\#$ and $\mathcal{D}^\#$ of V,1 it is clear that $\delta_x = n+1$ for all $x \in \Omega$, since $A^\#$ and $\mathcal{D}^\#$ contain all C_0^∞-functions and folpde's.

We now arrive at the following characterization of the secondary symbol space (rather, of the space $\mathbb{M}\setminus\mathbb{W}$ of all points of $\mathbb{M}$ over infinite x). Assume given an orthonormal base $D_1,\dots,D_\delta$ of S_{x_0} at $x_0 \in M_{A^\#}$, and a set of δ_{x_0} folpde's $D_j \in \mathcal{D}^\#$ with $\{D_j, D_l\} = \delta_{jl}$. For a maximal ideal $m \in \iota^{-1}(x_0)$ let us define

$$(2.3) \qquad \eta(m) = (\sigma_{D_1\Lambda}(m),\dots,\sigma_{D_\delta\Lambda}(m)) = (\eta_1(m),\dots,\eta_\delta(m)) .$$

Theorem 2.2. The map $\eta\colon \iota^{-1}(x_0) \to \mathbb{C}^\delta$, $\delta = \delta_{x_0}$ is continuous and injective from $\iota^{-1}(x_0) \subset \mathbb{M}\setminus\mathbb{W}$ onto a compact subset of $\mathbb{C}^\delta$. Moreover, for $m \in \iota^{-1}(x_0)$ we have the symbol of the generators defined by

$$(2.4) \qquad \begin{aligned} &\sigma_a(m) = a(x_0) , \ \sigma_{D_j\Lambda}(m) = \eta_j(m) , \ \sigma_{(D_j\Lambda)^*}(m) = \bar{\eta}_j(m) , \\ &\sigma_{D\Lambda}(m) = \sigma_{(D\Lambda)^*}(m) = 0 \text{ for all } D \in \mathcal{D}^\#_{x_0} . \end{aligned}$$

Proof. The continuity of $\eta_j(m) = \sigma_{D\Lambda_j}(m)$ is evident, since every $\sigma_A\colon \mathbb{M}\to\mathbb{C}$, for fixed $A \in C$, is continuous over $\mathbb{M}$, hence $\sigma_A|\iota^{-1}(x_0)$ continuous. The injectivity of η is proven just like Herman's lemma in [C1], IV,2 . Suppose for two maximal ideals $m, m^\sim \in \iota^{-1}(x_0)$ we have $\eta(m) = \eta(m^\sim)$. This means that $\sigma_a(m) = \sigma_a(m^\sim)$ for all $a \in A^\#$ (due to $\sigma_a(m) = a(\iota(m)) = a(x_0) = a(\iota(m^\sim)) = \sigma_a(m^\sim)$) , and also $\sigma_{D_j\Lambda}(m) = \eta_j(m) = \eta_j(m^\sim) = \sigma_{D_j\Lambda}(m^\sim)$, by (2.3) . Now let us first assume the following.

Proposition 2.3. We have

$$(2.5) \qquad \sigma_{D\Lambda}(m) = 0 \text{ for all } D \in \mathcal{D}^\#_{x_0} \text{ and } m \in \iota^{-1}(x_0) .$$

Assuming prop.2.3 it follows that $\sigma_{D\Lambda}(m) = \sigma_{D\Lambda}(m^\sim) = 0$ for all $D \in \mathcal{D}^\#_{x_0}$. Hence we conclude that the symbols of all the generators of C coincide at m and $m^\sim$ (including the symbols of the adjoints $(D\Lambda)^*$, which are just the complex conjugates of $\sigma_{D\Lambda}$).

In particular we find formula (2.4) confirmed. Also, the two homomorphisms $C/E \to \mathbb{C}$,defined by $A \to \sigma_A(m)$,and $A \to \sigma_A(m^\sim)$, respectively, for $A \in C$, must coincide, since they coincide for the generators. But these homomorphisms determine their corresponding maximal ideals uniquely, so that we must have $m = m^\sim$, proving the theorem.

<u>Proof of prop.2.3</u>. For $D \in \mathcal{D}^{\#}_{x_0}$ we have $\{D,D\}(x_0) = 0$. Thus there exists a neighbourhood N_ε of x_0 (in $M_{A^\#}$) such that $|\{D,D\}(x)| \leq \varepsilon^2/4$, $x \in N_\varepsilon$, for every $\varepsilon>0$, by continuity of $\{D,D\}$ on $M_{A^\#}$. There exists a function $a \in C(M_{A^\#})$ with $0 \leq a \leq 1$, $a(x_0)=1$, $a=0$ outside of $N_{\varepsilon/2}$. Let $M=(\sup\{1,|\{D,D\}(x)| : x \in \Omega\})^{1/2}$. Let b be in the algebra, finitely generated by $A^\#$ with $\|b-a\|_{L^\infty(\Omega)} < \varepsilon/2M$. Then we get

$$(2.6) \qquad \{bD,bD\}^{1/2} \leq \{(b-a)D,(b-a)D\}^{1/2} + \{aD,aD\}^{1/2} ,$$

at each $x \in \Omega$. The first term at right is bounded by $\varepsilon/2$, at each x, as follows from (2.1). But the second term also is bounded by $\varepsilon/2$, since $a(x)=0$, hence $\{aD,aD\}(x)=0$ outside N_ε , while $|\{aD,aD\}(x)| \leq \|a\|^2_{L^\infty}\{D,D\}(x) \leq \varepsilon^2/4$ there.

Accordingly we conclude that

$$(2.7) \qquad \sup\{|\{bD,bD\}(x)| : x \in \Omega\} \leq \varepsilon .$$

Then, using (2.6) we conclude that also

$$(2.8) \qquad \|bD\Lambda\| \leq \varepsilon .$$

Finally we observe that

$$(2.9) \sup\{|\sigma_{bD\Lambda}(m)| : m \in \mathbf{M}\} = \inf\{\|bD\Lambda+E\| : E \in \mathcal{E}\} \leq \|bD\Lambda\| \leq \varepsilon ,$$

using that the Gel'fand isomorphism of a C^*-algebra is an isometry. Also, $\sigma_b(m)= b(x_0)$, by construction, and $|b(x_0)-1| = |b(x_0)-a(x_0)| \leq \|b-a\|_{L^\infty} \leq \varepsilon/2M$, which implies $\sigma_b(m) \geq 1/2$,for sufficiently small ε . Thus (2.9) implies $\varepsilon \geq |\sigma_b(m)||\sigma_{D\Lambda}(m)| \geq |\sigma_{D\Lambda}(m)|/2$, or, $|\sigma_{D\Lambda}(m)| \leq 2\varepsilon$. Since $\varepsilon>0$ was arbitrary, we conclude that $\sigma_{D\Lambda}(m) = 0$, q.e.d.

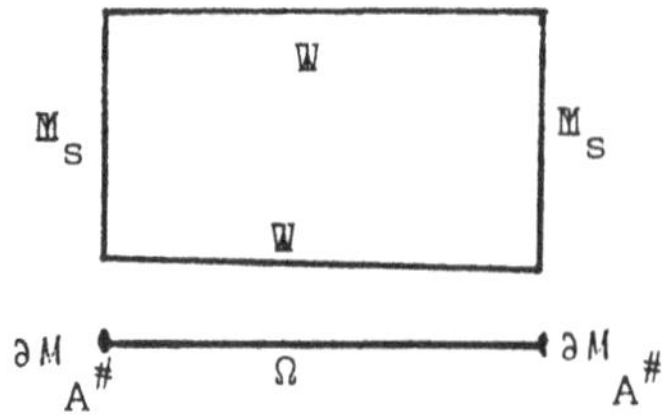

Fig. 2.1.

With thm.2.2 we now have the following model of the symbol space $\mathbf{M}$, from the structure of the inverse images $\iota^{-1}(x_0)$, $x_0 \in M_{A^\#}$(fig.2.1). $M_{A^\#}$ is a certain compactification of Ω . All functions in $A^\#$ extend continuously to $M_{A^{\#}}$. Ω is dense in $M_{A^{\#}}$, we write $\partial M_{A^{\#}} = M_{A^{\#}} \setminus \Omega$. Then $\iota: \mathbb{M} \to M_{A^{\#}}$ is a continuous surjective map .

The inverse image of the manifold Ω , i.e., the set $\mathbb{W}=\iota^{-1}\Omega$, called the wave front space, is (homeomorphic to) the bundle of unit spheres in the cotangent space (or better the spheres of radius ∞). In particular the sets $\iota^{-1}(x_0)$, for $x_0 \in \Omega$, are n-1-spheres. On the other hand, over any $x_0 \in \partial M_{A^{\#}}$ the sets $\iota^{-1}(x_0)$ are compact subsets of $\mathbb{C}^{\delta}$, with $\delta=\delta(x_0)$ the dimension of the local space S_{x_0} .

Since the topology of $\mathbb{M}$ is the weak topology, determined by the set $C(\mathbb{M})$, an explicit knowledge of the sets $\eta(\iota^{-1}(x_0))$, for all $x_0 \in \partial M_{A^{\#}}$, and of the functions η , will characterize the topology of $\mathbb{M}$ as well. In particular we have the following.

<u>Corollary 2.4</u>. For a point $x_0 \in \partial M_{A^{\#}}$ assume the dimension δ_{x_0} locally constant (i.e., $\delta_x=\delta_{x_0}$ for all $x \in \partial M_{A^{\#}} \cap N_{x_0}$, with some neighbourhood N_{x_0} of x_0). There exists a compact neighbourhood $O_{x_0} \subset N_{x_0} \cap \partial M_{A^{\#}}$ of x_0 in $\partial M_{A^{\#}}$ such that the map $\iota^{-1}(O_{x_0}) \to O_{x_0} \times \mathbb{C}^{\delta}$, $\delta=\delta_{x_0}$, defined by

$$(2.10) \qquad m \to (\iota(m), \eta(m)) ,$$

with the map (2.3) , defines a continuous injection of $\iota^{-1}(O_{x_0})$ onto a compact subset of $O_{x_0} \times \mathbb{C}^{\delta}$. Here the map $\eta(m)$ is defined with an orthonormal base of S_{x_0} , which still is a base at $x \in O_x$, for $x \in O_{x_0}$, although not necessarily an orthonormal one.

<u>Proof</u>. The continuity again is evident. For the injectivity we observe that $\iota(m)=\iota(m^{\sim})$ implies that $m, m^{\sim} \in \iota^{-1}(x)$, where $x=\iota(m)$. Then if also $\eta(m)=\eta(m^{\sim})$ we may repeat the conclusion of thm.2.2, to show that $m=m^{\sim}$. In particular, $D_x^{\#}$ is independent of the base $\{D_{\nu}\}$, and the orthogonality of the base is not essential. Q.E.D.

<u>Remark 2.5</u>. We observe that $\delta_x=n+1$ whenever $x \in \Omega$. Therefore the

assumption that, for some $x_0 \in \partial M_{A^{\#}}$ the dimension δ_x is constant in a neighbourhood of x_0 relative to $M_{A^{\#}}$, not only to $\partial M_{A^{\#}}$, implies that $\delta_x = n+1$, in such a neighbourhood. In this case an analogue of cor.2.4 for a neighbourhood O_{x_0} relative to $M_{A^{\#}}$ is valid.

However, in this case we can make a stronger statement under slightly stronger conditions, as will be discussed in sec.3.

Remark 2.5. If we have $\delta_x = \delta = \text{const.}$ in some subdomain $B \subset \partial M_{A^{\#}}$, then every point in B has a compact neighbourhood N such that $\iota^{-1}(N)$ is homeomorphic to a compact subset of $N \times \mathbb{C}^\delta$. In many concrete examples this property may be translated into a manifold-like structure of the subset $\iota^{-1}(B) \subset \mathbb{M} \backslash \mathbb{W}$.

Let us now examine some explicit examples.

Example (A) of V,4. We obtain the algebra $\mathfrak{S}_0$ of [CHe$_1$]. with the classes $A_s^{\#}$, $\mathcal{D}_s^{\#}$, as follows. $A_s^{\#}$ is the function algebra finitely generated by $\lambda(x) = \langle x \rangle^{-1}$, $s_j(x) = x_j \lambda(x)$, $j = 1, \ldots n$. $\mathcal{D}_s^{\#}$ is the class of all folpde's with coefficients in $A_s^{\#}$. Clearly $A_s^{\#}$ and $\mathcal{D}_s^{\#}$ satisfy (m_1) and (m_2). We get

$$\{D,D\} = \sum_{j=1}^n \bar{b}^j b^j + \bar{p}p \quad , \quad D = \sum_{j=1}^n b^j \partial_{x^j} + p \, , b^j, p \in A_s^{\#} \, . \tag{2.11}$$

In [C$_1$],IV we have seen that $M_{A_s^{\#}} = \mathbb{B}^n$ is the 'directional compactification' of $\mathbb{R}^n$, corresponding to the closure in $\mathbb{R}^n$ of the unit ball $s(\mathbb{R}^n)$, with the map $s: x \to x/\langle x \rangle$ taking $\mathbb{R}^n$ onto $\{|x|<1\} \subset \mathbb{R}^n$. Accordingly, $\partial M_{A_s^{\#}} = \partial \mathbb{B}^n$ is an n-1-sphere.

Clearly an orthonormal base of S_x is globally given by

$$D_0 = 1 \, , \, D_\nu = -i\partial_{x^\nu} \, , \, \nu = 1, \ldots, n \, , \text{ for all } x \in \mathbb{B}^n \, . \tag{2.12}$$

Thus the dimension $\delta_x = n+1$ is constant, we get

$$A_0 = D_0 \Lambda = \Lambda \, , \, A_j = S_j = D_j \Lambda \, , \, \eta_j(m) = \sigma_{A_j}(m) \, , \, j = 0, \ldots, n \, . \tag{2.13}$$

The A_j are self-adjoint operators, hence η_j are real-valued. Also A_0 is a positive operator, hence $\eta_0 \geq 0$. A calculation shows that

$$\sum_{\nu=0}^n A_j^* A_j = 1 \, , \tag{2.14}$$

so that $0 \le \eta_0 = (1-\sum_{\nu=1}^{n} |\eta_j|^2)^{1/2}$. Thus we have the following.

Proposition 2.7. The space $\mathbb{M}\backslash\mathbb{W}$ is homeomorphic to a compact subset of the set

(2.15) $\qquad \partial\mathbb{B}^n \times \mathbb{H}^n$, where $\mathbb{H}^n = \{\eta \in \mathbb{R}^{n+1}: \sum_{\nu=0}^{n} |\eta_\nu|^2 = 1 , \eta_0 \ge 0\}$.

It is clear that $\mathbb{H}^n$ is topologically equivalent to the ball $\mathbb{B}^n$ (of infinite radius). In [CHe_1] is was seen that

(2.16) $$\mathbb{M}\backslash\mathbb{W} = \partial\mathbb{B}^n \times \mathbb{B}^n \simeq \partial\mathbb{B}^n \times \mathbb{H}^n .$$

From V,4 we know that the algebra $\mathfrak{S}_0$ has compact commutators, as a subalgebra of the algebra obtained in example (A) there.

Similarly, for the algebra $\mathfrak{A}_0$ of V,4, example (A) we get

(2.17) $$\mathbb{M}\backslash\mathbb{W} = \partial M_{A_c^\#} \times \mathbb{H}^n .$$

Example (B) of V,4. Let us use the function algebra $A_c^\#$ and folpde-class $\mathcal{D}^\#$ introduced in V,4,(B). Write $x \in \Omega = \mathbb{G}^{jk} = \mathbb{R}^j \times \mathbb{T}^k$ in the form $x = (x', x")$, with $x' \in \mathbb{R}^j$, $x" \in \mathbb{T}^k$. A function $a(x', x") \in A^\#$ must also have its x"-derivarives decaying, as $x \to \infty$ (i.e., as $|x'| \to \infty$). This implies that functions in $A^\#$ must be nearly constant on the k-tori x'=const., $|x'|$ large. In VIII,2 we will consider a larger class $A^\#$ which, however, will involve non-compact commutators for the corresponding algebra C. The compactification $M_{A_c^\#} = \mathbb{P}^{jk}$ will look similar to the Stone-Cech compactification of $\mathbb{G}^{jk}$. One may use exactly the same basis D_ν , $\nu = 0, \dots, n$,as defined in (2.12) (with D_ν , $\nu > j$, now acting on a periodic function). Again the $A_j = D_j \Lambda$ are self-adjoint, and $A_0 \ge 0$. Again we get (2.14). Accordingly it follows again that $\mathbb{M}\backslash\mathbb{W} \subset \partial M_{A_c^\#} \times \mathbb{H}^n$. However, in the present case, we do not have equality. Rather, the secondary symbol space $(\mathbb{M}\backslash\mathbb{W})^{\text{int}}$ is homeomorphic to $\mathbb{R}^j \times \mathbb{Z}^k$, i.e., to the character group of the group $\mathbb{G}^{j,k}$ (cf. thm.3.1 of [CS] ,ch.5) .

We postpone the discussion of other examples of V,4. Here let us first discuss an example of different type.

Example (H): Let Ω be an open subdomain of $\mathbb{R}^n$, with smooth smooth boundary, compact or not. Assume the boundary $\partial\Omega$ to be a smooth n-1-dimensional submanifold of $\mathbb{R}^n$. We will focus on two special case only: either $\Omega\cup\partial\Omega$ is compact, or $\Omega = \mathbb{R}^n_+ = \{x\in\mathbb{R}^n : x_1 > 0\}$, referred to as "cases (a) and (b)" respectively.

Consider the Euclidean Laplace-triple, in either case, with the classes $A^\#$ and $\mathcal{D}^\#$ of restrictions to Ω of the functions (folpde's) of $A^\#_s$ and $\mathcal{D}^\#_s$ of example (A), above. Let C be the cor-esponding comparison algebra.

Again we may use the orthonormal base D_ν of (2.12), for every $x\in\partial M_{A^\#}$. (We have $M_{A^\#}=\Omega\cup\partial\Omega$ in case (a), and $M_{A^\#}=\mathbb{B}^n_+=(\mathbb{R}^n_+)^{clos}$, with the closure of $\mathbb{R}^n_+$ in $\mathbb{B}^n$.) The dimension is $\delta=n+1$ everywhere.

However, an investigation shows that the operators A_j no longer are self-adjoint, although we still have (2.14). In particular, focusing on case (b), the operator A_1, involving the derivative normal to the boundary, is not even self-adjoint modulo E. Accordingly the set $\eta(\iota^{-1}(x_0))$, for a point x_0 with $x_0^1=0$, only is a subset of the complex unit sphere $\sum_{\nu=0}^n |\eta_\nu|^2=1$, $\eta\in\mathbb{C}^{n+1}$.

For a precise description of the space $\mathbb{M}$ cf. [C_1], V, thm. 10.3. The description there is in a slightly different representation, to be translated into the present form. Over points of the finite boundary of Ω one only gets the (still well-defined) cosphere bundle, and, in addition certain 1-dimensional 'filaments'. Over points $|x|=\infty$, $x_0^1=0$, one obtains the ball $|\xi|\le\infty$ (i.e. η real, $|\eta|=1$, $c_0\ge 0$, and again some 1-dimensional filaments). For other points $|x|=\infty$ we of course get the same as for $\Omega=\mathbb{R}^n$, $A^\#_s$, $\mathcal{D}^\#_s$ again.

In [C_1], V we also investigate the commutator ideal E of case (b) in detail, and thus obtain a complete Fredholm theory for the algebra C. We have not analyzed the space $\mathbb{M}\setminus\mathbb{W}$ in case (a), but expect a very similar result.

Note that the last example pertains to the case of a genuine boundary problem, where cdn.(w) is never true. In [C_1] and [CE] we also show that the elliptic boundary problem with 'Lopatinsky-Shapiro'-condition is solvable with our algebra C.

3. Stronger conditions and more detail on $\mathbb{M}\setminus\mathbb{W}$.

It should be realized that our assumptions on $A^\#$ and $\mathcal{D}^\#$ as well as on the triple $\{\Omega, d\mu, H\}$ are minimal, so far. In sec. 2 we have studied some examples which allowed a more complete description of the space $\mathbb{M}\setminus\mathbb{W}$. We now first will look at conditions to allow a description of the sets $\iota^{-1}(x_0) \subset \mathbb{M}\backslash\mathbb{W}$ as compact subsets of a real upper hemisphere $\mathbb{H}^{\delta-1}$, just as in the examples of sec.2, although only locally.

For the $\mathbb{R}^n$-related examples of sec. 2 the following observation may be useful.

Remark 3.1. Suppose near a point $x_0 \in \partial M_{A^\#}$ the manifold Ω 'looks like $\mathbb{R}^n$', in the following sense: x_0 has a neighbourhood N_{x_0} relative to $M_{A^\#}$, such that $\Omega \cap N_{x_0} = U$ is a chart. In particular, assume that, if U is considered a subset of $\mathbb{R}^n$, then either x_0 is a point of a smooth boundary piece of U, or $|x_0|=\infty$, and $|x|=\infty$ for all $x \in N_{x_0} \backslash \Omega$. Suppose then that, in the coordinates of U, the coefficients $h^{jk}(x)$, $q(x)$ extend continuously onto N_{x_0}, and still define a positive definite $(n+1)\times(n+1)$-matrix, at all points of N_{x_0}. Suppose also that $\mathcal{D}^\#$ contains folpdes D_ν such that

$$(3.1) \qquad D_0 = 1 \ , \ D_\nu = -i\partial_{x^\nu} \ , \ \nu=1,\dots,n \ , \ \text{for } x \in U \ .$$

Then we have $\delta_x = n+1$ in N_{x_0}, and it is possible to use the folpde's (3.1) as a base of S_x, although not necessarily orthonormal. Accordingly cor.2.4 applies in the entire N_{x_0}, with η defined by (3.1). For a generator $A = a(x)$, or $=D\Lambda$, or $=(D\Lambda)^*$ of C we then have the value of $\sigma_A(m)$ at some $m \in \iota^{-1}(x_0)$, $x_0 \in \partial M_{A^\#}$, given by

$$(3.2) \quad \sigma_a(m) = a(x_0) \ , \ \sigma_{D\Lambda}(m) = b^j(x_0)\eta_j \ , \ \sigma_{(D\Lambda)^*}(m) = \bar{b}^j(x_0)\bar{\eta}_j .$$

for $D=\sum_0^n b_\nu D_\nu = b^j D_j + p$, summation convention used from 1 to n, only for non-greek indices.

In thm.3.2, below, we will show that, in this case, we get

$$(3.3) \quad h^{jk}(x_0)\bar{\eta}_j\eta_k + q(x_0)|\eta_0|^2 = 1 \ , \ \eta_0 \geq 0, \ \text{for } \eta_j = \eta_j(m) \ .$$

Therefore we may write (3.2) in the form

$$(3.4) \ \sigma_a(m) = a(x_0), \ \sigma_{D\Lambda}(m) = (b^j(x_0)\xi_j + p(x_0))/(h^{jk}(x_0)\bar{\xi}_j\xi_k + q(x_0))^{1/2},$$

using the coordinates $\xi_j = \eta_j/\eta_0$, with a $\xi \in \mathbb{E}^n$.

Here the vector $\xi = (\xi_1,\ldots,\xi_n)$ is not always finite. It is assumed to be a point in the directional compactification $\mathbb{E}^n$ of C^n , defined similarly as $\mathbb{B}^n$, above. If, in addition, it can be shown that all η_j are real, then we have $\xi \in \mathbb{B}^n$ (but we should recall example (H) of sec.2, of a comparison algebra on a manifold with boundary, where we do not get all η_j real).

In this representation, the space $N_{x_0} \times \mathbb{B}^n$ appears as a part of the compactification of the cotangent space $T^*\Omega$ of Ω , generated by the bounded continuous functions (3.4) over Ω . Moreover, the representations of $\iota^{-1}(x_0) \subset \mathbb{M}$ of thm.2.2 are obtained by a change to 'projective coordinates'.

In the following we will seek to recover these features. While we return to the general case, assuming only $(a_0),(a_1),(d_0)$, and $(m_1),(m_2)$, (and not (w)) we impose (some or all of) the following additional conditions.

Condition (m_3): For each $x_0 \in M_{A^\#}$ and open neighbourhood N of x_0 there exists a function $a \in A^\#$ with $0 \le a \le 1$ and $a=1$ near x_0, $a=0$ outside of N , and such that $D_a = a|^j \partial_{x^j} \in \mathcal{D}^\#$.

Condition (m_4): There exists an $\varepsilon > 0$ such that

$$(3.5) \quad \Lambda^{-1-\varepsilon}[a,\Lambda] \in C \ , \ \Lambda^{-\varepsilon}[D,\Lambda] \in C \ , \text{ for all } a \in A^\# \ , \ D \in \mathcal{D}^\# \ ,$$

where the last expression should be interpreted as $\Lambda^{-\varepsilon}(D\Lambda-(\Lambda D)^{**})$, as a product of unbounded operators, to be continuously extended.

Condition (m_5): We have $\Lambda \in C$.

Condition $(m_6)_x$: The dimension δ_x of $S^\#_x = \mathcal{D}^\#/\mathcal{D}^\#_x$ is maximal (i.e., we have $\delta_x = n+1$).

We define

$$(m_6) = \cup\{(m_6)_x : x \in \partial M_{A^\#}\} = \cup\{(m_6)_x : x \in M_{A^\#}\},$$

i.e., cdn.(m_6) (without subscript x) means that (m_6) holds for all $x \in M_{A^\#}$ (it is always true for $x \in \Omega$) .

Note that (m_3) is a separation condition comparable to the condition describing completely regular algebras (cf.[C_1],AII,5). On the other hand, (3.5) may be compared to the statements of V, lemmas 3.3 and 3.6. Using the techniques of V,3 one may reduce (m_4) to explicit conditions on $a \in A^\#$ or the coefficients of $D \in \mathcal{D}^\#$, but we will not make a systematic discussion here. Note that

(m_5) holds if $1 \in \mathcal{D}^{\#}$. For a weaker condition cf. IX,5 .

Theorem 3.2. Assume condition (m_j) , $j=1,2,3,4,5,6$, and let D_j, $j=1,\ldots,\delta$, of thm.2.2 be formally self-adjoint. Then the image set of the map $\eta : \iota^{-1}(x_0) \to \mathbb{C}^{\delta}$ is contained in the sphere

$$(3.6) \qquad \{\eta \in \mathbb{R}^{\delta} : \textstyle\sum_{\nu=1}^{\delta} |\eta_j|^2 = 1\} \subset \mathbb{R}^n .$$

Proof. Since the dimension $\delta(x)$ is lower semi-continuous, it has to be constant near x_0 if it is maximal at x_0 . Thus we conclude that $\delta_x = n+1$ near x_0 .

We have Parseval's relation in S_{x_0} for the orthonormal set $D_1,\ldots,D_\delta$. For every $D \in \mathcal{D}^{\#}$ we get

$$(3.7) \qquad \{D,D\} = \textstyle\sum_{\nu=1}^{\delta} |\{D_\nu,D\}|^2 \ , \text{ at } x=x_0 .$$

Due to continuity of $\{.,.\}$ as function of x this implies that

$$(3.8) \qquad |\{D,D\}(x) - \textstyle\sum_{\nu=1}^{\delta} |\{D_\nu,D\}(x)|^2| \le \varepsilon\{D,D\}(x) \ , \text{ as } x \in N_{x_0} ,$$

for a suitable neighbourhood N_{x_0} of x_0 . Indeed, we already used in prop.2.1 that $D^{\sim}_\nu(x) = \sum w_{\nu\mu}(x) D_m$, for suitable $W=((w_{\nu\mu}))$ close to $((\delta_{\nu\mu}))$, is orthonormal at x, for every $x \in N_{x_0}$. Since $\delta=n+1$, we will have (3.7) at $x \in N_{x_0}$ for $D^{\sim}_\nu$ instead of D_ν . This implies (3.8) , by a simple estimate, as N_{x_0} is chosen sufficiently small. (For more details cf. proof of IX, thm.2.1) .

Proposition 3.3. Let $a \in A^{\#}$ equal 1 and 0 , as x near x_0 ,and x outside $N_{x_0} = N_{x_0,\varepsilon}$, respectively, and let $0 \le a \le 1$. Let $A_\nu = D_\nu \Lambda$, $\nu=1,\ldots,\delta$. Then for all $u \in \Lambda C_0^\infty(\Omega)$ we have the estimate

$$(3.9) \qquad \left|\int_\Omega d\mu(a^2(h^{jk}(\Lambda u)_{x^j}(\overline{\Lambda u})_{x^k} + q|\Lambda u|^2) - \sum_\nu |aA_\nu u|^2)\right| \le \varepsilon \|u\|^2 .$$

This proposition follows from (3.8) by a calculation. Actually, (3.8) implies a corresponding local estimate, and then we may use that $\|u\|_1^{\,2} = (u,Hu) = \|\Lambda^{-1}u\|$ (cf. V, (1.4)) .

Using (m_3) we now may bring the factor a^2 of the first term in (3.9) under one of the derivatives$(\Lambda u)_{|j}$. In detail we get

$J = \int_\Omega d\mu a^2(h^{jk}(\overline{\Lambda u})_{|j}(\Lambda u)_{|k}+q|u|^2)$ into the form

(3.10) $2J = (a^2\Lambda u,\Lambda^{-1}u)+(\Lambda^{-1}u,a^2\Lambda u)-2((\Lambda a(D_a\Lambda)+(D_a\Lambda)^* a\Lambda)u,u)$,

with the inner product (.,.). Apply the first relation (3.5) for

(3.11) $a^2\Lambda = \Lambda(a^2+aC+Ca+C^2)$, $C = \Lambda^{-1}[a,\Lambda] \in C$.

It follows that the first two terms can be written as

(3.12) $(a^2\Lambda u,\Lambda^{-1}u)+(\Lambda^{-1}u,a^2\Lambda u) = 2(u,(a^2+C')u)$, $C'=aC+Ca+C^2$.

Finally (3.9) with (3.10), (3.12) may be polarized for

(3.13) $|((a^2+C'-(\Lambda a(D_a\Lambda)+(D_a\Lambda)^* a\Lambda)- \sum_\nu A_\nu^* a^2 A_\nu)u,v)| \le c\varepsilon(\|u\|^2+\|v\|^2)$,

valid for all $u,v \in \Lambda C_0^\infty$. All operators in (3.13) are continuous, hence we may extend to all of H . Denote the self-adjoint operator at right by Z ,and introduce Zu=v , for

(3.14) $\|Zu\|^2 \le\varepsilon(\|u\|^2+\|Zu\|^2)$, $u \in H$,

which results in

(3.15) $a^2- \sum_\nu (aA_\nu)^*(aA_\nu)-C'-(\Lambda a(D_a\Lambda)^*+(D_a\Lambda)^* a\Lambda)=Z$, $\|Z\|\le\varepsilon/(1-\varepsilon)$.

Accordingly we also get $|\sigma_Z(m)| \le \varepsilon/(1-\varepsilon)$. However, for a maximal ideal $m \in \iota^{-1}(x_0)$ we get $\sigma_a(m) = 1$, and $\sigma_{aA_\nu}(m) = \sigma_{A_\nu}(m)$ $= \eta_\nu(m)$. Also the symbol of all other terms of Z vanishes at m . Indeed, we get $D_a \in \mathcal{D}^\#$, by (m_3), while $D_a=0$ near x_0 , by (m_3) . Hence one may find a function $\chi \in A^\#$ equal to 1 near x_0 and 0 on supp ∇a , and set $\omega=1-\chi \in A^\#$, $\omega D_a = D_a$. It follows that $\sigma_{D_a\Lambda}(m)$ $= \omega(x_0)\sigma_{D_a\Lambda}(m) = 0$, which makes the symbols of the last two terms vanish. The same may be used for $C' = \Lambda^{-1}[a,\Lambda]$. With a resolvent integral we get

(3.16) $C' = \pi^{-1}\int_0^\infty \Lambda^{-1}(H+\lambda)^{-1}[H,a](H+\lambda)^{-1}\lambda^{-1/2}d\lambda$.

Using V,(2.24) we get $[H,a] = (D_{\bar a})^* - D_a$, where we again use ωD_a $= D_a$. (Note that $(H+\lambda)^{-1}$, $\Lambda^{-1}(H+\lambda)^{-1}$, $\Lambda^{-2}(H+\lambda)^{-2} \in C$, so that the integrand is in C and has symbol vanishing at m . The integral exists, by the standard estimates. Thus $\sigma_{C'}(m) = 0$.

As a consequence of (3.15) we therefore get

$$|1 - \sum_{\nu=1}^{\delta} |\eta_\nu(m)|^2| \leq \varepsilon \quad . \tag{3.17}$$

This estimate holds for all $\varepsilon>0$, so that we get

$$\sum_{\nu=1}^{\delta} |\eta_\nu(m)|^2 = 1 \quad . \tag{3.18}$$

Note that self-adjointness of the D_j so far has not been used. This condition will be essential if we try to show that $\eta_j(m)$ is real. Clearly the complex conjugate of $\eta_j(m)=\sigma_{A_j}(m)$ is $\sigma_{A_j^*}(m)$. Also $A_j^* = (D_j\Lambda)^* = (\Lambda D_j)^{**}$, using that $D^*=D$. By the second relation (3.5) we get

$$(D_j\Lambda)^* = D_j\Lambda + \Lambda^\varepsilon C \quad , \quad C \in C \ . \tag{3.19}$$

If the maximal ideal m considered has the property that $\sigma_\Lambda(m)=0$, then (3.19) implies at once that $\eta_j(m)$ is real, since the symbol of the last term vanishes. On the other hand, we get

$$[D_j\Lambda,\Lambda] = ((D_j\Lambda - (\Lambda D_j)^{**})\Lambda \in E \ , \tag{3.20}$$

so that $Y = D_j\Lambda - (\Lambda D_j)^{**}$ satisfies $\sigma_\Lambda(m)\sigma_Y(m)=0$, which again gives $\sigma_Y(m) = 0$ (,i.e., $\eta(m)$ real) whenever $\sigma_\Lambda(m) \neq 0$. Thus we indeed have proven our theorem.

Remark 3.4. Note that we have shown above that relation $(3.5)_2$ and (m_5) imply relation $(3.21)_1$ below. Similarly, (m_4) and (m_5) imply $(3.21)_2$:

$$[D,\Lambda] \in E \ , \ \Lambda^{-1}[a,\Lambda] \in E \ , \text{ for all } D \in \mathcal{D}^\# \ , \ a \in A^\# \ . \tag{3.21}$$

Indeed we showed that the symbol of $[D,\Lambda]$ vanishes for all m . Similarly, $\Lambda(\Lambda^{-1}[a,\Lambda]) = [a,\Lambda] \in E$. so that σ_C , with $C=\Lambda^{-1}[a,\Lambda]$, vanishes whenever $\sigma_\Lambda \neq 0$. But $C=\Lambda^\varepsilon C'$, $C' \in C$, which implies $\sigma_C=0$ also if $\sigma_\Lambda=(\sigma_{L^\varepsilon})^{1/\varepsilon}=0$, or, again $\sigma_C=0$ for all $m \notin \mathbb{M}$, i.e., $C \in E$.

Corollary 3.5. Under the general assumptions of thm.3.2, if we only require (m_j), j=1,2,3,4,5, but not (m_6) then we still have

$$\eta(i^{-1}(x_0)) \subset B_\delta = \{x \in \mathbb{R}^\delta : |x| \leq 1\} \ . \tag{3.22}$$

Moreover, if also the assumption of selfadjointness of the expressions D_j is dropped, then we still have

$$(3.23)\qquad \eta(\iota^{-1}(x_0)) \subset \{x\in\mathbb{C}^\delta : |x|\le 1\} .$$

Proof. With the present assumptions we can substitute (3.8), (3.9) by the weaker estimates $(1+\varepsilon)\{D,D\}\geq\sum_{\nu=1}^{\delta}|\{D_\nu,D\}|^2$, and

$$(3.24)\qquad (1+\varepsilon)\int_\Omega d\mu(a^2(h^{jk}(\overline{\Lambda u})_{|x^j}(\Lambda u)_{|x^k}+q|\Lambda u|^2) \geq \sum_{\nu=1}^{\delta}\|aA_\nu u\|^2 .$$

Here the left hand side may be treated exactly as above. One finds that it is of the form $(1+\varepsilon)(u,Vu)$ where $V\in C$ has symbol $\sigma_V(m)=1$.

Write $Z=\sum_{\nu=1}^{\delta}A_\nu^*a^2A_\nu$, and conclude from (3.24) that $(1+\varepsilon)V-Z\geq 0$, hence $(1+\varepsilon)V-Z=T^2$ with a positive self-adjoint $T \in C$. From this we conclude that $(1+\varepsilon)\sigma_V-\sigma_Z=(\sigma_T)^2\geq 0$, which yields $|\eta(m)|^2=\sigma_Z(m) \leq (1+\varepsilon)\sigma_V(m)=1+\varepsilon$. Since this holds for every ε we get the statement (3.22) . Now (3.23) is trivial, q.e.d.

Let us now come back to a representation of the symbol space $\mathbb{M}$ as a compact subset of a certain compactification of the cotangent space $T^*\Omega$ of Ω , as discussed for examples earlier in this section. For our generators a, $D\Lambda$ we define formal symbols similar to (3.2) by setting

$$(3.25)\quad \tau_a(x,\xi)=a(x),\ \tau_{D\Lambda}(x,\xi)=(b^j(x)\xi_j+p(x))/(h^{jk}(x)\xi_j\xi_k+q(x))^{1/2},$$

for all $(x,\xi)\in T^*\Omega$, where $D = -ib^j\partial_{x^j}+p \in \mathcal{D}^\#$. We require (m_4) and (m_5), hence can be assured that $(\Lambda D)^{**}=D\Lambda+E$, $E\in E$, by remark 3.4, so that $(\Lambda D)^{**}$ need not be taken as generators, mod E .

Note that the formal symbols are bounded complex-valued $C^\infty(\Omega)$-functions over Ω. Hence the class of functions (3.25), for all $a\in A^\#$, $D\in\mathcal{D}^\#$ determines a smallest compactification of Ω, called $\mathbb{P}^*\Omega$, with the property that each function (3.25) admits a continuous extension to $\mathbb{P}^*\Omega$ (cf. Kelley [Ke_1]) . This is just the maximal ideal space of the algebra of bounded continuous functions over $T^*\Omega$ generated by (3.25). The injection of the algebra generated by $A^\#$ into that algebra induces a surjective associate dual map π: $\mathbb{P}^*\Omega \to M_{A^\#}$ again, defined just as the map ι . Let $\partial\mathbb{P}^*\Omega = \mathbb{P}^*\Omega\setminus T^*\Omega$.

For the theorem, below, we will require a somewhat stronger form of cdn.(m_1), as well as cdn.(m_7), below.

Condition (m_1') : We have cdn.(m_1) , and, in addition, $p^2/q \in A^\#$, for all $D=b^j\partial_{x^j}+p \in \mathcal{D}^\#$.

Condition (m_7) : At each $x \in \partial M_{A^\#}$ there exists a folpde $D \in \mathcal{D}^\#$, such that $\{D,D\}_0(x)\neq 0$, but $\{D,D\}_1(x)=0$, with $\{D,D\}_j$ of (3.26).

Under cdn.(m_1') we do not only have $\{D,D\}(x)$ well defined for points $x\in\partial M_{A^\#}$, but even the decomposition

$$\{D,D\} = \{D,D\}_0+ \{D,D\}_1 \ , \ \{D,D\}_0=|p|^2/q \ , \ \{D,D\}_1=h_{jk}\bar{b}^j b^k \ , \tag{3.26}$$

extends to $\partial M_{A^\#}$. Then (m_7) requests a nontrivial $D\in S_x$ which 'emphasizes the 0-order component'.

Clearly it makes sense to request (m_1') or (m_7) only at a specific point $x_0\in \partial M_{A^\#}$. Then we will speak of $(m_1')_{x_0}$ or $(m_7)_{x_0}$ again.

Theorem 3.6. Assume that (m_1'), and (m_j) , $j=2,3,4,5,7$, hold. Then the maximal ideal space $\mathbb{M}$ of the comparison algebra C is homeomorphic to a compact subset of the space $\partial \mathbb{P}^*\Omega$. Under this homeomorphism the wave front space $\mathbb{W}$ is mapped onto the bundle $\pi^{-1}(\Omega)$ of co-spheres of infinite radius. Moreover, if we regard $\mathbb{M}$ as a compact subset of $\partial \mathbb{P}^*\Omega$, by virtue of this homeomorphism, then for each $m\in M$, and every generator $a \in A^\#$, $D\Lambda$, $D\in\mathcal{D}^\#$, the symbol is the restriction of the formal symbol . That is, $\sigma_a(m)$ and $\sigma_{D\Lambda}(m)$ equal the values of the continuous extension of a and $\tau_{D\Lambda}$ (of (3.25)) to $\mathbb{P}^*\Omega$ at m , respectively.

Thm.3.6 was first proven in [CS],ch.6. (cf. also [CM_2]) under stronger assumptions (cdn.(l_4),of IX,3). The idea of the present proof is due to McOwen [M_1], where the result was established for the Laplace comparison triple of a complete Riemannian manifold, and the sets $A^\#$ and $\mathcal{D}^\#$ of functions (folpdes) with covariant derivatives (of the coefficients) vanishing at ∞ . McOwen assumes the additional condition that the dimension δ_{x_0} is constant at each end of Ω, and that $E = K(H)$.

Proof. We only must provide an injection $\mathbb{M}\to\partial\mathbb{P}^*\Omega$ with the properties of the theorem. This will define a continuous map since the topology of $\mathbb{M}$ as well as that of $\partial\mathbb{P}^*\Omega$ is determined as the weak topology of the same function class (3.25).

One finds that, for $x_0\in\Omega$, the set $\pi^{-1}(x_0)\cap\partial\mathbb{P}^*\Omega$ is a sphere - there is one boundary point of $\partial\mathbb{P}^*\Omega$ in each direction $\tau\xi$, $\tau\to\infty$, $|\xi|=1$. By VI,thm.2.2 we may define the injection over x_0 as the map $\xi\to\infty\xi$ from the co-unit-sphere to $\pi^{-1}(x_0)\cap\partial\mathbb{P}^*\Omega$. Therefore we only must focus on points $x_0\in\partial M_{A^\#}$. First assume cdn.$(m_6)_{x_0}$. Then, for a formally self-adjoint orthonormal base D_ν, as in thm.3.2, we may define the injection map $\eta : \iota^{-1}(x_0)\to S^{\delta-1}$ of that thm., with the unit sphere $S^{\delta-1}$ of (3.6). Under the present assumptions we may choose the base D_ν such that $D_\delta=D_{n+1}$ is a folpde with $p/\sqrt{q}(x_0) = 1$, just by using a multiple of the assumed $D\neq 0$ with $\{D,D\}_1=0$. Then a conclusion like that in cor.3.5 may be used to show that $\eta_{n+1}(m) = \sigma_{D_\delta\Lambda}(m)\geq 0$ (where, of course (m_3) is essential). Accordingly, the map η now takes $\iota^{-1}(x_0)$ into the hemisphere $H^{\delta-1} = \{\eta\in S^{\delta-1}:\eta_\delta\geq 0\}$.

Now we claim, that also $\pi^{-1}(x_0)$ is homeomorphic to (the entire) $H^{\delta-1}$. For $p\in\pi^{-1}(x_0)$ let

$$\zeta(p)=(\zeta_1(p),\ldots,\zeta_\delta(p)) \text{ , with } \zeta_\nu(p)=\tau_{D_\nu\Lambda}(p) \text{ .} \tag{3.27}$$

Observe that $\tau_{D_\nu\Lambda}$ is real-valued, by self-adjointness of D_ν, and that $\sum_{\nu=1}^{\delta}\zeta_\nu^2(p) = 1$, again follows from Parseval's relation. Thus we get $\zeta:\pi^{-1}(x_0)\to S^{\delta-1}$ again, and the injectivity of ζ may be proven as in thm.2.2 again. In fact, we also have $\zeta_\delta\geq 0$, so that ζ again maps into $H^{\delta-1}$.

To show the surjectivity, let us introduce 'projective coordinates', for $(x,\xi)\in T^*\Omega$, with x near the point x_0, by setting $\xi_j=\eta_j/\eta_0$, $\eta_0>0$. We get

$$\tau_{D\Lambda}(x,\xi) = (b^j\eta_j+p\eta_0)/(h^{jk}\eta_j\eta_k+q\eta_0^2)^{1/2} \text{ .} \tag{3.28}$$

Here we introduce $\eta_0=r(x)/q(x)$, $\eta_j=h_{jk}(x)c^k(x)$, with the (real) coefficients of some folpde $F=-ic^j\partial_{x^j}+r$, into the right hand side. Then it will assume the form $\{F,D\}/\{F,F\}^{1/2}$. For $\xi_j=\eta j(x)/|r(x)|$

we thus have

(3.29) $$\tau_{D\Lambda}(x,\xi) = \pm\{F,D\}/\{F,F\}^{1/2} .$$

In (3.29) set $D=D_\nu$. Also assume D_ν orthonormal in some neighbourhood of x_0 , and let $\{F,F\}(x)=1$, near x_0 , without loss of generality. It follows that

(3.30) $$\tau_{D_\nu\Lambda}(x,\xi) = \pm\{F,D_\nu\} ,$$

where the same sign holds for all ν , since the sign is determined as the sign of the coefficient $r(x)$. If F runs through all unit vectors of the δ-dimensional space, at some x , then the δ-tuple $(\{F,D_\nu\})_{\nu=1,\dots,\delta}$ runs through the entire unit sphere. Thus, for any point $\lambda\in S^{\delta-1}$ we have either λ or $-\lambda$ assumed by $(\tau_{D_\nu\Lambda}(x,\xi))$.

The same property then must hold for the values $\zeta(p)$, as p runs through $\pi^{-1}(x_0)$. But we already know that $\zeta_\delta(p) \in H^{\delta-1}$. Thus it follows that all points of $(H^{\delta-1})^{int}$ must be assumed. Since im ζ is closed, we then conclude that im $\zeta = H^{\delta-1}$, i.e., ζ is surjective.

It then follows that the map $\zeta^{-1}\circ\eta=\upsilon$ is well-defined. Let $p = \upsilon(m)$, for $m\in\iota^{-1}(x_0)$. We get

(3.31) $$\tau_a(p)=a(x_0)=\sigma_a(m) , \quad \tau_{D_\nu\Lambda}(p)=\zeta_\nu(p)=\eta_\nu(m)=\sigma_{D_\nu\Lambda}(m) ,$$

so that indeed formal symbol and symbol coincide.

On the other hand, if $(m_6)_{x_0}$ is false, i.e., $\delta_{x_0}<n+1$, then we have seen that η maps into the ball $B^{\delta-1}=\{\xi\in\mathbb{R}^\delta : |\xi|\le 1\}$ (cor.3.5)

Again we may assume that D^δ satisfies $p/\sqrt{q}=1$ at x_0, and then conclude that $\eta_\delta\ge 0$, so that the map η really goes into the half-ball $B_+^{\delta-1}=\{\eta\in B^{\delta-1}:\eta_\delta\ge 0\}$.

But if $\delta_{x_0}<n+1$, then we get $\delta_x = \dim S_x = n+1>\dim S_{x_0} = \delta_{x_0}$, for all $x\in\Omega$, and every neighbourhood of x_0 contains such points. Repeating the above conclusion we find that the vector $(\tau_{D_\nu\Lambda}(x,\xi))$ now assumes either ζ or $-\zeta$, for every $\zeta\in B^{\delta-1}$. The same again is true for the vectors $\zeta(p)$, as p runs through $\pi^{-1}(x_0)$. Thus, in this case, the map ζ also goes onto $B_+^{\delta-1}$, and we again may define the map $\upsilon(m)$. Thus we have the same statement, and thm.3.6 is established.

4. More structure of $\mathbb{M}_s$, and more on examples.

Remark 4.1. Before we attempt application of the results of sec.3 let us point out that thm.3.2, cor.3.5, and thm.3.6 all have 'local' generalizations, in the following sense:

Under cdn's $(m_3)_{x_0}$ and $(m_6)_{x_0}$ only, for some given point $x_0 \in \partial M_{A^\#}$, we still have the statement of thm.3.2 true for that point, regardless of all other points, assuming the other (m_j), of course. Under $(m_3)_{x_0}$ alone we still get cor.3.5, at x_0 . Again, under $(m_1')_{x_0}$, $(m_3)_{x_0}$, with the other (m_j) required for thm.3.6 we get the homeomorphism of thm.3.6 at least between $\iota^{-1}(x_0)$ and $\pi^{-1}(x_0)$, although perhaps not between $\mathbb{M}$ and $\partial \mathbb{P}^*\Omega$.

Here, by $(m_3)_{x_0}$ we mean that the function a of (m_3) exists for that x_0, and its neighbourhoods. By cdn.$(m_1')_{x_0}$ we mean that (m_1') holds 'near x_0', i.e. $\{D,D\}$ and $p/\sqrt{q}$ each coincide with some function in $A^\#$ near x_0 . Similarly $(m_1)_{x_0}$, which may be substituted for (m_1) , above.

We shall refer to these local generalizations as thm.3.2_{x_0}, etc. Note that there also is a thm.2.2_{x_0} .

Let us point out again, that, speaking in terms of thm.3.6, examples have been given which show that the set $\iota^{-1}(x_0)$, for some $x_0 \in \partial M_{A^\#}$, may range from the minimum (the set $\mathbb{W}^{clos} \cap \pi^{-1}(x_0)$) to the maximum (all of $\pi^{-1}(x_0)$) . In particular, example (B) of V,4 shows that cases between these two extremes occur: For the Laplace comparison algebras of the polycylinders $\mathbb{G}^{jk}$ we may get more than $\mathbb{W}^{clos} \cap \pi^{-1}(x_0)$,but not all of $\pi^{-1}(x_0)$, rather, only some sub-surfaces of the hemisphere $H^{\delta-1}$ are contained in $\mathbb{M}$.

As a partial answer to the question for the set $\iota^{-1}(x_0)$ we prove the result, below.

Theorem 4.2. Under the assumptions of thm.3.6_{x_0}, if there exists a function $a\in A^{\#}$ with $a(x_0)\neq 0$, and $aD_0\Lambda\in E$, where D_0 denotes the folpde assumed in thm.3.6, with $\{D_0,D_0\}=\{D_0,D_0\}_0\neq 0$, at x_0 , then we have

$$(4.1)\qquad \iota^{-1}(x_0) = \iota^{-1}(x_0)\cap\mathbb{W}^{clos} .$$

Proof. Let $m\in\mathbb{M}^s = \mathbb{M}\setminus\mathbb{W}^{clos} = \mathbb{M}\setminus\mathbb{M}^s$, and let $\iota(m)=x_0$. Since $aD_0\Lambda\in E$, we conclude that $0 = \sigma_{aD_0\Lambda}(m) = a(x_0)\sigma_{D_0\Lambda}(m)$, or $\sigma_{D_0\Lambda}(m)=0$, since $a(x_0)\neq 0$, by assumption. Therefore the statement is an immediate consequence of cor.4.3, below, q.e.d.

Corollary 4.3. Under the assumptions of thm.3.6_{x_0}, with the operator $D=D_0$ of thm.3.6, we have

$$(4.2)\quad \iota^{-1}(x_0)\cap\mathbb{W}^{clos}=\{m\in\iota^{-1}(x_0):\sigma_{D_0\Lambda}(m)=0\}=\{m\in\iota^{-1}(x_0):\tau_{D_0\Lambda}(m)=0\} .$$

Proof. First notice that $\tau_{D_0\Lambda}(m)=0$ for points of $\mathbb{W}^{clos}$. Indeed, we know that, for an $(x,\xi)\in\mathbb{W}$ we have $|\xi|=\infty$, so that the formal symbol assumes the form

$$(4.3)\qquad \tau_{D\Lambda}(x,\xi) = b^j\xi_j^0/(h^{jk}\xi_j^0\xi_k^0)^{1/2} , \text{ for } \xi=\infty\cdot\xi^0 , |\xi^0|=1 .$$

With the technique of the proof of thm.3.6 we express this as

$$(4.4)\qquad \tau_{D\Lambda}(x,\xi) = (\{D,F\}_1/(\{F,F\}_1)^{1/2})(x) ,$$

where F is a suitable folpde. (Note that the sign of the 0-order term no longer is essential.) Since we have $\{D_0,D_0\}_1(x_0)=0$, we conclude from (4.4), that $\tau_{D_0\Lambda}(p) = 0$, for all $p\in\pi^{-1}(x_0)$, whenever $p\in\mathbb{W}^{clos}$, which proves the above.

Now, again in the terminology of thm.3.6, we must show that the restriction of the map η to the set $\{m\in\iota^{-1}(x_0):\eta_\delta(m)=0\}$ maps onto the sphere (or ball) $\eta_\delta=0$. However, for $x\in\Omega$ near x_0 the values of the vector $(\sigma_{D_\nu\Lambda}(m))$, $m\in\iota^{-1}(x)$, are exactly the values of $(\tau_{D_\nu\Lambda}(x,\xi)) = ((\{D_\nu,F\}/(\{F,F\}_1)^{1/2})(x))$, taken over all folpdes F, defined near x. From this one concludes that also in the limit $x\to x_0$ every vector of the sphere or ball is assumed. This completes the proof.

Remark 4.4. Let us emphasize again, that thm.3.6, and cor.4.3

require conditions (m_4) and (m_5), which imply that, for a basis (mod $A^{\#}$) of self-adjoint folpdes D_ν the corresponding generators $D_\nu\Lambda$ are self-adjoint mod E . If these conditions are violated, as in the case of a comparison algebra associated to a boundary problem, (cf. example (H) of section 2), then $\mathbb{M}$ no longer must be a subset of the space $\partial\mathbb{P}^*\Omega$. In $[C_1]$ and $[CC_1]$ we discussed such an example, where it turns out that certain additional 'filaments' of $\mathbb{M}$ occur, which cannot be found from the values of the formal symbol.

Remark 4.5. As another feature, illuminating the role of the formal symbol and the compactification $\partial\mathbb{P}^*\Omega$, let us observe that, in all of the results of sec's 2, 3, 4, involving the sesqui-linear form $\{D,F\}$, $D,F\in\mathcal{D}^{\#}$, we may use the form $\{D,F\}^\wedge(x)$ of any triple $\{\Omega^\wedge,d\mu^\wedge,H^\wedge\}$ (c> $\{\Omega,d\mu,H\}$, defined as

$$(4.5)\qquad \{D,D\}^\wedge = h^\wedge_{jk}\bar{b}^j b^k+|p|^2/q^\wedge ,$$

and by polarization for $\{D,F\}^\wedge$. Actually, one may use just any similar sesqui-linear form, positive definite over each space $\mathbb{C}^{n+1}$ of n+1-tuples $(\nabla u(x),u(x))$, at every $x\in\Omega$, as long as the estimate V,(1.6) remains satisfied with $\{D,D\}^\wedge$ instead of $\{D,D\}$. It is clear then that we get corresponding conditions $(m_j)^\wedge$, j=1,6,7, as well as $(m_1{}')^\wedge$, and the corresponding local conditions $(m_j)^\wedge_x$,etc. (Note that cdn's (m_j), j=2,3,4,5, do not depend on the form $\{D,F\}$). Here it is essential that $(m_1)^\wedge$ implies that $\{D,D\}^\wedge<\infty$ for all $D\in\mathcal{D}^{\#}$. With the modified conditions substituted for the ordinary ones we then again get thm.2.2, thm.3.2, cor.3.5, thm.3.6, thm.4.2, cor.4.3 true again, with exactly the same proofs, and including all the local versions.

However, the formal symbol changes with the form $\{D,F\}$. We now must take

$$(4.6)\qquad \tau^\wedge_a(x,\xi)=a(x)\ ,\ \tau^\wedge_{D\Lambda}(x,\xi)=(b^j\xi_j+p)/(h^{\wedge jk}\xi_j\xi_k+q^\wedge)^{1/2}$$

instead of (3.25). Accordingly, the boundary $\partial\mathbb{P}^*\Omega$ of the compactification $\mathbb{P}^*\Omega$ of $T^*\Omega$ also changes. In fact, one easily verifies, that there exists a surjective continuous map $\partial\mathbb{P}^{\wedge *}\Omega\to\partial\mathbb{P}^*\Omega$, which also is injective in $\mathbb{W}\subset\partial\mathbb{P}^{\wedge *}\Omega$ and acts as a homeomorphism between the two images of $\mathbb{W}$. Over the complement of $\mathbb{W}$, on the other hand, points may split into several points. In other words, the representation of $\mathbb{M}$ as a subset of $\partial\mathbb{P}^*\Omega$ changes with the form $\{D,F\}$.

Now let us return to the discussion of explicit examples again. First we look at the comparison algebra of V, thm.4.1. In details, we have Ω given as a subdomain with smooth compact boundary $\partial\Omega$ of a complete Riemannian manifold $\Omega^\wedge$ without boundary with metric tensor $h^{\wedge jk}$ and measure $d\mu^\wedge$. $\partial\Omega$ is an n-1-dimensional submanifold of $\Omega^\wedge$. We have $d\mu = d\mu^\wedge|\Omega$, and the tensor h^{jk} is the restriction of $h^{\wedge jk}$ to Ω , but q is not necessarily a restriction of $q^\wedge$. We have $q \geq q^\wedge$, and $\{\Omega^\wedge, d\mu^\wedge, H^\wedge\}$ (c> $\{\Omega, d\mu, H\}$ in the sense of V,3. For formal reasons it is convenient, however, to impose the following.

Condition (q_1) : We have $q(x) = q^\wedge(x)$ in $\Omega^\wedge\backslash N$, with some neighbourhood N of $\partial\Omega$.
To satisfy V,(3.13) and (w) we again require (as in V,4) that

$$(4.7) \quad q(x) \geq \gamma_1(\mathrm{dist}(x,\partial\Omega))^{-2} \text{ , and } \nabla(\log q^\wedge) = o(1) \text{ (in } \Omega^\wedge) .$$

Now it turns out that the classes $A_c^\#$, $\mathcal{D}_c^\#$ of V,4 in general will not satisfy conditions (m_j), j=1,2. To discuss an example only let us work with the form $\{D,F\}^\wedge$ of the above $h^{\wedge jk}$ and $q^\wedge$ and the following more restricted classes called $A_C^\#$ and $\mathcal{D}_C^\#$.

$A_C^\#$ is the class of all bounded functions $a \in C^\infty(\Omega)$ such that

$$(4.8) \quad \nabla a = o_{\Omega^\wedge}(q^{\wedge\varepsilon/2}) \text{ , for some } \varepsilon,\ 0 \leq \varepsilon < 1 .$$

(The difference to the class $A_c^\#$ is that the functions $a \in A_c^\#$ must have their derivative bounded near $\partial\Omega$, while ∇a , for $a \in A_C^\#$ may not remain bounded, but only must be $o(q^{\varepsilon/2})$, $0 \leq \varepsilon < 1$, near $\partial\Omega$.)

$\mathcal{D}_C^\#$ is the class of all $D = b^j\partial_{x^j} + p$ with bounded $\{D,D\}^\wedge$ and $\{D^*,D^*\}^\wedge$ (bounded only over Ω) satisfying the following stronger version of V, (3.16): There exists an ε , $0 \leq \varepsilon < 1$, such that

$$(4.9) \quad \begin{gathered} (b^i{}_{|k}) = o_{\Omega^\wedge}(q^{\wedge\varepsilon/2}) \text{ , } b^j(\log q)_{|j} = o_{\Omega^\wedge}(1) \text{ ,} \\ \nabla(p/\sqrt{q^\wedge}) = o_{\Omega^\wedge}(q^{\wedge\varepsilon/2}) \text{ , } \nabla((b^j\zeta)_{|j}/(\zeta\sqrt{q^\wedge})) = O_{\Omega^\wedge}(q^{\wedge\varepsilon/2}) . \end{gathered}$$

Again the requirements are distinctly different near ∞ of $\Omega^\wedge$, and near $\partial\Omega$: Near ∞ the covariant derivative $b^j{}_{|k}$ stays bounded, near ∞ it must be $o(q^{\varepsilon/2})$. Near $\partial\Omega$ the last two conditions just amount to

$$(4.10) \quad \nabla p = O(1) \text{ , } \nabla((b^j\zeta)_{|j}/\zeta) = O(1) .$$

Clearly we have $A^{\#}_{C} \subset A^{\#}_{c}$, $\mathcal{D}^{\#}_{C} \subset \mathcal{D}^{\#}_{c}$. A calculation, not discussed in detail will confirm the following.

Proposition 4.6. Let cdn.(q_1) hold for a triple as described above satisfying (4.7). Then the classes $A^{\#}_{C}$ and $\mathcal{D}^{\#}_{C}$ satisfy (m_1)^, (m_j), j=2,4,5,(as well as (m_1')^). Moreover, $A^{\#}_{C}$ even is an algebra, and $D^{\#}_{C}$ satisfies the condition of thm.3.6^, for every $x \in \partial M_{A^{\#}}$, with the expression D=1, at every point.

In particular (m_4)^ is an immediate consequence of V, lemma 3.3 , and V, lemma 3.6, since $K \subset C$.

We now consider the comparison algebra $C = C(A^{\#}_{C}, \mathcal{D}^{\#}_{C})$. This algebra clearly has compact commutators, and contains $K(H)$. Moreover,it also satisfies the conditions of thm.2.2, so that we get the sets $\iota^{-1}(x_0)$ imbedded in $\mathbb{C}^{\delta}$, for each $x_0 \in \partial M_{A^{\#}}$.

Note that the functions of $A^{\#}_{C}$ must have continuous extensions to $\Omega\cup\partial\Omega$, since they have their first derivative bounded near $\partial\Omega$, by (4.6). Also $A^{\#}_{C}$ contains all restrictions to Ω of functions in $C^{\infty}_{0}(\Omega\cup\partial\Omega)$. This shows that $\partial\Omega\subset\partial M_{A^{\#}}$, In fact, we find that $\partial M_{A^{\#}}$ is a disjoint union of $\partial\Omega$ and a set $\partial M^{\infty}_{A^{\#}}$ 'at ∞ ' , where both sets are separated, i.e., have disjoint closures. Again, it is easily verified that $(m_3)_{x_0}$ holds for all points $x_0 \in \partial\Omega$, since $\partial\Omega$ is smooth. The corresponding property $(m_3)_{x_0}$, for points in $\partial M_{A^{\#}}$, will be verifiable only under additional conditions on the Riemanian space Ω , as well as on the potential q . For a more detailed investigation of $\mathbb{M}\setminus\mathbb{W}$ in case of the Laplace comparison algebra of a complete locally symmetric Riemannian manifold cf. McOwen $[M_1]$. For a complete characterization of $\mathbb{M}$ in case of a Schroedinger expression on $\mathbb{R}^n$, for certain (spherically symmetric) potentials q , cf. Sohrab $[S_1]$, $[S_2]$,$[S_3]$.

Theorem 4.7. Under the above assumptions, with the decomposition $\partial M_{A^{\#}} = \partial\Omega \cup \partial M^{\infty}_{A^{\#}}$, we have

$$\iota^{-1}(\partial\Omega) \cap \mathbb{M}_s = \emptyset \ . \tag{4.11}$$

That is, any possible points of the secondary symbol space $\mathbb{M}_s = \mathbb{M}\setminus(\mathbb{W}^{clos})$ are over infinity, not over $\partial\Omega$.

Proof. We just apply III,thm.3.7 to conclude that $a\Lambda \in K(H)$, for every $a\in C_0^\infty(\Omega\cup\partial\Omega)$. For a point $x_0\in\partial\Omega$ one may find such a function with $a(x_0)\neq 0$. Since $E=K(H)$, and $(m_3)_{x_0}$ holds, we have the conditions of thm.4.2 satisfied for $\{.,.\}^\wedge$. It follows that there indeed is no point of $\mathbb{M}_s$ over x_0 , q.e.d.

The above discussion translates literally to the language used in V,4(E) . Details are left to the reader. On the other hand, just to illustrate the variety of problems encountered, let us shortly look at example V,4(D). For the classes $A_c^\#$ and $\mathcal{D}^\#$ used there one finds that the form $\{D,D\}$ is well defined, but $\{D,D\}^\wedge$ is not generally defined for all $D\in\mathcal{D}^\#$. Moreover one finds that $\mathcal{D}^\#$ is not an $A_c^\#$-module . Accordingly, none of the results of sec's 2, 3, 4 applies for these sets $A_c^\#$, $D^\#$, although the algebra $C(A_c^\#,\mathcal{D}^\#)$ has compact commutators and a well-defined symbol. For example, one now might well expect nontrivial parts of the secondary symbol space over $\partial\Omega$. We have not tried to apply the techniques described, although this should be easy and only a formal task.

Similarly for the examples V,4(C) and V,4(F), which will not be discussed in detail.

Finally let us discuss some 1-dimensional examples, extending an investigation by Sohrab [S_2] , where a complete knowledge of the symbol space will be obtained. These are examples of a Sturm-Liouville-triple V,(5.1).

Applying the Sturm-Liouville transform (V,prop.5.1(a)) we may assume V,(5.1) in the form V,(5.3). In that form we assume the interval $I^\sim$ to be $\mathbb{R}=(-\infty,\infty)$, which amounts to the condition that

$$\int_\alpha^\beta (\kappa/p)^{1/2}dx \text{ does not exist on both ends } \alpha,\beta \ . \tag{4.12}$$

We assume the triple in the normal form $\{\mathbb{R},dx,q-\partial_x^2\}$, where $q(x)\geq 1$ on $\mathbb{R}$. In order to reach the generality of [S_2] all conditions imposed below must be translated back to the general case.

In order to have only one point of $\partial M_{A^\#}$ at each end of the interval we select $A^\#$=algebra finitely generated by $C_0^\infty(\mathbb{R})$ and $s(x)=x/\langle x\rangle$, and $\mathcal{D}^\#$=span $\{D_0=\sqrt{q},\ D_1=-i\partial_x\}$(mod $A^\#$). We want $E=K(H)$, hence we require V,(3.13) and V,(3.15), V,(3.16), with $q=q^\wedge$. A calculation shows that this requires V,(3.13) , i.e.,

$$q' = o(q) \tag{4.13}$$

as only additional condition. (Actually, we could allow the larger function class $A_c^{\#}\supset A^{\#}$ of V,4, and still would get $E=K$ under (4.13) only, but this would give us many points in $\partial M_{A^{\#}}$ over $\pm\infty$.)

Note that (4.13) will not allow potentials of exponential growth, while potentials of arbitrary polynomial growth often are acceptable. In particular, the potentials $q(x)=1+x^2$ and $q(x)=1+x^2+x^4$ of the harmonic and the anharmonic oscillator in quantum mechanics satisfy (4.13).

All conditions (m_j), (m_1') hold, as well as the assumption of thm.3.6 (with $D=D_0=\sqrt{q}$). We will show that

$$\mathbb{M} = \partial \mathbb{P}^*(\mathbb{R}) \ , \tag{4.14}$$

where the form $\{D,D\}$ of $h^{11}=1$ and $q=q\hat{}$ is used. We trivially have $\mathbb{W}= \mathbb{R}\times\{-\infty\} \cup \mathbb{R}\times\{\infty\} \subset\mathbb{M}$. For the secondary symbol space Sohrab looks at the operator $T = \sqrt{q}\Lambda^2\sqrt{q}$, observing that the range of $\eta_2(m)$, in the set $\iota^{-1}x_0$, for either one of the infinite points $x_0=\pm\infty$, coincides with the essential spectrum of $\chi\sqrt{T}$, with a cut-off function $\chi=1$ near x_0 , $=0$ near $-x_0$.

Formally we get $T = L^{-1}$, with a self-adjoint realization L associated to the expression (in $\mathbb{R} = (-\infty,\infty)$)

$$L =-q^{-1/2}\partial_x^2 q^{-1/2}+1 = - \partial_x q^{-1}\partial_x +1-q^{-1/2}(q^{-1/2})'' \ . \tag{4.15}$$

Transforming L onto Sturm-Liouville normal form again we get

$$L^{\sim} = -\partial_y^2 + 1+q''/(4q^2) -5q'^2/(16q^3) \ , \ -\infty<y<\infty, \ y= \int_0^x \sqrt{q} \ . \tag{4.16}$$

Let us assume that the potential in (4.16) has the limits $1+\beta_\pm$ at $\pm\infty$. Then we have limit point case and the self-adjoint realization is unique. The essential spectrum of $L^{\sim -1/2}$ is equal to the interval $[0,(1+\beta)^{-1/2}]$, where $\beta=\mathrm{Min}\{\beta_\pm\}$ (cf.[We_1], [Ne_2], [DS_3]). Moreover, even the essential spectrum for a suitable realization over one of the half-lines $R^\pm$ of an expression modified near 0 is $[0,(1+\beta_\pm)^{-1/2}]$. Here we refer to $L^{\pm -1/2}$ where $L^+=L^\sim+c_0\omega/x^2$, for example, with a cut-off function $\omega=1$ near 0, $\omega\in C_0^\infty(\mathbb{R})$, and a sufficiently large constant c_0 .

It follows easily that the essential spectra of $\chi\sqrt{T}$ and of $\chi L^{\pm\wedge -1/2}$ (with the back-transform $L^{\pm\wedge}$ of $L^{\pm}$) coincide.

On the other hand $\beta_{\pm}=0$ follows if $\beta_{\pm}$ exists. Indeed, (4.13) implies $q'^2/q^3 \to 0$, and $\lim(q^{-1})'' = -4\beta_{\pm}$. If $\beta_{\pm}\neq 0$ this implies $|q^{-1}|\geq\gamma x^2$, with $\gamma > 0$, for every large (small) x . Or, $q=O(x^{-2})$, at that end. This contradicts $q\geq 1$.

This argument shows that $\eta_2(m)$ has range $[0,1]$, over each end $\pm\infty$ of $\mathbb{R}$. Accordingly $|\eta_1(m)|=(1-\eta_2(m))^{1/2}$ also has range $[0,1]$ This may be used to show that, at each end $\pm\infty$, the half-circle

$$H^1 = \{\eta_1{}^2+\eta_2{}^2=1 \ , \ \eta_2\geq 0\} \tag{4.17}$$

is in im η , so that we get (4.14). Indeed, if $\eta_2(m)+i\eta_1(m) = z$, where z is any value of the circle $|z|=1$, Re $z \geq 0$, then the operator $T_z = \chi(D_0+iD_1)\Lambda - z = \chi(\sqrt{q}+\partial_x)\Lambda - z$ is not Fredholm, since its essential spectrum contains zero. Here χ denotes any real-valued cut-off function: $\chi\in C^\infty(\mathbb{R})$, $0\leq\chi\leq 1$, $\chi=0$ in $(-\infty,x_0)$, $\chi=1$ in (x_1,∞). This means that there exists a sequence $u_j\in H$, $\|u_j\|=1$, with $\|T_z u_j\|\to 0$, $u_j\in (\ker T_z)^{\perp}$. But then the conjugate-complex sequence $\bar{u}_j$ will satisfy the same, with respect to $\bar{z}$, since $D_0+iD_1 = \partial_x+\sqrt{q}$ and Λ both are real operators, left invariant by complex conjugation. It follows that $T_{\bar{z}}$ is not Fredholm as well, so that $(-\eta_1(m),\eta_2(m))$ also must be assumed by the map η , as another maximal ideal of course. We have proven the following.

<u>Theorem 4.8.</u> Assume that $q\geq 1$, that (4.13) holds, and that $\lim_{x\to\pm\infty} q''/q^2$ exist. Then the comparison algebra C (with compact commutator) has symbol space $\mathbb{M}$ given by (4.14) . Topologically this symbol space is a circle.

CHAPTER 8. COMPARISON ALGEBRAS WITH NONCOMPACT COMMUTATORS.

In this chapter we consider some examples of comparison algebras with noncompact commutators. The most important example, perhaps, has been discussed in detail in [C_1],ch.5., in connection with the half space $R_+^{n+1} = \{(y,x):x\in\mathbb{R}^n,\ y\in\mathbb{R},\ y>0\}$. If the manifold Ω has a regular boundary, onto which the expression H, and measure $d\mu$, as well as $A^\#$ and $\mathcal{D}^\#$ may be extended, meeting similar conditions as for interior points, then commutators are no longer compact. However, both quotients, $\mathcal{C}/\mathcal{E}$ as well as $\mathcal{E}/\mathcal{K}(\mathcal{H})$, turn out to be function algebras. The maximal ideal space $\mathbb{M}$ of $\mathcal{C}/\mathcal{E}$ looks similar to the spaces encountered for complete manifolds like $\mathbb{R}^n$, except that it has certain one dimensional 'filaments' over the boundary of Ω .

On the other hand we get

$$(0.1)\qquad \mathcal{E}/\mathcal{K}(\mathcal{H}) \equiv C(\mathbb{N},\mathcal{K}(h)) ,$$

with the Hilbert space $h=L^2(\mathbb{R}^+)$, and the symbol space $\mathbb{N}$ of a well investigated Laplace comparison algebra over $\mathbb{R}^n$. If the $\mathcal{E}$-symbol of an operator $A\in\mathcal{L}(\mathcal{H})$ does not vanish, then A may be inverted mod $\mathcal{E}$, which corresponds to the first step of the common approach of solving a boundary problem: That of using a fundamental solution (or parametrix) of the differential equation to solve the differential equation, though not yet the boundary condition.

This then must be followed in common theory by solving another system of equations- for example a singular integral equation over the boundary, in case one uses the well known boundary layer approach. The latter corresponds to the inversion mod $\mathcal{K}(\mathcal{H})$ of a (system) $1+E$, with $E\in\mathcal{E}$.

Presently we do not study such manifolds with boundary, but rather turn to a pair of different examples. First (sec.1) we consider an algebra over $\mathbb{R}$, where some of the generators are periodic functions. (It was seen in V,5 that a commutator $[e^{ix},D\Lambda]$ is not compact) . It is found that again a 2-link chain results. Now we get

$$(0.2)\qquad E/K(H) \cong C(\mathbb{N},K(\ell^2))\ ,\ \mathbb{N} = S^1_+ \cup S^1_-\ ,$$

with the Hilbert space $\ell^2 = L^2(\mathbb{Z})$, and two disjoint copies $S^1_\pm$ of the circle S^1 .

In direct relation to the above special importance of $L^2(\mathbb{Z})$ this algebra is invariant under translations by $2k\pi$. Again there is a tensor decomposition of H involved, together with a certain unitary map of H. Results are related to those discussed in Gohberg-Krupnik [GK], ch.11, (cf. also Sarason [Sa_1] , and [CMe_1], where we expand on the present discussion).

Next, in sec.2, we look at a product manifold $\Omega = \mathbb{R}^{n'} \times B$, where B is a given compact manifold. The simplest example would be the circular cylinder $\mathbb{R}\times S^1$, with the circle S^1 . We studied a comparison algebra on $\mathbb{R}\times S^1$ in example (B) of sec.4, using an algebra $A^\#$ of bounded C^∞-functions having <u>all</u> derivatives tending to 0, at $\pm\infty$, including the derivatives in the 'S^1-variables' . That algebra had a compact commutator. In sec.2 we admit more general generators, and obtain noncompact commutators. Again a 2-link ideal chain is found. Again the second quotient is of a similar form:

$$(0.3)\qquad E/K(H) \cong C(\mathbb{N},K(h))\ ,$$

where $\mathbb{E}$ is the symbol space of a certain algebra of singular integral operators over $\mathbb{R}^{n'}$. In case of n=1 we get $\mathbb{E} = \mathbb{R} \cup \mathbb{R}$, a disjoint union of 2 copies of $\mathbb{R}$. In each case we have $h = L^2(B)$. Again a tensor decomposition of H is involved, after a unitary transformation. The ideal E is identified as a certain algebra of singular integral operators with compact operator valued symbol.

In section 4 the corresponding is done for a manifold with finitely many cylindrical ends. To make this discussion possible we introduce the technique of 'algebra surgery' in sec.3. This simply means that we cut out the portion of a comparison algebra over some manifold Ω_1 , corresponding to an open subset $U \subset \Omega$, and compare it with the corresponding portion of another algebra C_2 belonging to another manifolds Ω_2 , but with the same subset U . Of course algebra surgery also may be performed if commutators are compact.

It should be noticed that sec.3 only discusses the basic

method of surgery, at the example of a 'compact cut', - i.e. the boundary of the cut-out portion is assumed to be compact (not the cut-out portion itself). It is not hard to remove this assumption, at the expense of additional conditions near the infinite parts of the cut. We find that these additional conditions are too complicated to state, except for very specific geometrical models. For a discussion of surgery on a manifold with polycylindrical ends we refer to a [CDg_1] , sec.5.

In all three examples considered here the symbol space of C (i.e. the maximal ideal space of C/E) is a compactification of the wave front space. In other words, the secondary symbol space is void, fon each of these examples.

The theory of sec.4. has an application to a problem on $\mathbb{R}^n$, studied first by Nirenberg-Walker[NW_1], then by Cantor [Ct_1], Lockhart [Lk] and McOwen [M_1] (cf. also [LM]). This involves differential operators with coefficients constant at infinity and homogeneous symbol in certain weighted Sobolev spaces, where derivatives of different orders have different weights. Details of this application will be discussed elsewhere.

We note that there are numerous other approaches to singular boundary problems on manifolds with cylindrical or conical ends (cf. Agmon-Nirenberg [AN_1] , Bruening-Seeley [$BS_{1,2}$] , Melrose-Mendoza [MM] , Schulze [$Schu_1$]). Also let us point to a direct extension of our present theory discussed in [CPo] and [CMe_1] . It turns out that the E-symbol can be extended to the entire algebra C . One thus obtains Fredholm criteria without involving a chain of two inversions.

1. An algebra invariant under a discrete translation group.

Let us consider the Laplace comparison algebra C on $\mathbb{R}$ of example (G) of V,5 with generators V,(5.27). We have discussed the commutator ideal E in V,prop.5.5. Now we want to obtain more details about the ideal chain $C \supset E \supset K(H)$ and the quotients C/E and $E/K(H)$. First we study the symbol of the algebra C .

Proposition 1.1. The compactification $M_{A^\#}$ of the function class $A^\#$ described by V,(5.27), as a point set, is given by collapsing each circle $\{(x_0,y):y\in S^1\}$, for a fixed $x_0\in\mathbb{R}$, in the compact infinite cylinder $[-\infty,\infty]\times S^1$, into a single point. The points of the two circular 'caps' $\{(-\infty,y):y\in S^1\}$ and $\{(\infty,y):y\in S^1\}$ correspond to the maximal ideals $\{a(x): \lim_{k\to-\infty} a(2\pi y+2k\pi)=0\}$ (and with the limit $k\to+\infty$, respectively, for integers k) of the algebra $C_{A^\#}$.

The topology is the weak topology induced by $C_{A^\#}$. In particular, the relative topology induced on the two caps is equivalent to the Euclidean topology of the circle S^1. Every neighbourhood of one of the points θ of the circle caps contains an open neighbourhood of the point $\theta+2k\pi$, for all sufficiently large (sufficiently small) integers k, corresponding to the cap at $\pm\infty$, respectively.

The proof is left as an exercise.

Theorem 1.2. The symbol space $\mathbb{M}$ of the algebra $\mathcal{C}$ generated by $A^\#$ and $\mathcal{D}^\#$ of V, (5.27) is homeomorphic to a compact subset of $M_{A^\#}\times[-\infty,+\infty]$. Using the homeomorphism as an identification we have

$$\mathbb{M} = M_{A^\#}\times\{-\infty,\infty\} \quad (1.1)$$

That is, $\mathbb{M}$ is the closure of the wave front space in the above product. The symbols of the generators are given by

$$\sigma_\Lambda(x,\xi)=\lambda(\xi)\ ,\ \sigma_{D\Lambda}(x,\xi)=s(\xi)\ ,\ \sigma_a(x,\xi)=a(x), \quad (1.2)$$

where the functions λ, s, and a must be continuously extended.

Proof. We use the argument of Herman's Lemma again (cf.[C_1],IV,2): The algebra $\mathcal{C}$ has the two commutative subalgebras $\mathcal{C}_0 = C_{A^\#}$ and $\mathcal{C}^\#$ = algebra span of the operators $S=s(D)$ and $\Lambda=\lambda(D)$. These algebras have the maximal ideal spaces $M_{A^\#}$, and (the interval) $[-\infty,+\infty]$, respectively. Clearly both of them together generate the algebra $\mathcal{C}$.

Let $\pi_0:\mathbb{M}\to M_{A^\#}$ and $\pi^\#:\mathbb{M}\to[-\infty,\infty]$ be the duals of the injections $\mathcal{C}_0\to\mathcal{C}$ and $\mathcal{C}^\#\to\mathcal{C}$. Then $\upsilon:\mathbb{M}\to M_{A^\#}\times[-\infty,+\infty]$, defined as $\upsilon=\pi_0\times\pi^\#$, defines a continuous map onto a compact subset of the space at right. This map must be 1-1, hence a homeomorphism, because $\pi_0(m) = \pi_0(m')$ and $\pi^\#(m) = \pi^\#(m')$, for $m,m'\in\mathbb{M}$ implies that the homomorphisms $h_m:\mathcal{C}\to\mathbb{C}$ and $h_{m'}:\mathcal{C}\to\mathbb{C}$ corresponding to m and m' coincide on $\mathcal{C}_0$ and $\mathcal{C}^\#$, hence on $\mathcal{C}$. Then $h_m=h_{m'}$ implies $m=m'$.

Now $\mathbb{M}$ must contain the set $\mathbb{R}\times\{-\infty,+\infty\}$, identified as the wave front space, by VII,thm.1.5. Indeed, the map π_0 is identical with ι of the proof of VI,thm.2.2 over interior points of $\Omega=\mathbb{R}$. The cosphere at each $x_0\in\mathbb{R}$ contains exactly two points which must agree with the two points of $\pi_0^{-1}(x_0)$. Since $\mathbb{M}$ is closed, it must contain the closure of $\mathbb{W}$, i.e., the space (1.1). On the other hand, $\mathbb{M}$ cannot contain any (x_0,ξ_0) in the product set having ξ_0 finite. Indeed, let $\phi\in CO(\mathbb{R})$ be such that $\phi(\xi_0)\neq 0$. Then $\phi(D)\in E$, and also $\in C^{\#}$. Using the associate dual map one finds that the symbol of $A=\phi(D)$ assumes the value $\sigma_A=\phi(\xi_0)\neq 0$ at $m=(x_0,\xi_0)$. On the other hand $\sigma_A\equiv 0$ since $A\in E$ (by V,prop.5.5), a contradiction. This shows that $\mathbb{M}$ coincides with the set (1.1). Then formula (1.2) is a consequence of the fact that $\sigma_a(m) = a(\pi_0(m))$, $a\in C_0$, and $\sigma_{f(D)}(m) = \phi(\pi^{\#}(m))$, $\phi(D)\in C^{\#}$, by virtue of the properties of the associate dual map (cf.[C_1],AII,5). Q.E.D.

Next we examine the quotient $E/K(H)$. In fact it will be possible to obtain a much more precise control of the ideal E. First, in that respect, it is useful to look at the C^*-subalgebra F of E with generators

$$(1.3)\qquad E = \sum_{j=-N}^{N} a_j(D)e^{ijx} \in E \ , \ a_j\in CO(\mathbb{R}) \ , \ N=0,1,2,\dots \ .$$

Note that the equation $(1+E)u=f$, for such an E, is equivalent to the linear 2N-th order finite differences equation (for the inverse Fourier transforms $u^{\vee}$ and $f^{\vee}$)

$$(1.4)\qquad u^{\vee}(x) + \sum_{j=-N}^{N} a_j(-x)u^{\vee}(x+j) = f^{\vee}(x) \ , \ x\in\mathbb{R} \ .$$

Here (1.2) relates only the values of the vectors

$$(1.5)\qquad u^{\wedge}=(u_j)=(u^{\vee}(x-j))_{j=0,\pm1,\dots} \ , \ f^{\wedge}=(f_j)=(f^{\vee}(x-j))_{j=0,\pm1,\dots} \ ,$$

for each fixed x , $0\le x<1$. In fact, for each such x we get an infinite system of equations

$$(1.6)\qquad u_j+\sum_{k=j-N}^{j+N} a_{jk}u_k = f_j \ , \ j=0,\pm1,\dots \ , \text{ with } a_{jk}=a_{j-k}(j-x) \ .$$

Note that the infinite matrix $A=((a_{jk}))_{j,k=0,\pm1,\dots}$, where we set $a_{jk}=0$, as $|j-k|>N$, defines a compact operator on the Hilbert space $\ell^2= L^2(\mathbb{Z})$. (Such a matrix, with the property that all entries outside a finite number of diagonals vanish, is called a Toeplitz matrix.) The present Toeplitz matrix has the additional property that the elements in each diagonal have limit zero, as

$|j|\to\infty$, since $a_j \in CO(\mathbb{R})$. This implies compactness of A . (To prove this, observe that the matrix obtained from A by replacing all entries a_{jk} , with $|j|,|k|<M$, by 0 , has norm $\leq \varepsilon$, as M is large, by Schurs Lemma $[C_1]$,I,1. This means that A differs from an operator with finite rank by an arbitrarily small amount,in norm. Thus it must be compact.)

Notice that the vectors $u^\wedge$, $f^\wedge$ of (1.3) vary with x , as $x \in \mathbb{R}$, and that we have the 'periodicity condition'

$$(1.7) \qquad u^\wedge(x+1) = Y_{-1}u^\wedge(x) \ , \ f^\wedge(x+1) = Y_{-1}f^\wedge(x) \ ,$$

with the shift operator $Y_{-1}:\ell^2\to\ell^2$ defined by $Y_{-1}(u_j) = (u_{j-1})$. Correspondingly the matrix A is a function of x and we have

$$(1.8) \qquad A(x+1) = Y_{-1}A(x)Y_1 \ , \ x \in \mathbb{R} \ .$$

In order to convert (1.7) and (1.8) into true periodicity one may introduce a unitary deformation Y_t , $0\leq t\leq 1$, of the shift matrix Y_1 into the identity operator. Note that Y_1 is a unitary operator of ℓ^2 . For example one may define

$$(1.9) \qquad Y_t = ((\upsilon(j-k+t)))_{j,k=0,\pm1,\dots} \ , \ t\in\mathbb{R} \ ,$$

with the entire analytic function $\upsilon(\tau)=e^{i\pi\tau}\sin\pi\tau/\pi\tau$, $\tau\neq 0$, $\upsilon(0)=1$. It follows easily that $Y_m u^\wedge = (u_{j+m})$, so that indeed Y_m shifts the components of $u^\wedge=(u_j)$ by m units. Moreover we get $Y_{t+\tau}=Y_tY_\tau$, as $t,\tau\in\mathbb{R}$ (i.e.,Y_t is a 1-parameter group). Also Y_t is norm continuous and even $C^\infty(\mathbb{R},L(\ell^2))$ and unitary. All this follows from the fact that Y_t, in essence, may be obtained as the conjugation by the Fourier transform of $\mathbb{Z}$ of the family of multiplications (of a function in $L^2([0,1]))$ by $e^{2\pi it\theta}$. The Fourier transform of $\mathbb{Z}$ is defined by

$$(1.10) \qquad u^\wedge = (u_j) \to u^\sim(\theta) = \textstyle\sum_{j=-\infty}^{\infty} u_j e^{2\pi ij\theta} \ , \ \theta \in [0,1] \ .$$

Introducing $u^\wedge(x)=Y_{-x}v^\wedge(x)$, $f^\wedge(x)=Y_{-x}g^\wedge(x)$, $A^\wedge(x)=Y_xA(x)Y_{-x}$, we find that (1.7) and (1.8) go into

$$(1.11) \qquad v^\wedge(x+1)=v^\wedge(x) \ , \ g^\wedge(x+1)=g^\wedge(x) \ , \ A^\wedge(x+1) = A^\wedge(x) \ .$$

In other words, $v^\wedge(x)$ and $g^\wedge(x)$ now may be regarded as functions on S^1 , the one-dimensional circle obtained by identifying the endpoints of the unit interval $[0,1]$. In fact we get $v^\wedge$, $g^\wedge \in L^2(S^1,\ell^2)$, and, moreover, using that $Y_x:\ell^2\to\ell^2$ is unitary,

(1.12) $$\|u\|^2 = \int_R |u(x)|^2 dx = \int_{S^1} dt \|v^\blacktriangle\|^2 .$$

Thus we have proven:

Proposition 1.3. The map $u \to u^\blacktriangle \to v^\blacktriangle$ defines a unitary map $Z: H \to H^\blacktriangle$, with $H = L^2(\mathbb{R})$, $H^\blacktriangle = L^2(S^1, \ell^2)$. Moreover, for any $E \in E$ of the form (1.3) the operator $E^\blacktriangle = ZEZ^{-1} \in L(H^\blacktriangle)$ is a 'multiplication operator'

(1.13) $$(E^\blacktriangle v^\blacktriangle)(t) = A^\blacktriangle(t) v^\blacktriangle(t) \ , \ v^\blacktriangle(t) \in H^\blacktriangle \ ,$$

with a continuous function $A^\blacktriangle(t) \in C(S^1, K(\ell^2))$.

Proposition 1.4. Conjugation with the operator $Z^{-1}: H^\blacktriangle \to H$ takes the algebra $F \subset E$ onto the class $C(S^1, K(\ell^2))$ of all multiplications of the form (1.13) .

Proof. This is an immediate consequence of the Stone-Weierstrass theorem for algebras of matrix functions: (i) At each point $t \in S^1$ the algebra of all 'values' of functions $A^\blacktriangle(t)$ obtained from E of the form (1.3) is dense in $K(\ell^2)$, since by a special choice of the functions $a_j(x)$ it always can be arranged that only the coefficient $a^\blacktriangle_{jk}$, for a given fixed index pair j,k, is different from zero, while all other coefficients of the matrix $A^\blacktriangle(t)$, for the given t, vanish. (Note that conjugation by Y_t does not change this since Y_t is unitary.) Also, (ii) F is invariant under multiplication by $b(D)$, with $b \in C^\infty(\mathbb{R})$, b periodic with period 1. Such operation translates into a multiplication of $A^\blacktriangle(t)$ with the function $b \in C(S^1)$. This proves the statement.

Proposition 1.5. A multiplication $u^\blacktriangle(t) \to A^\blacktriangle(t) u^\blacktriangle(t)$ by an $A^\blacktriangle(t) \in C(S^1, K(\ell^2))$ is in $K(H^\blacktriangle)$ if and only if $A^\blacktriangle(t) \equiv 0$.

Proof. Let $A^\blacktriangle(t_0) \neq 0$. There exists $w \in \ell^2$ with $\|w\|=1$ and $\|A^\blacktriangle(t)w\| \geq p > 0$ in $|t-t_0| \leq \varepsilon$, for some $p>0$, $\varepsilon>0$. Then the class of all $u^\blacktriangle = \phi w$, $\phi \in C_0^\infty((t_0-\varepsilon, t_0+\varepsilon))$ defines an infinite dimensional subspace of $H^\blacktriangle$ on which we get $\|A^\blacktriangle u^\blacktriangle\| \geq p \|u^\blacktriangle\|$. Hence $A^\blacktriangle(t)$ cannot be a compact operator, q.e.d.

In order to obtain full control on the ideal E it now appears natural to investigate operators of the form $B^\blacktriangle = Za(x)Z^{-1}$, where $a \in CS(\mathbb{R})$. These operators will only appear in the form $B^\blacktriangle E^\blacktriangle$ or $E^\blacktriangle B^\blacktriangle$, with an $E^\blacktriangle \in ZFZ^{-1}$. Getting an explicit form for $B^\blacktriangle$ implies that we can carry out all investigations on E by looking at $E^\blacktriangle = ZEZ^{-1}$ only.

Note that $a(x)E \in K(H)$ for all $a \in CO(\mathbb{R})$. Also we may get restricted to the special case $a(x)=s(x)=x/\langle x\rangle \in CS(\mathbb{R})$. Then we have $a(D)=(s^{\vee}*)$ equal to the convolution operator with kernel $s^{\vee}=F^{-1}s$. The properties of $s^{\vee}$ are well known: Using [C_1],II, for example, one finds that $s^{\vee}\in S^0_{ad}$. The singularity at 0 has a homogeneous expansion mod D with leading term c_1/x. We have $s^{\vee}\in C^{\infty}(\mathbb{R}\setminus\{0\})$, and it equals a function in S outside $|x|\leq 1$. More directly, $s^{\vee}$ may be expressed by modified Hankel functions, showing the same feature, and even that it decays exponentially with all derivatives, as $|x|\to\infty$. Using a cutoff function $\chi\in C_0^{\infty}(\mathbb{R})$, $\chi=1$ near 0, $\chi=0$ for $|x|\geq 1/4$ we write $s^{\vee} = \psi + \omega^{\vee}$, where $\psi=\chi s^{\vee}$, $\omega^{\vee}=(1-\chi)\sigma^{\vee}$. It follows that $\omega^{\vee}\in S$, hence $\omega\in S\subset CO$. Accordingly $s(D)=\psi^{\wedge}(D)+\omega(D)$. where one may neglect the second term for reason mentionned above. In fact one may split again $\psi(x) = c_1\xi(x)/x + \psi^{\sim}(x)$, where $\psi^{\sim\wedge}\in CO$, using the asymptotic expansion of $s^{\vee}$.

This discussion shows that it suffices to assume

$$(1.14)\qquad \Psi u^{\vee}(x) = a(D)u^{\vee}(x) = \int\psi(x-y)u^{\vee}(y)dy\ ,\ \psi(z) = \chi(z)/z\ ,$$

with a cut-off function χ as described above.

<u>Proposition 1.6</u>. For the operator Ψ of (1.14) and $\Psi^{\wedge}=(ZF)\Psi(ZF)^{-1}$ we have

$$(1.15)\qquad \Psi^{\wedge}v^{\wedge}(x) = \int\psi(x-y)Y_{x-y}v^{\wedge}(y)dy\ ,\ v^{\wedge}=(v_j(x)) \in C^{\infty}(S^1,\ell^2),$$

where $Y_t:\ell^2\to\ell^2$ denotes the operator of (1.9), and where we interpret $v^{\wedge}$ and $\Psi^{\wedge}v^{\wedge}$ as periodic functions (1) defined on $\mathbb{R}$.

<u>Proof</u>. Given $v^{\wedge}(x) = (v_j(x))$, where the v_j are assumed defined over $\mathbb{R}$ and periodic (1). Let $u = Z^{-1}v^{\wedge}$. We then have $u^{\wedge}(x) = (u^{\vee}(x-j)) = Y_{-x}v^{\wedge}(x)$. Or, using the periodicity of v_k

$$(1.16)\qquad u^{\vee}(x-j) = \textstyle\sum_k \upsilon(j-k-x)v_k(x) = \sum_k \upsilon(-(x-j)-k)v_k(x-j)\ .$$

We replace x-j by x and then apply the operator Ψ of (1.14), for

$$(1.17)\qquad f^{\vee}(x) = \Psi u^{\vee}(x) = \int dy\textstyle\sum_k \psi(x-yW\upsilon(-y-k)v_k(y).$$

Finally let $g^{\wedge} = (g_j) = Zf$. We get $g^{\wedge}(x) = Y_x(f^{\vee}(x-j))$. Or,

$$(1.18)\quad g_j(x)=\textstyle\sum_l\upsilon(j-l+x)f(x-l)=\int dy\sum_{kl}\psi(x-y-l)\upsilon(j-l+x)\upsilon(-y-k)v_k(y).$$

Note that near each x,y only one term of the infinite series $\sum_l$ can be $\neq 0$, since supp $\psi\subset\{|x|\leq 1/4\}$, by construction of ψ, hence interchange of sum and integral in (1.18) is justified. Using once more the periodicity of v_k we shift the integration variable, for

$$(1.19)\qquad g_j(x) = \int dy\psi(x-y)\sum_{kl}\upsilon(j-l+x)\upsilon(l-k-y)\psi_k(y) \ .$$

Note that $\sum_l \upsilon(j-l+x)\upsilon(l-k-y)$ is the (j,l)-matrix element of the product $Y_x Y_{-y} = Y_{x-y}$, using the properties of Y_t of (1.9). Thus,

$$(1.20)\qquad g_j(x) = \int dy \ \psi(x-y)\sum_k \upsilon(j-k+x-y)v_k(y) \ ,$$

where the integral extends over $\mathbb{R}$, but may be extended from $y=x-1/2$ to $y=x+1/2$ only, due to the properties of supp ψ . Q.E.D.

It is convenient to introduce the periodic matrix kernel

$$(1.21)\qquad \Psi^{\blacktriangle}(z) = \psi(z)Y_z \ , \text{ as } |z|\leq 1/2 \ , \text{ periodic (1) elsewhere.}$$

Then (1.15) assumes the form

$$(1.22)\qquad \Psi^{\blacktriangle} v^{\blacktriangle}(x) = \int_{S^1} \Psi^{\blacktriangle}(x-y)v^{\blacktriangle}(y)dy \ .$$

Moreover, using that the group Y_t is not only continuous but even $C^\infty(\mathbb{R})$, in norm convergence, and that $Y_0=1$, we conclude that

$$(1.23)\qquad \Psi^{\blacktriangle}(z) - \pi\cot\pi z = \Phi(z) \in C(S^1,L(\ell^2)).$$

<u>Proposition 1.7</u>. For every $E^{\blacktriangle} \in F^{\blacktriangle}$ we have

$$(1.24)\qquad \Psi^{\blacktriangle} E^{\blacktriangle} - \pi H^{\blacktriangle} E^{\blacktriangle} \in K(H^{\blacktriangle}) \ ,$$

where

$$(1.25)\qquad H^{\blacktriangle} u(x) = \int_{S^1}\cot\pi(x-y)u(y)dy$$

denotes the Hilbert transform of S^1 .

Indeed, an operator $E^{\blacktriangle}$ is a multiplication by a compact operator valued function $E^{\blacktriangle}(x)\in C(S^1,K(\ell^2))$, by prop.1.4. Thus the right hand side of (1.24) is an integral operator with norm continuous kernel taking values in $K(\ell^2)$. Such an operator must be in $K(H^{\blacktriangle})$, q.e.d.

At this point it will be useful to recall the topological tensor structure of the Hilbert space $H^{\blacktriangle} = L^2(S^1,\ell^2)$. Defining $h = L^2(S^1)$ we find that

$$(1.26)\qquad H^{\blacktriangle} = h\hat{\otimes}\ell^2 \ .$$

with the topological tensor product of the two Hilbert spaces, as defined in [C_1],V,8 (or [BC_2]). On S^1 we introduce the C^*-algebra SL of singular integral operators, i.e. just the unique Laplace comparison algebra of the compact space S^1. Clearly SL has compact commutators and its symbol space $\mathbb{M}_{SL}$ coincides with wave front space, since S^1 is compact. Note also that the space $\mathbb{M}_{SL}$ consists of two disjoint copies of S^1, since the unit sphere in $\mathbb{R}$ consists of two points only.

<u>Theorem 1.8</u>. The C^*-subalgebra $E^\blacktriangle = ZEZ^{-1}$ of $L(H^\blacktriangle)$ coincides with the algebra $SL_\infty = SL\hat{\otimes}K(\ell^2)$. Moreover, the quotient $E^\blacktriangle/K(H^\blacktriangle)$, and hence also the quotient $E/K(H)$ is isometrically isomorphic to the function algebra $C(\mathbb{M}_{SL},K(\ell^2))$. This isometry is such that an operator $E^\blacktriangle$ = (multipliation by) $x^\blacktriangle(x) = ((e_{jk}(x)))$ has the symbol $((e_{jk}(x))$ on both copies of S^1. For an operator E of the form (1.3) the symbol on $\mathbb{M}_{SL} = S^1_+ \cup S^1_-$ is given as

$$(1.27)\qquad \gamma_E(x) = E^\blacktriangle(x)\ ,\ x \in S^1_{\pm}\ ,\ E^\blacktriangle = ZEZ^{-1}\ .$$

Moreover, for an operator $A = a(x)E$, $E\in F$, $a\in CS(\mathbb{R})$ we get

$$(1.28)\qquad \gamma_A(x) = a(\pm\infty)E^\blacktriangle(x)\ ,\ x\in S^1_{\pm}\ .$$

We shall offer only a skeleton proof. For details cf.[CMe_1].

It may be shown that the generators V,(5.33) transform into terms of the form

$$(1.29)\qquad X_+E^{\blacktriangle +} + X_-E^{\blacktriangle -} + K\ ,\ X_\pm = 1/2(1\pm iH^\blacktriangle)\ ,$$

with the Z^{-1}-conjugates $E^\blacktriangle_\pm$ of the two terms $E_\pm \in F$ occurring in V,(5.33). Here it must be verified by a calculation that the Z^{-1}-conjugates of $\xi_\pm(x)$ equal the terms $X_\pm$, modulo a term $W_\pm$ such that $W_\pm E^\blacktriangle_\pm \in K(H)$. This calculation was prepared by prop's 1.6 and 1.7, but its details are left to the reader.

Let e^j be the unit vector of ℓ^2 with j-th component =1. From V,lemma 1.1 we get $K(H^\blacktriangle) \subset E^\blacktriangle$. Using prop.1.4 we then get all multiplication operators $a(x)\otimes(e^j\rangle\langle e^j)$ contained in $E^\blacktriangle$, where $j\in\mathbb{Z}$ and $a\in C^\infty(S^1)$ are arbitrary. Then also $H^\blacktriangle a(x)\otimes(e^j\rangle\langle e^j) \in E^\blacktriangle$, by prop.1.7. Now it follows that all operators $A\otimes(e^j\rangle\langle e^j)$, for $A \in SL$ are contained in $E^\blacktriangle$. Repeating the same conclusion with e^j replaced by e^j+e^l we find that also $A\otimes(e^j\rangle\langle e^l)$, for $A\in SL$, $j,l\in\mathbb{Z}$, is in $E^\blacktriangle$. Thus, letting K_N denote the set of all infinite matrices $C\in K(\ell^2)$ with components zero outside $|j|,|k|\le N$, we find that

$A\otimes K\in E^{\blacktriangle}$ for all $A\in SL$, $K\in K_N$. In other words, $SL\otimes K_N \subset E^{\blacktriangle}$. In [C_1], V,8 (or [BC_2]) we have shown that $SL_\infty = SL\hat{\otimes}K(\ell^2)$ is the closure of $SL\otimes(UK_N)$. Since $E^{\blacktriangle}$ is closed we get $E^{\blacktriangle}\supset SL_\infty$, and equality follows since the opposite inclusion is trivial.

Finally, the statements about the quotients are immediate consequences of an investigation for abstract algebras with symbol in [BC_2], sec.6. Q.E.D.

2. A C^*-algebra on a poly-cylinder.

Next we look at a Laplace comparison algebra with noncompact commutator on a poly-cylinder $\Omega=\mathbb{R}^{n'}\times B$. Here B denotes a compact Riemannian space of dimension n" with metric $d\rho^2=g_{jk}dx^j dx^k$. Accordingly, for the metric and the Laplace operator of Ω we get

$$(2.1)\quad ds^2= dt^2+ d\rho^2 \quad , \quad \Delta = \Delta_t + \Delta_x \quad , \quad \Delta_x=(\sqrt{g})^{-1}\partial_{x^j}\sqrt{g}g^{jk}\partial_{x^k} ,$$

where Δ_x is the Laplace operator on B . In (2.1) we are using the Euclidean metric $dt^2={dt^1}^2+\ldots+{dt^{n'}}^2$ of $\mathbb{R}^{n'}=\{t=(t^1,\ldots,t^{n'}):t_j\in\mathbb{R}\}$, and the Euclidean Laplace operator $\Delta_t=\sum\partial_{t^j}^2$. The summation convention often will be used from 1 to n", as will be clear from the context. We set n=n'+n", so that Ω is n-dimensional.

Let H be the Hilbert space $L^2(\Omega) = L^2(\Omega,dS)$, with the surface measure $dS=dS'dS''=\sqrt{g}dtdx$ of the metric (2.1). Let $C \subset L(H)$ be the smallest C^*-subalgebra containing the (5 types of)operators

$$(2.2)\, a\in A^\#_B,\ s_j(t)=t_j/\langle t\rangle\ ,\ \Lambda=(1-\Delta)^{-1/2},\ \partial_t^{\,j}\Lambda,\ D_x\Lambda,\ D_x\in\mathcal{D}^\#_B,\ j=1,..,n'.$$

Here $\langle t\rangle=(1+t^2)^{1/2}$; we write $A^\#_B=C^\infty(B)$ while $\mathcal{D}^\#_B$ denotes the collection of all C^∞-vector fields on B . Also Λ is the unique positive inverse square root of the (unique) self-adjoint realization $(1-\Delta)$ of the Laplace differential expression Δ . There is a unique such self-adjoint realization because Ω is a complete Riemannian space (cf.IV, cor.1.7). The functions $a\in A^\#_B$ and $s_j(t)$ in in (2.2) represent the corresponding multiplication operators on functions $u(t,x) \in H$. Correspondingly, we denote by $A^\#$ and $\mathcal{D}^\#$ the algebra finitely generated by C^∞_0 and the first two kinds of generators, and the left module of $A^\#$ spanned by $\{1, \partial_{t^j} , \mathcal{D}_x\}$, respectively. The operations $\partial_{t^j}\Lambda u$ and $D_x\Lambda u$ are well defined for $u\in\Lambda^{-1}C_0^\infty(\Omega)$, a dense subspace of H (cf.V,1) and have continuous extensions to H (cf. also (2.6), below).

We notice that $H = h_t\hat{\otimes}h_x$ is the topological tensor product of the Hilbert spaces $h_t = L^2(\mathbb{R}^{n'})$, and $h_x = L^2(B)$ (cf.[C_1],V,8). Let us write $L_x = L(h_x)$, $k_x = K(h_x)$, $L_t = L(h_t)$, $k_t = K(h_t)$. It may be observed that the topological tensor products $L_t\hat{\otimes}_* L_x = L_{tx}$, $k_t\hat{\otimes}k_x = K(H)$, $L_t\hat{\otimes}k_x = K_x$, $k_t\hat{\otimes}L_x = K_t$ are all well-defined C^*-subalgebras of $L(H)$, where $K(H)\subset K_x$, $K_t \subset L_{xt} \subset L(H)$ all are proper inclusions. In fact, K_x and K_t are proper closed two-sided ideals of L_{tx} , and, of course, $K(H)$ is a proper closed ideal of all the others. Evidently, L_{tx} (but not K_x or K_t) contains the identity operator I .

Note that we may write $H = L^2(\mathbb{R}^{n'},h_x)$ as the space of all functions over $\mathbb{R}^{n'}$ with values in $h_x = L^2(B)$ such that

$$\|u\|^2 = \int_{\mathbb{R}^{n'}} dt\|u(t,.)\|^2 < \infty , \tag{2.3}$$

by Fubini's theorem. Correspondingly, if $CB(\mathbb{R}^{n'},k_x)$ and $CB(\mathbb{R}^{n'},L_x)$ denote the classes of bounded norm continuous functions over $\mathbb{R}^{n'}$, taking values in k_x and L_x, respectively, then a function $\Phi\in CB(\mathbb{R}^{n'},L_x)$ has a natural interpretation as an operator in $L(H)$, defined by $u(t,.)\to\Phi(t)u(t)$. Moreover, this operator is in K_x whenever $\Phi\in CO(\mathbb{R},k_x)$ (cf.[C_1],V,8). (We indicate a proof of this fact in cor.2.4.) This establishes an isometric *-isomorphism of $CB(\mathbb{R};L_x)$ (as a Banach algebra with norm

$$\sup\{\|A(\tau)\| : \tau\in\mathbb{R}\} , \tag{2.4}$$

where $\|A(\tau)\|$ is the norm in L_x) into $L(H)$. In the following we will use this interpretation of functions as operators in $L(H)$.

In order to find the commutator ideal E of the unital C^*-algebra C with generators (2.2) we conjugate the generators with the Fourier transform in the t-direction. In detail we have

$$F_t u(\tau) = (2\pi)^{-n'/2}\int_{\mathbb{R}^{n'}} e^{-it\tau}u(t)dt , \quad \tau\in\mathbb{R}^{n'} , \tag{2.5}$$

which defines a unitary operator of h_t . By conjugation with F_t we, of course, mean conjugation with $F_t\otimes I_x$, where I_x denotes the identity operator. (However,we will write this as $F_t^{-1}AF_t$,for $A\in L(H)$.) First consider the F_t-conjugations of Λ, $D_{t^j}\Lambda$, $D_x\Lambda$:

(2.6) $\Lambda^{\sim}(\tau) = (\langle\tau\rangle^2 - \Delta_x)^{-1/2}$, $\tau_j\Lambda^{\sim}(\tau)$, $D_x\Lambda^{\sim}(\tau)$,

with $D_{t^j}=-i\partial_{t^j}$ and above interpretation of a function as operator.

Since B is compact we have $\Lambda^{\sim}(\tau) \in K(h_x)$, for all τ (cf.III, cor.3.9). Moreover, we even get $\Lambda^{\sim}(\tau) \in CO(\mathbb{R}^{n'},k_x)$, where $CO \subset CB$ denotes the class of functions with limit 0 at infinity. In fact, $\Lambda^{\sim}(\tau)$ even is analytic, in norm topology of $L(h_x)$, and we have $\|\Lambda^{\sim}(t)\| \leq \langle\tau\rangle^{-1}$ as well as $\|\Lambda^{\sim}(0)^{-1}\Lambda^{\sim}(\tau)\|\leq 1$, so that $\tau_j\Lambda^{\sim}(\tau)$, $D_x\Lambda^{\sim}(\tau)=(D_x\Lambda_x)(\Lambda_x^{-1}\Lambda^{\sim}(\tau))$ are functions in $CB(\mathbb{R}^{n'},L_x)\subset L(H)$, confirming the boundedness of the last two types (2.2). Also, writing $\Lambda_x=\Lambda^{\sim}(0)$, $T(\tau)=(1+\tau^2\Lambda_x^2)^{-1/2}$, the generators (2.2) correspond to

(2.7) $a(x)$, μ_j* , $\Lambda_xT(\tau)$, $(i\tau_j\Lambda_x)T(\tau)$, $(D_x\Lambda_x)T(\tau)$,

with $\mu_j=F_t^{-1}s_j$ and $\mu_t*=$ convolution in h_t .

<u>Proposition 2.1</u>. The operator functions $\Lambda_x^{\varepsilon}T(\tau)$ and $|\tau|^{1-\varepsilon}\Lambda_xT(\tau)$, for each fixed ε , $0\leq\varepsilon<1$, are in $CO(\mathbb{R},k_x)$.

<u>Proof</u>. Just note that $\Lambda^{\sim}=\Lambda_xT(\tau)\in CO(\mathbb{R},k_x)$, while $T(\tau)$ and $|\tau|\Lambda_xT(\tau)$ belong to CB . All operators are positive self-adjoint, so that $\Lambda_x^{\varepsilon}T(\tau) = \Lambda^{\varepsilon}T(\tau)^{1-\varepsilon} \in CO$, $|\tau|^{1-\varepsilon}\Lambda_xT(\tau) = \Lambda^{\varepsilon}(|\tau|\Lambda_xT(\tau))^{1-\varepsilon}\in CO$, as products of a bounded function and a function with limit 0, q.e.d.

Now we first will describe the algebra generated by the commutators of the F_t-conjugated generators (2.7) .

<u>Proposition 2.2</u>. All commutators of the F_t-conjugated generators and their adjoints are contained in the algebra

(2.8) $G^{\wedge} = CO(\mathbb{R}^{n'},K(h_x)) + K(H) \subset K_x$.

Here a function $C(\tau) \in CO(\mathbb{R}^{n'},K(h_x))$ must be interpreted as an operator in K_x in the manner described above.

<u>Proof</u>. We will use the resolvent integral technique used in V,3. Let the operators (2.7) be denoted by $G_1,\dots,G_5$, in the order listed (we write $t=t^j$, $\partial_t=\partial_{t^j}$, $\tau=\tau_j$, etc). Clearly $[G_1,G_2]=[G_3,G_4]=0$.

We get $(D_x\Lambda^{\sim})^*-D_x^*\Lambda^{\sim}=[\Lambda^{\sim},D_x^*]\in CO(\mathbb{R}^{n'},k_x)$, by (2.14) below, and adjoint invariance of $\mathcal{D}_B^{\#}$). All other generator(classe)s are self-adjoint. Hence the adjoint generators need no special attention.

For the commutators $[G_j,G_l]$,$j\neq 2$, we use the well known resolvent integral representation of $\Lambda^{\sim}=G_3$: Let $R(s)=(s+\langle\tau\rangle^2-\Delta_x)^{-1}$. Then we have

$$G_3=\Lambda^{\sim}(\tau)=\Lambda_x T(\tau)=(\langle\tau\rangle^2-\Delta_x)^{-1/2}=1/\pi\int_0^\infty R(s)ds/\sqrt{s}, \tag{2.9}$$

(cf. VI,(1.6)), with a norm convergent improper Riemann integral, in the algebra $L(h_x)$. We get

$$[G_3,G_5]=[\Lambda^{\sim},D_x]\Lambda^{\sim},\ [G_4,G_5]=[\Lambda^{\sim},D_x](\tau\Lambda^{\sim}),\ \text{where } \Lambda^{\sim},\tau\Lambda^{\sim}\in CB, \tag{2.10}$$

so that it suffices to show that $[\Lambda^{\sim},D_x]\in CO(\mathbb{R},k_x)$.

Similarly,

$$\begin{aligned}&[G_1,G_3]=[a_x,\Lambda^{\sim}],\ [G_1,G_4]=(\tau\Lambda^{\sim})(\Lambda^{\sim-1}[a_x,\Lambda^{\sim}]),\\ &[G_1,G_5]=p_x\Lambda^{\sim}+(D_x\Lambda^{\sim})(\Lambda^{\sim-1}[a_x,\Lambda^{\sim}]),\ \text{with } p_x=[a_x,D_x]\in C^\infty(B).\end{aligned} \tag{2.11}$$

Accordingly we also must show that $\Lambda^{\sim-1}[a_x,\Lambda^{\sim}]\in CO(\mathbb{R},k_x)$. Both these facts are consequences of prop.1.3, below. Before we discuss it we turn to the commutators $[G_2,G_1]$. There we find it practical to work without the Fourier transform (2.5), writing

$$\Lambda = 1/\pi\int_0^\infty S(r)dr/\sqrt{r}\ ,\ S(r) = (r+1-\Delta_t-\Delta_x)^{-1} . \tag{2.12}$$

Instead of 'diagonalizing the t-variable by using the Fourier transform' we will consider the x-variable diagonalized later on. Let $A_j=F_tG_jF_t^{-1}$ be the generators (2.2). Then we have

$$[A_2,A_4]=p_t\Lambda+(\partial_t\Lambda)V,\ [A_2,A_5]=(D_x\Lambda)V,\ V=\Lambda^{-1}[A_2,A_3],\ p_t\in C^\infty . \tag{2.13}$$

Note that $p_t\in CO(\mathbb{R}^{n'})$ hence $p_t\Lambda\in K(H)$, by a well known result (cf. III,thm.3.7). We claim that all 3 commutators $[A_2,A_l]$, $l=3,4,5$, are in $K(H)$. Again this is a trivial consequence of prop. 1.3 below, so that all of prop.1.2 has been reduced to prop.1.3.

<u>Proposition 2.3</u>. For ε with $0\le\varepsilon<1$ we have

$$\Lambda^{\sim-1-\varepsilon}[a_x,\Lambda^{\sim}]\in CO(\mathbb{R}^{n'},k_x)\ ,\ \Lambda^{\sim-\varepsilon}[D_x,\Lambda^{\sim}]\in CO(\mathbb{R}^{n'},k_x)\ , \tag{2.14}$$

$$\Lambda^{-1-\varepsilon}[s_j(t),\Lambda]\in K(H)\ ,\ s_j(t)=t_j/\langle t\rangle\ . \tag{2.15}$$

<u>Remark</u>. As a consequence of prop.2.3 the algebra C satisfies condition (m_4) of VII,3 as required later on (cf.thm.2.8).

For the proof of prop.2.3 we use (2.9) for

$$\Lambda^{\sim-1-\varepsilon}[G_1,G_3]=1/\pi\int_0^\infty\Lambda^{\sim-1-\varepsilon}R(s)L_1R(s)ds/\sqrt{s}\ ,\ L_1=[a(x),\Delta_x]. \tag{2.16}$$

The integrand in (2.16) is a norm-continuous (even analytic) function $F(s,\tau)$, with values in k_x. Indeed, one may write

$$(2.17)\qquad F(s,\tau) = (\Lambda^{\sim -1-\varepsilon}R(s))(L_1\Lambda_x)(\Lambda_x^{-1}R(s)/\sqrt{s}) ,$$

and use the estimates

$$(2.18)\quad \|\Lambda_x^{-\eta}R(\sigma)\| \le (1+s+\tau^2)^{\eta/2-1},\ \|\Lambda^{\sim -\eta}R(s)\| \le (1+s)^{\eta/2-1},\ 0\le\eta<2 ,$$

easily derived from the spectral decomposition of the self-adjoint operator $-\Delta_x \ge 0$ of h_x. Note that L_1 is a folpde on B independent of s,τ, so that the second factor in (2.17) is a constant in $L(h_x)$. Analyticity of the first and third factor is a consequence of analyticity of the resolvent R(s). These factors are $O((1+s)^{(\varepsilon-1)/2})$ and $O(s^{-1/2}(1+s)^{-1/2})$, respectively. Thus $F(s,\tau)=O((1+s)^{-1+\varepsilon/2}/\sqrt{s})$ uniformly for all $\tau\in\mathbb{R}^{n'}$. Also both factors are in k_x since B is compact, insuring the compactness of the resolvent R(s) of the Laplace operator Δ_x (III,thm.3.1). This implies existence of the improper Riemann integral (2.16) in norm convergence of $L(h_x)$ and uniformly so, for $\tau\in\mathbb{R}^{n'}$. Thus the integral is in $CB(\mathbb{R}^{n'},k_x)$ For more detail in such a proof cf. V,3).Moreover, since $0<\varepsilon<1$ is arbitrary, one may use this for $\varepsilon+\delta<1$, with some $\delta>0$. This gives a factor $\Lambda^{\sim\delta}\in CO$, whence the first (2.14) is CO, not only CB.

Similarly we may use (2.9) for

$$(2.19)\quad \Lambda^{\sim -\varepsilon}[D_x,\Lambda^\sim] = 1/\pi \int_0^\infty (\Lambda^{\sim -\varepsilon}\Lambda_x^{-1}R(s))(\Lambda_x[D_x,\Delta_x]\Lambda_x)(\Lambda_x^{-1}R(s))/\sqrt{s},$$

where the integral exists for analogous reason. (Now the second factor contains the second order operator $[D_x,\Delta_x]$ over B so that again the factor is a constant in $L(h_x)$.)

For (2.15) we use the other resolvent integral (2.12), for

$$(2.20)\quad \Lambda^{-1-\varepsilon}[s(t),\Lambda] = 1/\pi \int_0^\infty \Lambda^{-1-\varepsilon}S(r)M_1S(r)dr/\sqrt{r},\ M_1=[s(t),\Delta_t] .$$

Now, with respect to an orthonormal base of the Hilbert space h_x consisting of eigen-functions of the self-adjoint operator Δ_x, the operator S(r) is diagonalized, with respect to h_x, in the tensor product decomposition of H. Its diagonal components are

$$(2.21)\qquad S_j(r) = (r+\langle\lambda_j\rangle^2-\Delta_t)^{-1} ,$$

with the eigenvalues λ_j^2 of the positive self-adjoint operator Δ_x on B. In analogy to (2.18) we conclude that, with $\Lambda_t=(1-\Delta_t)^{-1/2}$,

(2.22) $\|\Lambda_t^{-\eta}S_j(r)\| \le (1+r+\lambda_j^2)^{\eta/2-1}$, $\|\Lambda_t^{j-\eta}S_j(r)\| \le (1+r)^{\eta/2-1}$,

for $j=1,2,\ldots$, and $\Lambda_t=(1-\Delta_t)^{-1/2}$, $\Lambda_t^j=(1+\lambda_j^2+\Delta_t)^{-1/2}$.

Notice that the entire relation (2.20) is 'x-diagonalized', i.e., decomposes into a set of countably many relations, involving Λ_t^j and S_j instead of Λ , $j=1,2,\ldots,$. Again one may write the integrand $F_j(r)$ as a product of three factors:

(2.23) $$F_j(r) = (\Lambda_t^{j-1-\varepsilon}S_j(r))(M_1\Lambda_t)(\Lambda_t^{-1}S_j(r)/\sqrt{r}) ,$$

where the second term is a constant in $L(H)$. For the first and third term we get estimates $O((1+r)^{(\varepsilon-1)/2})$ and $O((1+r+\lambda_j^2)^{-1/2}/\sqrt{r})$ $= O(\langle\lambda_j\rangle^{-\delta}(1+r)^{(\delta-1)/2}r^{-1/2})$, for any $0<\delta<1$. Thus

(2.24) $$\|F_j(s)\| = O(\langle\lambda_j\rangle^{-\delta}r^{-1/2}(1+r)^{(\varepsilon+\delta)/2-1}) .$$

The right hand side is integrable, as long as $\varepsilon+\delta<1$. The integral $\int F_j(r)dr = C_j$ is an operator in $K(h_t)$, and we get

(2.25) $$\|C_j\| = O(\langle\lambda_j\rangle^{-\delta}) .$$

Accordingly the operator (2.20) corresponds to a diagonal matrix $((C_j\delta_{kl}))_{k,l=1,2,\ldots}$. with diagonal components converging to 0 in norm. This indeed implies that (2.20) belongs to $K(H)$. (It is limit in $L(H)$ of the sequence of diagonal matrices Tk obtained by setting all C_j , $j>k$, equal to zero, while the matrices T_k are in $K(h_t)\boxtimes F(h_x) \subset K(h_t)\boxtimes K(h_x) \subset K(H)$, with the class $F(h_x)$ of bounded operators of finite rank over h_x .) This completes the proof of prop.1.3.

Corollary 2.4. The C^*-algebra $G^\wedge$ and its F_t^{-1}-conjugate G are subalgebras of $K_x = L_t\hat{\boxtimes} k_x$.

Proof. It is sufficient to show that $CO(\mathbb{R}^{n'},k_x) \subset K_x$. For any orthonormal basis $\{\phi_j:j=1,2,\ldots\}$, and the orthogonal projection P_n onto the span of $\{\phi_1,\ldots,\phi_n\}$ we get uniform convergence $P_nC(\tau)P_n \to C(\tau)$, $\tau\in\mathbb{R}$,for every $C(\tau) \in CO(\mathbb{R}^{n'},K(h_x))$. But the operator $P_nC(\tau)P_n$ is a finite sum of operators $(\phi_j)\langle\phi_l)c_{jl}(\tau)$, $c_{jl}(\tau)$ $\in CO(\mathbb{R})$. Thus $P_nC(\tau)P_n \in k_x$,and the limit $C(\tau)$ as well, q.e.d.

Proposition.2.5. The commutator ideal GO of the C^*-subalgebra of $L(H)$ generated by the operators of the form G_1,G_3,G_5 contains the algebra $CO(\mathbb{R}^{n'},k_x)$.

Proof. For a moment consider only the C^*-algebra B generated by G_1,G_3, and G_5. By virtue of the isometric isomorphism mentionned initially in this section, the generators of B belong to the function algebra $CO(\mathbb{R}^{n'},L_x)$. Hence the algebra B may be interprete as a subalgebra of $CO(\mathbb{R}^{n'},L_x)$. For a fixed $\tau=\tau_0\in\mathbb{R}^{n'}$ the values $B_{\tau_0}=\{A(\tau_0):A(\tau)\in B\}$ form a *-subalgebra of L_x. It is clear that B_{τ_0} is a *-subalgebra of the C^*-subalgebra C_{τ_0} of L_x generated by the values of the functions G_j, $j=1,3,5$ at τ_0, i.e., by the operators

(2.26) $$a_x\ ,\ \Lambda^{\sim}(\tau_0)=(-\Delta_x+\langle\tau_0\rangle^2)^{-1/2}\ ,\ D_x\Lambda^{\sim}(\tau_0)\ ,$$

where a_x and D_x run through all the functions (folpdes) over B. Moreover, C_{τ_0} evidently is the closure of B_{τ_0}. Also C_{τ_0} is just the minimal comparison algebra ,in the sense of V,1 generated by the triple $\{B,-\Delta_x+\langle\tau_0\rangle^2,dS\}$, on the compact manifold B. By V,lemma 1.1, it follows that C_{τ_0} and even its commutator ideal contain all of k_x. But commutators in C_{τ_0} are compact, since B is compact. Therefore the commutator ideal of C_{τ_0} equals k_x. Since that commutator ideal is the closure of the commutator ideal $E^0_{\tau_0}$ of the finitely generated algebra, we must have $E^0_{\tau_0}$ dense in k_x. On the other hand $E^0_{\tau_0}$ clearly is contained in the commutator ideal of the algebra B_{τ_0}, and even in the localization GO_{τ_0} at τ_0 of the commutator ideal GO. Thus we conclude that the algebra GO_τ of 'values' of GO at τ is dense in k_x, for all $\tau\in\mathbb{R}^{n'}$.

We also find that the algebra B contains $\Lambda^{\sim}(\tau)$, hence also contains every $f(\tau,\Lambda_x)$, for a general $f\in CO(\mathbb{R}^{n'}\times[0,1])$, by the spectral theorem and the Stone-Weierstrass theorem. Hence B contains all functions $\psi(\tau)E_\lambda$, with $\psi\in CO(\mathbb{R}^{n'})$ and the projection operators E_λ of the spectral family of Λ_x^{-1}. Note that E_λ are of finite rank, and that $E^N\to 1$, strongly, as $N\to\infty$. Since GO is an ideal of B, it follows that GO contains

(2.27) $$GO_N=\{\psi(\tau)E_NA(\tau)E_N : A(\tau)\in GO,\ \psi(\tau)\in CO(\mathbb{R}^n)\}\ .$$

But GO_N is a self-adjoint algebra of (finite) $j_N \times j_N$-matrix-valued functions, separating points in the following sense. For every $\tau_1, \tau_2 \in \mathbb{R}^{n'}$, $\varepsilon>0$, and $j_N \times j_N$-matrix P there exists $K(\tau) \in GO_N$ such that $K(\tau_2)=0$ and $|K(\tau_1)-P|<\varepsilon$ (This follows from the above). By the matrix version of the Stone Weierstrass theorem this implies that $GO_N = CO(\mathbb{R}^{n'}, L(\mathbb{C}^{j_N}))$. Since this holds for all N we find that GO contains all these matrix algebras. But for a general $A(\tau) \in CO(\mathbb{R}^{n'}, k_x)$ we get $A_N(\tau)=E_N A(\tau) E_N \in CO(\mathbb{R}^{n'}, L(\mathbb{C}^{j_N}))$. Also $A_N(\tau)-A(\tau) \to 0$, in $CO(\mathbb{R}^{n'}, k_x)$ (with the norm (2.4)), since $A(\tau)$ is compact and E_N converges strongly to 1 . Since GO is closed we conclude that $A(\tau) \in GO$, so that indeed $CO(\mathbb{R}, k_x) \subset GO$, q.e.d.

Returning to our task of describing the commutator ideal E of the cylinder algebra C we conclude from prop.2.5 and V,lemma 1.1 that E contains the C^*-algebras $GO^\vee = F_t GO F_t^{-1}$ and $K(H)$. It is convenient again to work with the F_t-conjugated ideal $E^\wedge = F_t^{-1} E F_t$, containing the sum $GO+K(H)=SO$ (which in turn contains all the commutators of the generators G_j, $j=1,\dots,5$) .

Notice that SO is invariant under left and right multiplication with the (functions in $CB(\mathbb{R}, k_x)$) G_1, G_3, G_4, G_5, but not under multiplication with G_2 . Accordingly $E^\wedge$ must be properly larger than SO.

Specifically $G_2 = s_j(D_t)$ is a singular convolution operator with Cauchy-type singular integral and kernel $\mu_j = s_j^\vee$, so that a product $K(\tau)G_2$, for $K(\tau) \in CO(\mathbb{R}^{n'}, k_x)$, appears as an infinite matrix of singular integral operators on $\mathbb{R}^{n'}$, if we introduce some orthonormal base of h_x.

<u>Theorem 2.6</u>. Let SQ be the C^*-algebra of singular integral operators over h_t generated by the multiplications in $CO(\mathbb{R}^{n'})$, and the operators $a(M)s_j(D)$, $a \in CO(\mathbb{R}^{n'})$, $j=1,\dots,n'$. Then the ideal $E^\wedge$ coincides with the topological tensor product $SQ_\infty = SQ \hat{\otimes} k_x$.

<u>Proof</u>. Notice that SQ coincides with the minimal comparison algebra of the triple $\{\mathbb{R}^{n'}, dt, 1-\Delta_t\}$ (cf. ch.VI). Thus it contains the compact ideal k_t of $L(h_t)$, and SQ_∞ contains $k_t \hat{\otimes} k_x = K(H)$. Therefore it is trivial from the above that SQ_∞ contains all the commutators $[G_j, G_l]$, $j,l=1,\dots,5$ and that it is a closed *-ideal of C . Hence we have $E^\wedge \subset SQ_\infty$. To show equality we introduce a fixed orthonormal base $\phi_1, \phi_2, \dots$ of the space h_x and first consider $K(\tau) \in GO$

such that $K(\tau)$ takes span $\{\phi_1,\dots,\phi_N\}$ into itself and its orthogonal complement to 0, for all τ. Thus the infinite matrix vanishes outside its first N rows and columns. Let CO_N denote the subalgebra of $CO(\mathbb{R}^{n'},k_x)$ of all such finite matrices, for a given fixed N. Clearly CO_N is isometrically isomorphic to $CO(\mathbb{R}^{n'},L(\mathbb{C}^N))$.

Now we observe that $K(\tau)$ and $L(\tau)s(D_t)$, with $K,L\in CO_N$, belong to $E^\wedge$, and generate the algebra $SQ_N = SQ\otimes L(\mathbb{C}^N)$, for each $N=1,2,\dots$. Also we again find that SQ_∞ is the norm closure of $\cup SQ_N$. Therefore indeed $SQ_\infty\subset E^\wedge$, q.e.d.

We now come to our main task: The description of the symbol chain of the cylinder algebra C . First let us look at the ideal quotient $E/K(H)$. In that respect we observe that the algebra SQ is a subalgebra of the algebra SI of singular integral operators on $\mathbb{R}^{n'}$ generated by $S_j=s_j(D_t)=(\mu_j*)$, $j=1,\dots,n'$, and the multiplications with functions in $C(\mathbb{B}^{n'})$, with the 'ball compactification' $\mathbb{B}^{n'}$ of $\mathbb{R}^{n'}$, having one infinite point in each direction $\infty\cdot t_0$, $|t_0|=1$ (cf.[C_1],IV,1,problems). The special comparison algebra $SI\subset L(h_t)$ has symbol space

$$(2.28)\qquad \mathbb{M}_I = \{(t,\tau)\in C(\mathbb{B}^{n'}\times\mathbb{B}^{n'}) : |t|+|\tau|=\infty\} ,$$

(cf. ex.(A) of V,4, or [CHe_1]). The subalgebra SQ of SI consists precisely of all operators in SI with symbol vanishing at $|\tau|=\infty$, as follows from the Stone-Weierstrass theorem, looking at the symhols of the generators. (Note that the generators are written as multiplications by functions of the variable τ and convolutions by functions of τ as well, since we consider the F_t-conjugated ideal $E^\wedge$. Accordingly we have t and τ reversed, compared to the normal notation for space and momentum coordinates.) It follows that SQ/k_t is isometrically isomorphic to the function algebra $CO(\mathbb{E})$ with the locally compact space
compact space

$$(2.29)\qquad \mathbb{E} = \mathbb{M}_Q = \{(t,\tau)\in \mathbb{M}_I : |t|=\infty , |\tau|<\infty\} = \partial\mathbb{B}^{n'}\times\mathbb{R}^{n'} .$$

In case of $n'=1$ this space is a disjoint union of the two sets $\{\infty\}\times\mathbb{R}=\mathbb{E}^-$ and $\{-\infty\}\times\mathbb{R}=\mathbb{E}^+$. Both $\mathbb{E}^\pm$ are copies of $\mathbb{R}$, with the variable τ running over $\mathbb{R}$. In the general case $n'>1$ the space $\mathbb{E}$ is connected, and is a product of the infinite sphere $\partial B^{n'}=\mathbb{B}^{n'}\setminus\mathbb{R}^{n'}$ with $\mathbb{R}^n$.

Clearly this also is just the wave front space $\mathbb{R}^{n'}$, but with t and τ interchanged. We have proven the following result:

Theorem 2.7. The quotient algebra $E/K(H)$ is isometrically isomorphic to the function algebra $CO(\mathbb{E},k_x)$, so that E is a C^*-algebra with (compact operator valued) symbol, with symbol space $\mathbb{E}$. In the special case $n'=1$ we have

$$(2.30)\qquad CO(\mathbb{E},k_x) = CO(\mathbb{E}^-,k_x) \oplus CO(\mathbb{E}^+,k_x) .$$

The symbols of the generators of E are given as compact operator valued functions of (t,τ), for $t\in\partial\mathbb{B}^{n'}$ (i.e., $t=\infty\cdot t_0$, $t_0\in\mathbb{R}^{n'}$, $|t_0|=1$) and $\tau\in\mathbb{R}^{n'}$, as follows:

Let A_j, $j=1,\dots,5$, be the generators (2.2) of C, in the order listed (so that G_j are their Fourier transforms). Then $[A_1,A_2]=0=[A_3,A_4]$. The symbols $\gamma_{[A_3,A_5]}$, $\gamma_{[A_4,A_5]}$ are independent of t, as $\tau\in\partial\mathbb{B}^{n'}$, the value given by the terms of (2.10), respectively, where $\Lambda^\sim(\tau)=\Lambda_x T(\tau)$, while the commutator $[\Lambda^\sim,D_x]$ is obtained from the resolvent integral from (2.19). Similarly $\gamma_{[A_1,A_j]}$, $j=3,4,5$, are independent of t, as $|t|=\infty$, and, for $\tau\in\mathbb{R}^{n'}$, their values are given by (2.11) and the resolvent integral (2.16).

Also, for $j\neq 2$, the symbol of a product $A_j[A_k,A_l]$ or $[A_k,A_l]A_j$ is obtained by multiplying $\gamma_{[A_k,A_l]}$ with the corresponding function G_j of (2.6) , from the left or right, respectively. Also, for $j=2$, the symbols of these products equal the product of $\gamma_{[A_k,A_l]}(\tau)$ with the value of the function $s(t)$ (extended continuously to $\partial\mathbb{B}^{n'}$) at t. More generally, for every function $b\in C(\mathbb{B}^{n'})$ the operator $b(t)[A_k,A_l] \in E$ has the symbol

$$(2.31)\qquad \gamma = b(t)\gamma_{[A_k,A_l]}(\tau) \ , \ k,l\neq 2 .$$

Next we turn to the discussion of the quotient C/E , i.e., of the symbol and symbol space of C .

Theorem 2.8. The C^*-algebra C/E is isometrically isomorphic to the algebra $C(\mathbb{M})$ of continuous complex-valued functions over a compact space $\mathbb{M}$, called the symbol space of C. Here $\mathbb{M}$ is (homeomorphic to) the bundle of cospheres with infinite radius of the compactified poly-cylinder $\mathbb{B}^{n'}\times B$ considered as a compact C^∞-manifold with boundary (i.e., the product $B^{n'}\times B$ of the unit ball $B^{n'}=\{t\in\mathbb{R}^{n'}:|t|=1\}$ in $\mathbb{R}^{n'}$ with the compact manifold B).

Let $A_1,\dots,A_5$ be the generators (2.2) again. Then the C-symbols $\sigma_{A_j}=\sigma_{A_j}(t,x,\tau,\xi)$, i.e., the functions in $C(\mathbb{M})$, associated to A_j by the above isomorphism, are given as explicit functions of (t,x,τ,ξ) as follows:

$$(2.32)\qquad \sigma_{A_1} = a(x)\ ,\ \sigma_{A_2} = t/\langle t\rangle\ ,\ \sigma_{A_3} = 0\ .$$

$$\sigma_{A_4} = i\tau/(\tau^2+|\xi|^2)^{1/2}\ ,\ \sigma_{A_5} = (b^j\xi_j)/(\tau^2+|\xi|^2)^{1/2}\ ,$$

with

$$(2.33)\qquad |\xi| = (g^{jk}(x)\xi_j\xi_k)^{1/2}\ ,\ D_x = b^j(x)\partial_{x^j}+p(x)\ ,$$

in local coordinates of B, where $t\in\mathbb{B}^{n'}$, $x\in B$, while $(\tau,\xi)\in S^\infty_{t,x}$, the co-sphere at (t,x) with infinite radius. (Actually the last two symbols are the limits of the full symbol quotients

$$(2.34)\qquad \tau/(1+\tau^2+|\xi|^2)^{1/2}\ ,\ (b^j(x)\xi_j+p(x))/(1+\tau^2+|\xi|^2)^{1/2}\ ,$$

as (τ,ξ) is replaced by $(\rho\tau,\rho\xi)$, and $\rho\to\infty$).

Proof. We may just apply the general results of ch.VII, first verifying their assumptions. Let us shortly summarize these facts. First of all every symbol space of a comparison algebra contains the wave front space $\mathbb{W}$, normally identified with the bundle of unit spheres in the cotangent space $T^*\Omega$ of the underlying manifold Ω (cf.VI,thm.1.5). The space $\mathbb{W}$ is an open subset of $\mathbb{M}$. It precisely coincides with the set of points (x,t,τ,ξ), with $|t|<\infty$. Essentially this follows whenever C can be shown to contain all C_0^∞-functions and all operators $D\Lambda$, for a general first order differential expression with C^∞-cofficients and compact support.

In order to study the points at $|t|=\infty$ we require the compactification $\mathbb{P}^*\Omega$ of $T^*\Omega$ induced by the formal symbol quotients (2.34) together with the functions $a(x)$, $a\in A^\#_B$, and $t/\langle t\rangle$, and $(1+\tau^2+|\xi|^2)^{-1/2}$ (in other words, by the formal symbols of the 5 types of generators (2.2)). (That is, $\mathbb{P}^*\Omega$ is defined as the smallest compactification of $T^*\Omega$ onto which all above functions can be continuously extended.) It is readily verified that $\mathbb{P}^*\Omega$ is given as the compactification of $T^*\Omega$ obtained by adding the infinite sphere $\{\infty(\tau,\xi) : |\tau|^2+|\xi|^2=1\}$ to each fiber $T^*_{t,x}$ of the cotangent bundle $T^*\Omega^c$, of the compactification $\Omega^c=\mathbb{B}^n\times B$ of Ω. In other words, $\mathbb{P}^*\Omega$ is the disjoint union of all balls $\{(t,x)\}\times\mathbb{B}^n$ of infinite radius, as $(t,x)\in\Omega^c$.

Moreover, Ω^c coincides with the compactification $M_{A^\#}$ of Ω defined by the functions $a(x)$, $t/\langle t\rangle \in C^\infty(\Omega)$, introduced in ch.VI, and the algebra C satisfies conditions (m_1'), and (m_j), $j=2,3,4,5,7$. (We noted before that (m_4) is implied by prop.2.3. Cdn.(m_5) is trivial: We have $\Lambda\in C$. Cdn.(m_1'), first used by McOwen, requires that the functions

$$(2.35) \qquad |p(x)|^2 \text{ and } g_{jk}(x)\bar{b}^j(x)b^k(x) \ ,$$

are in $A^\#$, for every $D_x=b^j(x)\partial_{x^j}+p(x)\in D^\#$. This again is trivially true, since $A^\#$ contains all of $C^\infty(B)$. Furthermore the conditions (m_j) involve some separation conditions which can be satisfied by enlarging $A^\#$ and $D^\#$ in such a way that the generated algebra C remains the same. Details are left to the reader.

As a consequence we may apply VII, thm.3.6. The conclusion is that the symbol space $\mathbb{M}$ of C is a compact subset of the boundary $\partial\mathbb{P}^*\Omega = \mathbb{P}^*\Omega\setminus T^*\Omega$ of our compactification $\mathbb{P}^*\Omega$, containing the wave front space $\mathbb{W}$, i.e., the bundle of cospheres of infinite radius over Ω.

Since $\mathbb{M}$ is compact, it must contain all points of the infinite cosphere bundle over $\partial\Omega^c$ as well. Thus it remains to be shown that no other point of $\partial\mathbb{P}^*\Omega$ is contained in $\mathbb{M}$. In particular none of the points $|t|=\infty$, $\tau^2+|\xi|^2<\infty$ can be contained in $\mathbb{M}$.

This, on the other hand, is a consequence of VII,thm.4.2. To indicate at least the idea, we find that $\Lambda\in E$, for our present algebra, while the formal symbol of Λ is $(1+\tau^2+|\xi|^2)^{-1/2}$, i.e., is $\neq 0$ for the latter type of points. Hence such points can not be in $\mathbb{M}$, since the symbol of $\Lambda\in E$ must vanish, while VII, thm.3.6 implies that symbol and formal symbol coincide for the points of $\partial P^*\Omega$ which are in $\mathbb{M}$.

This completes the proof of thm.2.8.

3. Algebra surgery.

Let us assume cdn.(w) (at least) , in this section.

We already started discussing the relation between the minimal comparison algebras of different triples, although so far we only allowed a change of H and $d\mu$ over a given compact set $K\subset\Omega$.

In this section we will look at the comparison algebras of

triples over two different manifolds Ω_1 and Ω_2 . Suppose two triples $\{\Omega_1, d\mu_1, H_1\}$, and $\{\Omega_2, d\mu_2, H_2\}$ are given, and assume that an open subdomain U_1 of Ω_1 is diffeomorphic to a subdomain $U_2 \subset \Omega_2$, in such a way that the two triples correspond to each other over $U_1 \sim U_2$. In order to have simple notations we will regard the two sets U_j identified by the diffeomorphism.

This will bring a configuration as sketched in fig.3.1. We are given a manifold U , and two different extensions Ω_1 and Ω_2 of U. The two triples coincide over U . That is, we have

$$d\mu_1|U = d\mu_2|U \ , \ H_1|U = H_2|U.$$

Also we will assume that the boundary ∂U of U is compact, although U itself may be noncompact.

The last assumption may be removed, at the expense of additional conditions at the infinite ends of ∂U.

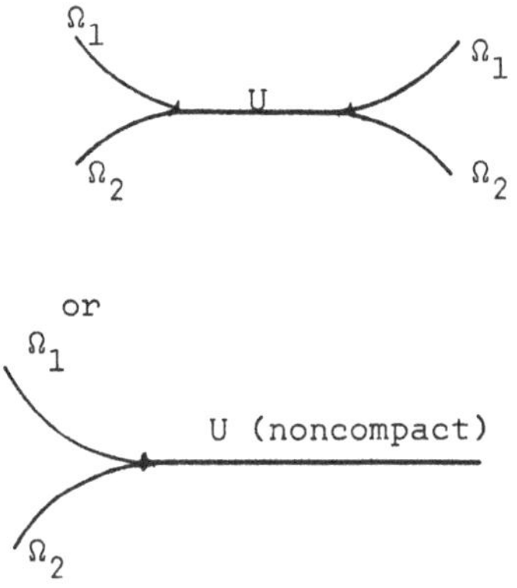

Fig.3.1

For a discussion of manifolds with polycylindrical ends involving such noncompact cuts cf.[CDg],sec.5.

We shall discover a 'common part' of the two comparison algebras $C_j = C(A_j^\#, \mathcal{D}_j^\#)$, induced by classes $A_j^\#$, $\mathcal{D}_j^\#$ over Ω_j , j=1,2, whenever the classes of restrictions of functions (folpde's) to U coincide, i.e.,

$$(3.1) \qquad A_U^\# = A_1^\#|U = A_2^\#|U \ , \ \mathcal{D}_U^\# = \mathcal{D}_1^\#|U = \mathcal{D}_2^\#|U \ .$$

We introduce the Hilbert spaces $H_1 = L^2(\Omega_1, d\mu_1)$, $H_2 = L^2(\Omega_2, d\mu_2)$, and the space $H = L^2(U, d\mu_j)$. Any $C_0^\infty(U)$ function will also be considered a $C_0^\infty(\Omega_j)$-function (with support in U), without change in notation. Similarly the functions $u \in H_1$ with support in U will also be considered functions in H or in H_2, and vice versa. Also, for $A \in L(H_1)$ and $\phi \in C_0^\infty(U)$ we may regard $A\phi$ as an operator in $L(H_2, H_1)$ and similar.

Clearly we will get corresponding maximal ideal spaces $M_{A_1^\#}$ and $M_{A_2^\#}$, and corresponding associate dual maps $\iota_1 : \mathbb{M}_1 \to M_{A_1^\#}$ and $\iota_2 : \mathbb{M}_2 \to M_{A_2^\#}$. Note that the restriction to U of an $a \in A^\#$ induces an algebra homomorphism $C(M_{A_1^\#}) \to C(M_{A_U^\#})$ which identifies the closure of $U \cup \partial U$ in $M_{A_1^\#}$ with the space $M_{A_U^\#}$, by virtue of its (injective) associated dual map. Similarly for $M_{A_2^\#}$, so that $M_{A_U^\#}$ is found as a closed subspace of both spaces $M_{A_1^\#}$ and $M_{A_2^\#}$, i.e., as the closure of $U \cup \partial U$.

Note that the common restriction

$$(3.2)\qquad \{U,d\mu,H\}\ ,\ d\mu = d\mu_j|U\ ,\quad H = H_j|U\ ,\ j=1,2\ ,$$

represents a comparison triple of its own. As we know, this triple cannot satisfy (w), unless ∂U is void. However, as long as we use sets $A^\#$, $\mathcal{D}^\#$ of functions and folpde's vanishing near ∂U, we may use the techniques of VI,1 and V,2 to achieve (w) or more by modifying the potential near $\partial\Omega$.

Now, for the remainder of sec.3, we assume the following separation condition.

<u>Condition</u> $(m_3)^U$: For every $a \in A^\#$ and $D \in \mathcal{D}^\#$ with $a=0$ near ∂U and $D=0$ near ∂U we have $\chi_U^j a \in A^\#$, $\chi_U^j D \in \mathcal{D}^\#$, with the characteristic function χ_U^j of $U \subset \Omega_j$.

Notice that $(m_3)^U$ is always true if $U \cup \partial U$ is compact, and that $(m_3)^U$ follows trivially from (m_3).

Accordingly we now define classes $A^\#$ and $\mathcal{D}^\#$ on U by setting

$$(3.3)\qquad A^\#_{U,0} = \{a \in A^\#_U : a=0 \text{ near } \partial U\}\ ,\ \mathcal{D}^\#_{U,0} = \{D \in \mathcal{D}^\#_U : D=0 \text{ near } \partial U\}\ ,$$

and then consider the comparison algebra $C = C(A^\#_{U,0}, \mathcal{D}^\#_{U,0}) \subset L(H)$.

We notice that $A^\#_{U,0}$ in general does not contain the constant functions. Therefore this is one of the exceptions to the general rules of V,1. In general C will not have a unit (except only in case where $\partial U = \emptyset$). However, as a consequence of $(m_3)^U$ it follows that $A^\#_{U,0}$ satisfies (a_1') of V,1. Also, we conclude that, for every $a \in A^\#_{U,0}$, $D \in \mathcal{D}^\#_{U,0}$, the trivial (i.e.=0) extension to Ω_j is contained in $A^\#_j$, or $\mathcal{D}^\#_j$, respectively.

We claim that C is the 'common part' of C_1 and C_2, in the following sense.

<u>Theorem 3.1</u>. Let $\mathcal{V}_j \subset C_j$ denote the closed two-sided ideal of all $A \in C_j$ with symbol σ_A having support in the set $\iota_j^{-1}(M_{A_U^\#})$. Then we have

$$(3.4)\qquad \mathcal{V}_j = C + E_j\ ,\quad C = \chi_U^j \mathcal{V}_j \chi_U^j\ ,$$

with the commutator ideal E_j of C_j , and the characteristic function χ_U^j of the subdomain $U \subset \Omega_j$. Here in the first relation (3.4) an operator $A \in C$ must be interpreted as $A\chi_U^j \in L(H_j)$, while in the second relation (3.4) we mean the restriction to its invariant subspace H of $\chi_U^j B \chi_U^j$, for $B \in V_j$.

Proof. First we notice that a statement like VI,prop.2.3 is valid, with exactly the same proof:

Proposition 3.2. The ideal $V_j \subset C_j$ is generated by E_j and all a, $D\Lambda_j$ with $a \in A^{\#}_{U,0}$, $D \in \mathcal{D}^{\#}_{U,0}$ (extended zero outside of U).

After this prop. we must use prop.3.3, below. From (3.6) it follows that $C^0\chi_U^j \subset V_j$, for the finitely generated algebra C^0 corresponding to C . Taking closure we even get $C\chi_U^j \subset V_j$, using that

$$(3.5) \qquad \|A\chi_U^j\|_{L(H_j)} = \|A\|_{L(H)} , \text{ for all } A \in L(H) .$$

Accordingly, $C + E_j \subset V_j$. Vice versa, for $A \in V_j$ let $A_k \in C_j^0$ be an approximating sequence, $\|A - A_k\|_{L(H_j)} \to 0$, where A_k is a finite sum of finite products of the generators of prop.3.2. Using prop.3.3 again we substitute each $D\Lambda_j$ in the products of A_k by $D\Lambda'$, and obtain an operator $A_k^0 \in C = C\chi_U^j$ with $A_k = A_0^k + K_k$, $K_k \in E_j$.

Therefore $\|A - A_k^0\chi_U^j - K_k\| \to 0$, $k \to \infty$. Applying χ_U^j from both sides we conclude that $A_k^0 + \chi_U^j K_k \chi_U^j \to \chi_U^j A \chi_U^j$ in $L(H)$, where $A^0 = \chi_U^j A \chi_U^j$ and $K_k^0 = \chi_U^j K \chi_U^j$ now stand for the restrictions of these operators to their invariant subspace H . We claim that $K_k^0 \in E$. Indeed, let $\gamma \in C_0^\infty(\Omega_j)$, $\gamma = 1$ near ∂U. From VII,thm.1.5 we know that γK_k^0, $K_k^0\gamma \in K(H_j)$. Thus, with $\delta = 1-\gamma$, $K_k^0 = \delta K_k^0 \delta + C_k$, $C_k \in K(H_j)$. Now $K_k^0 \in E$ follows from thm.3.4, below, and from the fact that $\chi_U^j K(H_j)\chi_U^j \subset K(H)$, since χ_U^j acts as a projection, hence a contractive map $H_j \to H$. In the limit $k \to \infty$ we get $A_0 = \chi_U^j A \chi_U^j \in C \subset C_j$, and the second (3.4) follows.

Finally consider $A - A_0 = K \in C_j$. For $\omega \in A_j^{\#}$ with supp $\omega \in \Omega_j \setminus \partial U$ we get $\omega K \omega = \omega A \omega - (\omega\chi_U^j)A(\omega\chi_U^j)$. Using that $A \in V_j$ has symbol with support in $U \cup \partial U$ we conclude that $\sigma_{\omega K \omega} = (\omega \circ \iota_j)^2 \sigma_K = 0$. Since this holds for all such ω we find that $\sigma_K(m) = 0$ for all $m \in \iota_j^{-1}(M_{A^{\#}} \setminus \partial U)$. However, the set $\iota^{-1}(\partial U) \subset \mathbb{M}_j$ is nowhere dense in $\mathbb{M}$, since this is just the co-sphere bundle over $\partial\Omega$. Hence the continuous function $\sigma_K:\mathbb{M}\to\mathbb{C}$, zero in $\iota^{-1}(M_{A^\#})$, must vanish identically. This proves that $A-A_0 \in E_j$, and completes the proof of thm.3.1.

The existence of the potential q' , in prop.3.3, below, is a consequence of IV,thm.5.1.

Proposition 3.3. Let $H'=H+q'$ satisfy (w), where $q'\geq 0, q'=0$ outside a given neighbourhood N' of ∂U , and let $\Lambda'=H'^{-1/2}$. Then we have

$$(3.6)\qquad D\Lambda_j - D\Lambda'\chi_U^j \in K(H_j) \subset E_j,\ j = 1,2,$$

whenever supp $D \subset U\setminus N'$.

The proof will again use resolvent integrals. Let $R(\lambda) =(H_j+\lambda)^{-1}$, $R'(\lambda) = (H'+\lambda)^{-1}$. For $u \in C_0^\infty(\Omega_j)$ write

$$(3.7)\qquad Iu = D\Lambda_j u - D\Lambda'\chi_U^j u = D(\Lambda_j-\Lambda')\chi u + Cu\ ,\ C \in K(H_j)\ ,$$

where V,lemma 3.3 was used, as well as the fact that $K(H)\chi_U^j\subset K(H_j)$ Also we wrote $D=D\chi$ with $\chi \in C^\infty(U)$, $0\leq\chi\leq 1$ $\chi=1$ in supp D, =0 in $N \supset N'^{\text{clos}}$. Then

$$(3.8)\qquad (I-C)u = \pi^{-1}\int_0^\infty \lambda^{-1/2}d\lambda D\chi(R(\lambda)-R'(\lambda))\chi u\ ,$$

where we write

$$(3.9)\qquad \chi(R-R')\chi = [\chi,R]\chi - (\chi R'-R\chi)\chi\ .$$

For $v \in \text{im}(H_0'+\lambda)$ we get

$$(3.10)\qquad \chi R'v - R\chi v = R(H+\lambda)\chi R'v-R\chi(H'+\lambda)R'v = R[H,\chi]Rv\ .$$

In view of (w) for H' we may extend continuously to H . Similarly for $[\chi,H]$, using (w) for H_j . Accordingly we get

$$(3.11)\qquad \chi(R-R')\chi = R[H,\chi](R-R')\chi\ ,$$

which may be used in (3.8) for

$$(3.12)\qquad I = C + \pi^{-1}\int_0^\infty \lambda^{-1/2}d\lambda DR(\lambda)[H,\chi](R(\lambda)-R'(\lambda))\chi\ .$$

The integral in (3.12) has a compact integrand, since the folpde $[H,\chi]$ has compact support. Also we use an estimate like V,(3.9) again, to show that we have a norm convergent improper Riemann integral. This shows that $I \in K(H)$, q.e.d.

Next let us compare the commutator ideals.

Theorem 3.4. The commutator ideal E of C is given as algebraic sum

(3.13) $$E = \chi E_j \chi + K(H) ,$$

with any cut-off function $\chi \in C_0^\infty(U)$, $0 \leq \chi \leq 1$, $\chi=1$ outside some compact subset of $U \cup \partial U$. Here again we consider χ extended zero to $\Omega_j \setminus U$, and, for $E \in E$, interpret $\chi E \chi$ as an operator in $L(H)$.

Proof. Note that the following operators form a set of generators (mod $K(H)$) for E_j ,(with the finitely generated algebras C_j^0) :

(3.14) $$A[B,C] \quad : A , B , C \in C_j^0 .$$

Similarly for E , where now A , B , $C \in C^0$, and where the Λ' of prop.3.3 may be used for C^0, instead of Λ . Now such a generator of E is contained in E_j , as a consequence of prop.3.3, so that $E \subset E_j$, all by virtue of the interpretation $E = E\chi_U^j \in L(H_j)$, for $E \in E$, of course. Also we get $K(H) \subset K(H_j) \subset E_j$, so that $E_j \supset \chi E \chi + K(H)$. For the converse we first note that trivially $\chi K \chi \in K(H)$, for every $K \in K(K_j)$. Also, for a generator E_j of the form (3.14), consider $\chi E_j \chi = \chi A[B,C]\chi$. Here we use the fact that there exist 'wider cut-off functions' $\chi' \in C^\infty(\Omega)$, $0 \leq \chi' \leq 1$, $\chi'=0$ near $\partial\Omega$, with $\chi\chi'^N = \chi$, for every $N=1,2,\ldots$. Note that χ' commutes mod $K(H_j)$ with every generator a , $D\Lambda_j$, $(\Lambda_j D)^{**}$ of C_j^0 . Indeed, the gradient $\nabla\chi'$ has compact support, so that V,thm.3.1 applies to a commutator $[\chi', D\Lambda_j]$, while trivially $[\chi', a] = 0$. It follows that we can write

(3.15) $$\chi E_j \chi = \chi(\chi'(A_0[B_0,C_0] + K)\chi')\chi ,$$

where A_0 has been constructed from A by replacing in the finite sum of finite products of generators a, $D\Lambda_j$, every a by $a_0 = \chi' a$, and every $D\Lambda_j$ by $D_0\Lambda'$, with $D_0 = \chi' D$, and Λ' of prop.3.3. Similarly for B and C , going into B_0 and C_0 . Then we have $\chi' K \chi' \in K(H)$, and $\chi' A_0[B_0,C_0]\chi' \in E^0$, proving that $\chi E_j \chi \subset C$. Q.E.D.

Corollary 3.5. We have

(3.16) $$E\chi_U^j = E_j \cap (C\chi_U^j) .$$

That is, an operator $A \in C$ has $A\chi_U^j \in E_j$ if and only if $A \in E$.

Proof. We know that $E\chi_U^j \in E_j$, by (3.13). Suppose $A\chi_U^j \in E_j$ for some $A \in C$. With the decomposition $1 = \gamma + \delta$ used in the proof of thm.3.1 we get $A = \delta A \delta + K$, where $K \in K(H) \subset E$. Also $\delta A \delta = (\delta\chi_U^j) A (\delta\chi_U^j) \in E$, by (3.13), since $A\in E_j$, q.e.d.

Theorem 3.6. The quotient algebras V_j/E_j and C/E are isometrically *-isomorphic under the canonical map $A+E \to A+E_j$, $A\in C$. The symbol space of C/E is homeomorphic to the set $\iota_j^{-1}(M_{A_U\#}) \subset \mathbb{M}_j$, with the map $m \to m^{\sim}$, for $m \in C/E$, $m^{\sim}\in V_j/E_j$ given by $m = m^{\sim}\cap C$, or $m^{\sim}= m \cup E_j$. For $A\in C$ the C-symbol equals the restriction $\sigma_A|\mathbb{M}$ of the C_j-symbol to $\mathbb{M}\simeq\iota_j^{-1}(M_{A_U\#})$.

The proof of thm.3.6 is evident, after the above discussion. In particular, every *-isomorphism is an isometry ([C_1],AII,thm. 7.17, or [R_1], [Dx_1]).

Remark 3.7. Let us give a special consideration to the case where $U \cup \partial U$ is compact. In that case, clearly, we have $A^{\#}_{U,0} = C_0^{\infty}(U)$, and $\mathcal{D}^{\#}_{U,0}$ is the class of all folpde's over U with compact support. Hence, in that case, the algebra $C = C_U$ coincides with the minimal comparison algebra J_0 of the restricted triple $\{\Omega,d\mu|U,H|U\}$. Its commutator ideal is $K(H)$, and its symbol space is the restriction of the co-sphere bundle to U . The homeomorphism of thm.3.6 is just the injection of that restriction into the co-sphere bundle of Ω_j .

Remark 3.8. Now consider the case where $U \cup \partial U$ is non-compact. Then either $\partial M_{A_1\#} = \partial M_{A_1 U\#}$, or $\partial M_{A_1\#} = \partial M_{A_1 U\#} \cup \partial M_{A_1 V\#}$, where $V = \Omega\setminus(U\cup\partial U)$ is the interior of the noncompact complement of U . A similar alternative holds for j=2 . Now the homeomorphism of thm.3.6 will put the cosphere bundle parts $\iota_j^{-1}(U)$ of $\mathbb{M}_j$, j=1,2, into correspondence. But it also will establish a homeomorphism between the spaces $\iota_1^{-1}(\partial M_{A_1 U\#})$ and $\iota_2^{-1}(\partial M_{A_2 U\#})$, under which the symbols of generators remain invariant.

Remark 3.9. We also get relation (3.17), below, as an equivalent of cor.3.5 for E , $K(H)$ and $K(H_j)$.

(3.17) $$K(H)\chi_U^j = E\chi_U^j \cap K(H_j) \quad .$$

(The proof is left to the reader ; in fact (3.17) remains true if we substitute $L(H)$ for E). This will be useful in the nextfollowing section, where the algebra C/E turns out to be a function algebra again.

4. Complete Riemannian spaces with cylindrical ends.

In the present section we consider a noncompact (infinitely differentiable) Riemannian manifold Ω of dimension n , with a disjoint decomposition into a finite number of subdomains (fig.4.1)

$$\Omega = \Omega_0 \cup \Omega_1 \cup \ldots \cup \Omega_N \cup \partial\Omega_1 \cup \ldots \cup \partial\Omega_N , \tag{4.1}$$

where the Ω_j are mutually disjoint with compact boundaries $\partial\Omega_j$,

$$\partial\Omega_0 = \partial\Omega_1 \cup \ldots \cup \partial\Omega_N \quad , \partial\Omega_j \cap \partial\Omega_l = \emptyset , \ j,l > 0 , \tag{4.2}$$

each $\partial\Omega_j$,j=1,...N being a (connected) n-1-dimensional submanifold of Ω . Moreover, $\Omega_0 \cup \partial\Omega_0$ is compact and,

$$\Omega_j \cup \partial\Omega_j = B_j \times \mathbb{R}^+ , \ \mathbb{R}^+ = [0,\infty) , \text{ with } \partial\Omega_j = B_j \times \{0\} . \tag{4.3}$$

In other words, using the terminology of [KG_1] , the manifold Ω has N ends, and is a half cylinder $B_j \times \mathbb{R}^+$ at each end , where B_j is a compact n-1-dimensional Riemannian space, and where $B_j \times \mathbb{R}^+$ carries the product metric

$$ds^2 = d\rho^2 + dt^2 , \ d\rho^2 = \text{metric of } B_j , \tag{4.4}$$

with Euclidean metric on $\mathbb{R}^+$. On such a manifold Ω we now just consider the Laplace comparison algebra C for generating sets $A^\#$ and $D^\#$ satisfying (a_0), (a_1), (d_0).

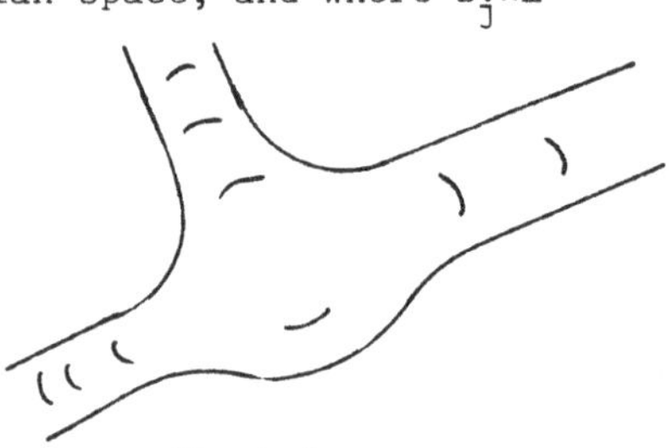

Fig.4.1

In general we cannot expect compact commutators. For the investigation of $C=C(A^\#,D^\#)$, and its commutator ideal $E=E(A^\#,D^\#)$ it appears natural to use algebra surgery of sec.3 to link the algebra C with corresponding C_j, j=1,...,N, where C_j live on the full cylinders $\mathbb{R} \times B_j = \Omega_j^0$.

Notice that for this type of surgery we indeed get 'compact cuts', i.e., the cylindrical end Ω_j is a common subdomain U_j of Ω and Ω_j^0 with compact boundary $\partial\Omega_j$. For the corresponding task involving a poly-cylinder of sec.2 , having n'>1, (i.e., some noncompact part of Ω is identified with a subdomain of a poly-cylinder) we get a noncompact boundary, however. Such type of algebra surgery is attempted in [CDg_1] .

First of all it is clear that the wave front space $\mathbb{W}$ of C is just the cosphere bundle of Ω , and the restrictions of the symbols of generators to $\mathbb{W}$ are the functions

$$(4.5) \qquad a(x)\ ,\ b^j(x)\xi_j\ ,\ a\in A^{\#}\ ,\ D=b^j\partial_{x^j}+p\in\mathcal{D}^{\#}\ ,$$

for a multiplication operator a , and for $D\Lambda$, respectively.

The important question to be answered is about the secondary symbol space $\mathbb{M}\setminus\mathbb{W}$, i.e., we ask for points of $\mathbb{M}$ over infinite points of the compactification $M_{A^{\#}}$. At each end the open subdomain Ω_j of the manifold Ω is identified with an open subdomain $U_j = B_j\times\mathbb{R}^+$, $\mathbb{R}^+=(0,\infty)$, of the cylinder Ω_j^0 .

Thus we have the configuration of thm.3.1 , assuming that the generating sets $A^{\#}$, $\mathcal{D}^{\#}$ are matched with corresponding sets $A_j^{\#}$, $\mathcal{D}_j^{\#}$ on Ω_j^0 . In that respect let us choose $A_j^{\#}$ and $\mathcal{D}_j^{\#}$ in accordance with (2.2) and the proposed sets $A^{\#}$, $\mathcal{D}^{\#}$ following (2.2), chosing n'=1 and $B=B_j$, j=1,...,N . This and (3.1) describe the sets $A^{\#}$, $\mathcal{D}^{\#}$ uniquely, since in any part of Ω away from its ends we just get all C^{∞}-functions (folpde's).

Next we look at the commutator ideals $E_j = E(A_j^{\#},\mathcal{D}_j^{\#})$ of the Laplace comparison algebras $C_j = C(A_j^{\#},\mathcal{D}_j^{\#})$, j=1,...,N . These were studied in sec.2. In fact, thm.2.6 gives a precise desciption of the F_t-conjugated ideal E_j^ of E_j , where F_t denotes the Fourier transform (2.3) in the t-direction of Ω_j^0 .

<u>Remark 4.1</u>. The ideal E_j contains the operator $\Lambda_j = (1-\Delta_j)^{-1/2}$, where $\Delta_j = \partial_t^2+\Delta_{x,j}$ denotes the Laplace operator on Ω_j^0 , with the Laplace operator $\Delta_{x,j}$ on B_j . Indeed, we get

$$(4.6) \qquad \Lambda_j\hat{}\ = F_t^{-1}\Lambda_jF_t = (1+\tau^2-\Delta_{x,j})^{-1/2} \le \langle\tau\rangle^{-1} \to 0\ ,\ |\tau|\to\infty\ ,$$

Also the operator valued function $S(\tau) = (1+\tau^2-\Delta_{x,j})$ takes values in $K(h_{x,j})$, $h_{x,j} = L^2(B_j)$, because the manifold B_j is compact. Thus we indeed get $\Lambda_j\hat{}\in CO(\mathbb{R},h_{x,j})\subset E_j\hat{}$, or, $\Lambda_j\in E_j$.

The above remark shows that VII,thm.4.7 may be applied to each of the cylinder algebras C_j , showing that the secondary symbol space is empty. (In particular we have $M_{A_j^{\#}} = B_j\times[-\infty,\infty]$, as easily seen. One thus confirms easily that all assumptions of VII, thm.4.7 hold.) In view of remark 4.8 we have the following result.

<u>Theorem 4.2</u>. The secondary symbol space of the Laplace comparison algebra $C(A^{\#},\mathcal{D}^{\#})$ is void. The symbol space $\mathbb{M}$ of this algebra coincides with the compactification of the wave front space $\mathbb{W}$

under the symbols (4.5) of the generators .

Next we look at the commutator ideal $E(A^{\#}, D^{\#})$. Here we may apply thm.3.4, which will link E with E_j . Again Ω and Ω_j^0 have the common subdomain $U_j=\Omega_j=B_j\times\mathbb{R}^+$. Using thm.2.4 we get a complete description of the commutator ideal E_j . Then we may generate the comparison algebra - here called $C_j^\sim$ - of the subdomain $U_j\subset\Omega_j^0$ with the classes (3.3) and call its commutator ideal $E_j^\sim$.

Note that $C_j^\sim$ and $E_j^\sim$ also live on the manifold Ω , where they again are generated by the common subdomain $U_j=\Omega_j$ of Ω , from restrictions of the classes $A^{\#}$ and $D^{\#}$ defined on Ω .

For each j=1,...,N choose a cut-off function $\psi(\tau)$, depending on t only, and meeting the requirements of thm.3.4. Then, we get $1-\sum_{j=1}^N\psi_j=\psi_0 \in C_0^\infty(\Omega)$. For any $E\in E$ we have $\psi_0 E$, $E\psi_0 \in K(H)$, by VII,thm.1.3, and $[\psi_j,E]\in K(H)$, by V,thm.3.1 and $E\subset C$. Accordingly

$$(4.7)\qquad E \equiv \sum_{j,k=1}^N\psi_j E\psi_k \equiv \sum_j \psi_j E\psi_j + \sum_{j\neq l} E\psi_j\psi_l = \sum_{j=1}^N\psi_j E\psi_j \pmod{K(H)} .$$

Using thm.3.4 this implies the proposition, below.

Proposition 4.3. The quotient algebra $E/K(H)$ allows the direct decomposition

$$(4.8)\qquad E/K(H) = E_1^\sim/K(H_1^\sim) \oplus \ldots\oplus E_N^\sim/K(H_N^\sim) \ , \ H_j^\sim = L^2(\Omega_j) .$$

On the other hand, we now look at $E_j^\sim$ as a subalgebra of E_j, again using thm.3.4 and remark 3.9:

Theorem 4.4. We have, with $k_{x,j}=K(h_j)$, $h_j=L^2(B_j)$,

$$(4.9)\qquad E_j^\sim/K(H_j^\sim) \simeq C(\mathbb{E}^+,k_{x,j}) \ , \ \mathbb{E}_j = \mathbb{R} \ , \ j=1,\ldots,N.$$

Proof. We apply thm.3.4 and conclude that, for $E\in E_j^\sim$, we have

$$(4.10)\qquad E + K(H_j) = \psi_j E_j\psi_j + K + K(H_j) = \psi_j E_j\psi_j + K(H_j) ,$$

with some $E_j\in E_j$, and $H_j=L^2(\Omega_j^0)$. Since $K(H_j^\sim)\subset K(H_j)$ with the imbedding of thm.3.4, we get a map $\upsilon:E_j^\sim/K(H_j^\sim)\to E_j/K(H_j)$, which is an isomorphism, by virtue of (3.17). On the other hand, thm.2.7 implies that $E_j/K(H_j) \simeq C(\mathbb{E}^+,k_{x,j})\oplus C(\mathbb{E}^-,k_{x,j})$. Also, from (2.31) we conclude that the symbols of operators of the form $\psi_j E_j\psi_j$ vanish identically on $\mathbb{E}^-$. Again, checking on the symbols of all the generators, as they are listed in thm.2.7, one finds that there are enough symbols of this form to generate the entire $C(\mathbb{E}^+, k_{x,j})$, by the Stone Weierstrass theorem. We still introduce the notation $\mathbb{E}_j = \mathbb{E}^+ = \mathbb{R}$, for each $j=1,\ldots,N$, and find (4.9) established, and our proof is complete.

We now summarize the results about the commutator ideal E , below. Remaining proofs are standard and will be left to the reader.

Theorem 4.5. For the quotient algebra $E/K(H)$ we have

$$(4.11) \quad E/K(H) = C(\mathbb{E}_1, k_{x,j}) \oplus C(\mathbb{E}_2, k_{x,j}) \oplus \ldots \oplus C(\mathbb{E}_N, k_{x,N}) ,$$

where each $\mathbb{E}_j$ is homeomorphic to $\mathbb{R}$, and where $k_{x,j}$ are the compact ideals of the spaces $L^2(B_j)=h_j$. Moreover, E (mod $K(H)$) is generated by

$$(4.12) \quad \psi_j a_{x,j}\Lambda_j\psi_j,\ \psi_j D_{x,j}\Lambda_j^2\psi_j \ :\ a_{x,j}\in A_j^{\#},\ D_{x,j}\in \mathcal{D}\#_j,\ j=1,\ldots,N,$$

with the classes $A_j^{\#}$ and $\mathcal{D}_j^{\#}$ of (2.2), for $B=B_j$. In (4.12) the support of the cut-off functions ψ_j may be chosen arbitrarily far out - i.e. within $t\geq R$, for arbitrarily large R. The generators (4.12) have symbol

$$(4.13) \quad a_{x,j}\Lambda_{x,j}T_j(\tau),\ D_{x,j}\Lambda_{x,j}^2 T_j^2(\tau),\ \tau\in\mathbb{E}_j=\mathbb{R}\ ,=0 \text{ on } \mathbb{E}_k,\ k\neq j ,$$

with

$$(4.14) \quad \Lambda_{x,j} = (1-\Delta_{x,j})^{-1/2} ,\ T_j(\tau) = (1+\tau^2\Lambda_{x,j}^2)^{-1/2} .$$

Thm.4.5 has an application to a singular elliptic boundary problem on $\mathbb{R}^n$ considered by Nirenberg and Walker [NW], which we shall not discuss in detail.

CHAPTER 9. H_s-ALGEBRAS; HIGHER ORDER OPERATORS WITHIN REACH.

In the present chapter we attend to two different, but related problems. First, we propose to also study comparison algebras in $L(H_s)$, where H_s is an L^2-Sobolev space on Ω of order s. Here the spaces H_s always are understood as spaces of the HS-chain induced by the (self-adjoint) Friedrichs extension H of H_0, for a a given triple $\{\Omega, d\mu, H\}$ (cf.I,6). In that respect we must assume cdn.(s) in order to be in agreement with the customary definition of Sobolev spaces.

Second, we will focus on more general N-th order expressions <u>within</u> <u>reach</u> of a given comparison algebra C (cf. V, def.6.2). Every compactly supported expression already is within reach of the minimal comparison algebra, as we know from V,6. However, for a general algebra C on a noncompact Ω the general expression L no longer is within reach. Thus we will ask for criteria to decide whether a given L is within reach.

The relation between these two tasks is discussed in sec.1 below. There we also discuss the organization of the present chapter. In particular we point out that some of the theorems have to be 'recycled', in the following sense. The first application only applies to the minimal comparison algebra J_0 in $L(H)$ (or in $L(H_s)$). That brings certain higher order differential expressions 'within reach' of J_0 , hence of larger algebras, so that, in turn, the same theorem now may be applied again, to get larger classes within reach, etc.

Unfortunately this makes it necessary to allow as assumptions, in many theorems, a variety of special cases, so that the theorems look complicated.

1. Higher order Sobolev spaces, and H_s-comparison algebras.

In this section, and in the following we assume cdn. (s): All operators H_0^m are essentially self-adjoint, for m=1,2,... . Then it becomes possible to work with general Sobolev spaces H_s, as already introduced in V,rem.6.3 (and rem.6.3'). We recall the definition:

For $s > 0$ we define $H_s = \mathrm{dom}\,\Lambda^{-s}$, and

$$(1.1)\qquad (u,v)_s = (\Lambda^{-s}u,\Lambda^{-s}v)\ ,\ \|u\|_s = (u,u)^{1/2}\ ,\ u,v \in H_s\ .$$

For $s < 0$ we define $\|u\|_s$ and $(u,v)_s$ by (1.1) for all $u,v \in H = \mathrm{dom}\,\Lambda^{-s}$ and then H_s as the completion of H under $\|u\|_s$.

The spaces H_∞ and $H_{-\infty}$ are defined as the intersection and the union of all Sobolev spaces:

$$(1.2)\qquad H_\infty = \cap\{H_s : s\in\mathbb{R}\}\ ,\ H_{-\infty} = \cup\{H_s : s \in \mathbb{R}\}\ .$$

In particular H_∞ is considered a Frechet space under the topology of the class $\{\|.\|_s : s \in \mathbb{R}\}$. The space $H_{-\infty}$ is provided with the inductive limit topology. For details cf. [C_1],III,(1.23) (but this will not be required in the sequel).

Notice that the collection $\{H_s : |s| \leq \infty\}$ defines an HS-chain in the sense of I,6, generated by the unbounded self-adjoint operator H -i.e., the Friedrichs extension of the minimal operator H_0 of our expression H .

By definition H_s consists of special L^2-functions, as $s \geq 0$. For s=1 and $u \in C_0^\infty$ the Sobolev norm $\|u\|_1$ coincides with a (weighted) L^2-norm of the (n+1)-vector-valued function $(u;\nabla u)$, by V,(1.3). Correspondingly, every norm $\|u\|_k$, k=2,3,..., may be expressed as a weighted L^2-norm of the 'jet' $(u^{(\alpha)} : |\alpha| \leq k)$ of local derivatives of order $\leq k$. We shall discuss this in sec.6, where covariant derivatives will be used, for a global representation, under some assumptions (cf. also [C_1],III,(2.7), (2.8)).

In VI,cor.4.3 we already discussed a weakened version of the Sobolev imbedding lemma, stating that $H_s \subset C^k(\Omega)$ for $s \geq [n/2]+k+1$. Accordingly H_s consists of smoother functions as s grows larger, and $H_\infty \subset C^\infty(\Omega)$.

For $s<0$ the general $u \in H_s$ no longer will be an L^2-function, but still may be interpreted as a distribution. Indeed, for $s<0$, and $u \in H$, $v \in H_{-s} = \mathrm{dom}\,\Lambda^s$, we get

(1.3) $|(u,v)| = |(u,v)_0| = |(\Lambda^{-s}u,\Lambda^{s}v)| \le \|u\|_s \|v\|_{-s}$.

Accordingly, the sesqui-linear functional (u,v) , extends continuously to a sesquilinear functional over $H_s \times H_{-s}$, which again will be denoted by (u,v) , for the moment. For every $u \in H_s$ the map $v \to (u,v)=l_u(v)$ defines a continuous linear functional over H_s , and we get $\|l_u\| = \|u\|_s$, by (1.3) ,and "=" for $v=\Lambda^{-2s}u \in H_s$.

This shows, that H_s coincides with the set of linear functionals over H_{-s} , under the induced pairing $H_s \times H_{-s} \to \mathbb{C}$. In other words, $u \in H_s$, $s<0$, may be characterized by the values of the linear functional (u,v) , $v \in H_{-s}$.

Note that the above pairing is identical to the pairing I,(6.13) introduced for abstract HS-chains. By lemma 1.1, below, the functional $v \to (u,v)$ is characterized by its values over $C_0^\infty(\Omega)$. Moreover, the restriction to C_0^∞ of the functional is a distribution over Ω , as an immediate consequence of VI, cor.4.3.

In this sense it is true that all the Sobolev spaces H_s , $|s| \le \infty$, are subsets of $\mathcal{D}'(\Omega)$, the space of all distributions over Ω . Note also that we have $H_s \subset H_t$, as $s > t$, (including $\pm\infty$). In fact the real-valued functions $s \to \|u\|_s$ for a fixed $u \in \mathcal{D}'(\Omega)$ are defined and nondecreasing on the half-line $(-\infty,s_0)$ if $u \in H_{s_0}$. This is a trivial consequence of the definition.

Unweighted L^2-Sobolev spaces over $\mathbb{R}^n$ have been introduced in [C_1],ch.III, using distributions and the Fourier transform. The same approach is used in [C_3] as well. The present definition is easily related to the earlier one if we observe that $1-\Delta = F^{-1}(1+|x|^2)F$, in case of the triple $\{\mathbb{R}^n,dx,1-\Delta\}$, with the Fourier transform F, and its inverse F^{-1} . Note that this triple satisfies cdn.(s) , and that (s) is quite essential in proving equivalence of the above definition with that of [C_1],V,(2.1) .

The lemma below is an easy consequence of I,prop.6.1 and the remarks on density of $\Lambda^{-N}C_0^\infty(\Omega)$ in H following V,def.6.1.

<u>Lemma 1.1</u>. The space $C_0^\infty(\Omega)$ is dense in H_s for $-\infty<s\le\infty$.

After the preliminary introduction into Sobolev spaces of I,6 (from an abstract view point) and VI,4 (mainly directed to compact manifolds) we will have to make a more systematical investigation for noncompact spaces now. In particular we will show independence of H_s of the choice of the comparison triple in a sense similar to that derived for comparison algebras: One may change $d\mu$ and H on a compact set without changing H_s . Also, if the manifold is changed outside a region $U \subset \Omega$ with compact boundary then H_s does not change over U , in a manner to be discussed (thm.1.2 and thm.1.3 below, to be proven in sec.5). However, our most important task will be the discussion of comparison algebras $C_s \subset L(H_s)$. Such algebras will have a similar application as the algebras $C \subset L(H)$ discussed in the previous chapters. We will use them to characterize the Fredholm properties of realizations of differential expressions within their reach.

For the concept of H_s-comparison algebra to be studied next we again require classes $A^{\#}$, $\mathcal{D}^{\#}$ satisfying (a_0), (a_1) (or (a_1')), and (d_0) of V,1. In addition to (s), and (m_2),(m_3) we now must impose stronger conditions on $A^{\#}$ and $\mathcal{D}^{\#}$. We choose to work with a subselection of the conditions (l_j), j=1,2,3,4,5, below. (Note that (l_1) implies (m_2) , and (l_3) implies (m_3).)

Condition (l_1): $A^{\#}$ is an algebra under the pointwise product of functions, and $\mathcal{D}^{\#}$ is a Lie-algebra under the commutator product [.,.] of folpde's. Also, $\mathcal{D}^{\#}$ is a two-sided $A^{\#}$-module, as under (m_2), and $A^{\#}$, interpreted as a set of zero order differential expressions, is a subset of $\mathcal{D}^{\#}$. Also, for $a \in A^{\#}$, $D \in \mathcal{D}^{\#}$ we have $[a,D] \in A^{\#}$.

Condition (l_2): $\mathcal{D}^{\#}$ contains the formal adjoints of its folpde's.

Condition (l_3): Condition (m_3) holds, and, in addition, for any $a \in A^{\#}$, satisfying $a \neq 0$ at a closed set $Q \subset M_{A^{\#}}$ there exists a function $b \in A^{\#}$ with ab=1 near Q .

Condition (l_4): The expression H can be written as a finite sum of terms of the form aDF , with $a \in A^{\#}$, $D,F \in \mathcal{D}^{\#}$.

Condition (l_5): For every $a \in A^{\#}$ and $D \in \mathcal{D}^{\#}$ we have $[a,H] \in \mathcal{D}^{\#}$, and $[D,H]$ is a sum of finitely many terms bEF, $b \in A^{\#}$, $E,F \in \mathcal{D}^{\#}$.

In addition (with one exception) we will either impose (m_5), or even the following formally stronger condition.

Condition (m_5') : We have $1 \in \mathcal{D}^{\#}$.

Clearly (m_5') implies (m_5), but (m_5') means that Λ even belongs to the generators of C , hence is in the finitely generated algebra C^0 . This will be formally useful for the H_s-comparison algebras, below.

The exception, mentioned above, is the minimal algebra J_0. It is clear that its generating classes $A^\#_m$, $D^\#_m$ satisfy (l_j), j=1,2,3,5, while, on the other hand, (m_5) will not generally hold. The case $A^\# = A^\#_m$, $D^\# = D^\#_m$, nevertheless, must be admitted in all results of sec's 3f. Actually, we know that J_0 is a compact commutator algebra, and it will be seen that (m_5) will not be required if the algebra C has compact commutators.

Note that the other exceptional case, not satisfying (a_1), and giving a non-unital C -i.e. the classes $A^\#_{U,0}$, $D^\#_{U,0}$ of VIII,3 - needs no special consideration here. For, in order to achieve cdn. (s) in spite of the truncation boundary ∂U , one will have to work with a potential q tending to infinity at ∂U. This implies (m_5), (cf. prop.2.3).

For a pair $A^\#$, $D^\#$ satisfying (m_5), and (l_j), j=1,2,5, (in addition to (a_0), (d_0) and either (a_1) or (a_1')), we introduce the new notation $A!=A^\#$, and revised family $D!=D^\#+\{\alpha 1\}$, where now $A!$, $D!$ generate exactly the same algebra C, but (m_5') is valid for $A!$ and $D!$. Clearly $A!$, $D!$ also satisfy (l_j), j=1,2, since the constants commute with all of $D!$ and $A!$. However, (a_1') might be violated for $A!$, $D!$, although valid for $A^\#$, $D^\#$. (In case where the minimal algebra J_0 is under consideration, we will use thm's 3.8, 4.2, and 4.3 with $A! = A^\#_m$, $D! = D^\#_m$, since then the adjunction of the constants to $D^\#_m$ might give a larger algebra.)

It is clear that (m_5') follows from (a_1) and (l_1). We will show in sec.2 that (a_1), and (l_j) : j=1,2, imply (m_4), and we will see that (m_5) even follows from (a_1') and (l_1) alone, if 'q grows large where $D^\#$ gets scarce', as to be specified (prop.2.3). It follows by a calculation that $\{(l_j):j=1,2,4\}$ implies (l_5) , while $\{(l_j):j=1,2,5\}$ does not necessarily imply (l_4) .

In thm.2.1 we will show that (l_4) is a consequence of (m_6) , assuming some of the other conditions (m_j) and (l_j) . Actually, we will show that (l_4) is equivalent to a 'global Parseval relation' of the form

$$(1.4) \qquad \{D,F\} = \sum_{\nu,\mu=1}^N a_{\nu\mu}\{D,D_\nu\}\{D_\mu,F\} \text{ for all } D,F \in D^\# ,$$

with a symmetric positive matrix $((a_{\nu\mu}))$ of functions $a_{\nu\mu} \in A^\#$, and a system $D_1,\dots,D_N \in \mathcal{D}^{\#}$. The same relation (1.4) also follows locally from $(m_6)_{x_0}$ for an orthonormal base $D_j, j=1,\dots$ $..,n+1$, at x_0, and then can be pieced together globally, using the compactness of $M_{A^{\#}}$.

It should be emphasized that condition (1_4) may amount to a restriction of the Riemannian space Ω at infinity. It implies that the metric tensor h^{jk} can be globally written as a finite sum $\sum a_\nu b^j_\nu c^l_\nu$, with bounded functions a_ν and tensors b^j_ν, c^j_ν still satisfying a set of conditions, as principal parts of $D,F \in \mathcal{D}!$. (cf.[CS],VI,L.2.5, [CM_1], [CM_2], and sec.3, below).

To say this again in different words: We presently depend on (1_5) which is difficult to achieve for classes $A!$, $\mathcal{D}!$, nontrivial near infinity, except through (1_4). But (1_4) often cannot be satisfied. However (1_5) holds trivially for the generators $A^{\#}_m$, $\mathcal{D}^{\#}_m$ of the minimal algebra J_0 , regardless of any other properties of the triple or the space Ω . The above difficulty will in part be removed later on, as follows: After proving our results for the algebra J_0 (and its H_s-equivalents) we will be able to substitute (1_5) with a more general condition, but will have to go several times through the same chain of arguments, under slightly varying conditions. This forces us to make split assumptions.

Also, all the results, below, concerning our above Sobolev spaces, will only require the higher order theory for the special case of the minimal algebra J_0 .

<u>Theorem 1.2</u>. Let two triples $\{\Omega,d\mu,H\}$, $\{\Omega,d\nu,K\}$ on the same manifold be given, where $H=K$, $d\mu = d\nu$ holds outside some compact set $M \subset \Omega$. Then we have cdn.(s) valid for $\{\Omega,d\mu,H\}$ if and only if it holds for $\{\Omega,d\nu,K\}$. For each $s\in[-\infty,\infty]$ the Sobolev spaces H_s of the two triples coincide as subsets of $\mathcal{D}'(\Omega)$. The corresponding Sobolev norms are equivalent (although the inner products and H_s-adjoints in general are different).

<u>Theorem 1.3</u>. Let two triples $\{\Omega_j,d\mu_j,H_j\}$ $j=1,2$, be given, where Ω_1 and Ω_2 both are extensions of a manifold U such that U appears as an open subdomain with compact boundary ∂U of Ω_1 and Ω_2 ,each. Let condition (s) be satisfied for both triples, and assume that the restrictions to U of the two triples coincide:

(1.5) $d\mu = d\mu_1|U = d\mu_2|U$, $H = H_1|U = H_2|U$.

Then, if the Sobolev spaces of the triples are denoted by H_s^j , then for every open set $V \subset U$ with $V^{clos} \subset U$ we have

(1.6) $H_s^1|V = H_s^2|V$, for all $s \in \mathbb{R}$,

where

(1.7) $H_s^j|V = \{u|V : u \in H_s^j\}$.

Moreover, if ∂U is a smooth n-1-dimensional (compact) submanifold of Ω_j , then the statement also holds for V=U . (Recall that the restriction $u|V$ of a distribution $u \in \mathcal{D}'(\Omega)$ to the open subset V is defined as the restriction of the linear functional to the testing functions with support in V.)

The proofs of thm.1.2 and thm.1.3 depend on techniques involving higher order commutators to be prepared in sec.3. We will discuss these proofs in sec.5.

For a given triple on a manifold Ω consider a chart U as a common subset with the manifold $\mathbb{R}^n$. Assume that U is a subset with compact closure of a larger chart V . Then all assumptions of thm.1.3 hold, and it follows that the subset of $u \in H_s$ with support in U coincides with the subset of $H_s(\mathbb{R}^n)$ with support in U . Here we first must use a measure $d\mu$ and expression H different from the surface measure and Laplace operator of $\mathbb{R}^n$, on some compact subset containing U , but we then may use thm.1.2 to remove this and obtain the Laplace comparison triple of $\mathbb{R}^n$. Thus we have the following.

Corollary 1.4. At each point $x_0 \in \Omega$ there exists a neighbourhood N of x_0 such that the set $H_s|N$ of restrictions $u|N$ of $u \in H_s$ to N if related to $\mathbb{R}^n$ by the local coordinate map, precisely consists of all restrictions of $u \in H_s(\mathbb{R}^n)$ to N .

Indeed it is a direct consequence of thm.1.3, that $H_s|N = H^\sim_s|N$, where $H^\sim_s$ is formed with the manifold $\mathbb{R}^n$ and a triple equal to the Laplace triple outside a compact set, containing N , but coinciding on N with the given triple on Ω. Then one uses thm. 1.2 to show that $H^\sim_s$ is just the ordinary Sobolev space $H_s(\mathbb{R}^n)$.

This corollary connects our present Sobolev spaces with those of $\mathbb{R}^n$, discussed in detail in [C_1] ,III. In particular we obtain the Sobolev imbedding, generalizing VI, cor.4.3:

Proposition 1.5. We have $H_s \subset C(\Omega)$, as $s>n/2$, and $H_s \subset C^k(\Omega)$, as $s>n/2 + k$. In particular, for any triple over Ω, we have

(1.8) $$C_0^\infty(\Omega) \subset H_\infty \subset C^\infty(\Omega) .$$

Note that we are not attempting an investigation of the spaces H_s at infinity. For example, for $\Omega=\mathbb{R}^n$ it is known that the functions of H_∞ have all derivatives vanishing at infinity. More generally one could ask how H_s will be influenced by a change of the metric tensor h^{jk} or the potential q. Discussions of this type will not be made here.

Assume now that we are given a function algebra $A!$ and a Lie-algebra $\mathcal{D}!$ satisfying (m_5'), and (l_j), j=1,2,5, as well as (a_0),(a_1)(or (a_1') for $\mathcal{D}^\#$ only, $\mathcal{D}!=\mathcal{D}^\#+\{\alpha 1\}$), (d_0), (or also, we admit the case of the minimal classes).

Under these assumptions we will show that the operators

(1.9) $$a\ ,\ D\Lambda\ ,\ \Lambda D\ ,\ \text{for } a \in A!\ ,\ D \in \mathcal{D}!$$

belong to $L(H_s)$ (cf. thm.4.2). Accordingly, for $s \in \mathbb{R}$, the Banach algebra $C_s = C_s(A!,\mathcal{D}!)$ obtained as norm closure in $L(H_s)$ of the finitely generated algebra C^0 of the operators (1.9) will be called the H_s-comparison algebra of the triple $\{\Omega,d\mu,H\}$, and the generating sets $A!$ and $\mathcal{D}!$. Similarly we introduce the algebra C_∞ as closure of C^0 in $O(0)$. For a precise definition of the operators (1.9) one may think of $D=b^j\partial_{x^j}+p$ in terms of distribution derivatives. Or else, one also may think of the closures (in H_s) of the unbounded operators $a|H_s$, $D_0(\Lambda|H_s)$, $(\Lambda|H_s)D_0$, respectively.

For technical reasons we introduce a second type of finitely generated algebra, called C_∞^0 , obtained from the generators

(1.10) $$a\ ,\ D\Lambda^j\ ,\ \Lambda^j D\ ,\ \text{for } a \in A!\ ,\ D \in \mathcal{D}!\ ,\ j=1,2,\ldots .$$

Clearly we get $C^0 \subset C_\infty^0$, while $C_\infty^0 = C^0$ whenever (m_5') holds, since then we have $\Lambda \in C^0$, hence $D\Lambda^j= (D\Lambda)\Lambda^{j-1}\in C^0$, and similarly $\Lambda^j D \in C^0$. If (m_5') does not hold, then (1.9) and (1.10) still generate the same Banach-subalgebra of $L(H)$, (and of $L(H_s)$ or $O(1)$ as well, as will be seen later on). Indeed, from (a_1') we get existence of a function $\chi \in A!$ such that $\chi D=D\chi=D$, for any given $D \in \mathcal{D}!$, and with $\nabla\chi$ of compact support. Then we get $D\Lambda^2=(D\Lambda)(\Lambda\chi)$

$+D[\chi,\Lambda^2]$, where the last term is in $K(H)$, by V,lemma 3.2. By (l_1) we have $\chi \in \mathcal{D}!$, so that the first term is contained in C^0 . Also, $C \supset K(H)$, by V,lemma 1,1, thus it follows that $D\Lambda^2 \in C$. A similar conclusion implies $D\Lambda^j \in C$, $j=3,4,\dots$, and $\Lambda^j D\in C$, $j=2,3,\dots$.

In thm.4.2 we also will show that, under above conditions, the algebra C_s is a C^*-subalgebra of $L(H_s)$, while the ideal chains C and C_s are in agreement, and the symbol spaces remain the same. In fact, under suitable additional conditions, such as (l_4), even the symbols of the generators remain independent of s.

2. Closer analysis of some of the conditions (l_j) and (m_j).

In this section we assume (a_0),(d_0),(a_1)(or (a_1')), and more cdn's as stated. In thm.2.1 we show that (m_6) implies (l_4), assuming some other (m_j) and (l_j) . Actually, we show that (l_4) and

$$(**) \qquad D=b^j\partial_{x^j}+p \in \mathcal{D}^\# \text{ implies } p \in \mathcal{D}^\#, \text{ and } D_0=b^j\partial_{x^j} \in \mathcal{D}^\# ,$$

imply a 'global Parseval relation' of the form

$$(2.1) \qquad \{D,F\} = \textstyle\sum_{\nu,\lambda=1}^N a_{\nu\lambda}\{D,D_\nu\}\{D_\lambda,F\} \text{ for all } D,F \in \mathcal{D}^\# ,$$

with a symmetric positive $((a_{\nu\lambda}))$, $a_{\nu\lambda}\in\mathbb{R}$, and a system $D_1,\dots,D_N\in\mathcal{D}^\#$ while (2.1), even with $a_{\nu\lambda}\in A^\#$, and without (**), but with invertible $((a_{\nu\lambda}))$ implies (l_4). Thus $(m_6)_{x_0}$ gives a local (2.1),i.e.,

$$(2.2) \qquad H = \textstyle\sum_{\nu,\lambda=1}^\delta D_\nu^* a_{\nu\lambda}D_\lambda \ , \ a_{\nu\lambda}=a_{\lambda\nu}\in A^\# \ , \ D_\nu\in\mathcal{D}^\# \ , \text{ near } x_0,$$

from which a global (l_4) may be pieced together, if (m_6) holds.

Theorem 2.1. Assume cdn's (m_1),(l_j),j=1,2,3. Then $(m_6)_{x_0}$ implies (2.2), while (m_6) implies (l_4), and even (with certain $D_\nu^j\in\mathcal{D}^\#$)

$$(2.3) \qquad H = \textstyle\sum_{j=1}^N\sum_{\nu,\lambda=1}^\delta D_\lambda^{j*}\chi_j a_{\nu\lambda}^j D_\nu^j \ , \ \delta=\delta(x)=n+1 \ ,$$

with a partition $\{\chi_j:j=1,\dots,N\}$ of $M_{A^\#}$, $\chi_j\in A^\#$, $a_{\nu\lambda}^j=a_{\lambda\mu}^j=\bar a_{\nu\lambda}^j\in A^\#$.

Vice versa, (l_4) and (**) imply (2.1) with certain $D_\nu\in\mathcal{D}^\#$, $a_{\nu\lambda}\in\mathbb{R}$.

Proof. Assume $(m_6)_{x_0}$ for $x_0\in\partial M_{A^\#}$. With some orthonormal base $\{D_j\}$ of S_{x_0} we get $((a_{jl}(x)))$ invertible for $x\in N=N_{x_0}$, where

$$(2.4) \qquad a_{jl}(x) = \{D_j,D_l\}(x)\in A^\# \ , \ a_{jl}(x_0) = \delta_{jl} \ .$$

Using (1_3) on $\det((a_{jl}))$ we get $\chi((a_{\nu\lambda}(x)))^{-1}=\chi((a^{\nu\lambda}(x)))$, with functions $a^{\nu\lambda}\in A^{\#}$, with a cut-off function $\chi\in A^{\#}$, in accordance with (m_3), for x_0 and N. We introduce $((b_{\nu\lambda}))=((a_{\nu\lambda}))^{1/2}$, and find that $D_\nu^{\sim}=\sum_\lambda b_{\nu\lambda}D_\lambda$, $\nu=1,\dots,\delta$, is an orthonormal base at every $x\in N$ (Perhaps $D_\nu^{\sim}\in \mathcal{D}^{\#}$, but still $D_\nu^{\sim}\in S_x$). Use Parseval's relation for the $D_\nu^{\sim}$. Define the local matrix $H^{\sim}=((h_{\nu\lambda}^{\sim}))_{\nu,\lambda=0,\dots n}$ by $h_{\nu\lambda}^{\sim}=h_{\nu\lambda}$ for $\nu,\lambda\geq 1$, $=q^{-1}$ for $\nu=\lambda=0$, $=0$ for $\nu=0$, $\lambda\geq 1$, and $\nu\geq 1$, $\mu=0$, for a moment. Also let $D_\nu^{\sim}=\sum b_\nu^j\partial_{x^j}+b_\nu^0$, and then define the matrix $B=((b_\lambda^\nu))_{\nu=0,\dots,n,\lambda=1,\dots\delta}$, with ν=row-index, μ=column-index. The above is valid in local coordinates, near points of Ω, dense in $M_{A^{\#}}$. Observe that then Parseval's relation can be written as

$$(2.5)\qquad b^*H^{\sim}b=b^*(H^{\sim}BB^*H^{\sim})b\ ,\ \text{for all } b\in\mathbb{C}^\delta\ ,\ x\in N\cap\Omega\ .$$

This relation may be polarized, and then yields $H^{\sim -1}=BB^*$. Or,

$$(2.6)\quad h^{jk}(x)\bar{u}_ju_k+q(x)\bar{u}_0u_0=\sum_{\nu=1}^{\delta}(\bar{b}_\nu^j(x)\bar{u}_j+\bar{b}_\nu^0(x)\bar{u}_0)(b_\nu^j(x)u_j+b_\nu^0(x)u_0),$$

which is valid for all $x\in N$, $u\in\mathbb{C}^\delta$. If we substitute $u_0=u(x)$, $u_j=u_{|x^j}(x)$, for some $u\in C_0^\infty(\Omega)$, and multiply by $\chi(x)$, we get

$$(2.7)\qquad \int_\Omega d\mu\chi^2(h^{jk}u_{|x^j}u_{|x^k}+q|u|^2)=\sum_{\nu=1}^{\delta}\|\chi D_\nu^{\sim}u\|^2\ ,\ u\in C_0^\infty\ .$$

Now we get $D_\nu=\sum c_{\nu\lambda}D_\lambda^{\sim}$, with the matrix $((c_{\nu\lambda}))=((b_{\nu\lambda}))^{-1}=((a_{\nu\lambda}))^{1/2}$. As a consequence the right hand side of (2.7) may be written in the form

$$(2.8)\qquad \sum_{\nu,\lambda=1}^{\delta}\int_\Omega d\mu\chi^2a_{\nu\lambda}D_\nu uD_\lambda u\quad .$$

Finally, assuming $(m_6)=\cup(m_6)_x$, use (1_3) to construct a partition of unity $\{\chi_1,\dots,\chi_N\}$ on the compact space $M_{A^{\#}}$, where we require that $\chi_j\in A^{\#}$, $j=1,\dots,N$, while $\sum\chi_j^2=1$ on $M_{A^{\#}}$, and such that (2.7) with (2.8) hold for every χ_j with suitable $D_\nu=D_{\nu j}$. Taking a sum over j we then get

$$(2.9)\quad (u,u)_1=\sum_{j=1}^{N}\sum_{\nu,\lambda=1}^{\delta}(\chi_j^2a_{\nu\lambda}^jD_{\nu j}u,D_{\lambda j}u)\ \text{for all } u\in C_0^\infty(\Omega)\ .$$

Here we finally may polarize and change notation $\chi_j^2 \to c_j$, for (2.3), i.e., the first half of the theorem.

Vice versa, let (l_4) and (**) hold. Accordingly we can write

$$(2.10) \qquad H = \textstyle\sum_{\nu=1}^{N} a_\nu D_\nu F_\nu \ , \ a_\nu \in A^{\#} \ , \ D_\nu \ , \ F_\nu \in \mathcal{D}^{\#} \ ,$$

where all coefficients may be assumed real-valued, since H has real coefficients. Using $(l_1),(l_2),(m_1)$ one may rewrite this as

$$(2.11) \qquad H = \textstyle\sum_{\nu=1}^{N} D_\nu^{\ *} F_\nu \quad ,$$

with changed, but still real-valued D_ν, F_ν . Let G_λ , $\lambda=1,\ldots,R$, be a basis of span $\{D_\nu, D_\nu^0, F_\nu, F_\nu^0 : \nu=1,\ldots,N\}$, with the principal parts D_ν^0 , F_ν^0 of D_ν , F_ν . We may express D_ν and F_ν as linear combinations of the G_ν , and (2.11) takes the form

$$(2.12) \qquad H = \textstyle\sum_{\nu,\lambda=1}^{R} a_{\nu\lambda} G_\nu^{\ *} G_\lambda \quad , \ a_{\nu\lambda} = a_{\nu\lambda} = a_{\nu\lambda} \in \mathbb{R} \ .$$

Now use that H is self-adjoint, so that $H = 1/2(H+H^*)$. We get

$$(2.13) \qquad H = \textstyle\sum_{\nu,\lambda=1}^{R} a_{\nu\lambda} G_\nu^{\ *} G_\lambda \quad ,$$

where again the $a_{\nu\lambda}$ have been changed. Now the matrix $((a_{\nu\lambda}))$ is real and symmetric.

Returning to our old notation we have proven

<u>Proposition 2.2</u>. If condition (l_4) is valid, then we have

$$(2.14) \qquad H = \textstyle\sum_{\nu,\lambda=1}^{N} a_{\nu\lambda} D_\nu^{\ *} D_\lambda \ , \text{ with } a_{\nu\lambda} = \bar{a}_{\nu\lambda} = a_{\lambda\nu} \in A^{\#} \ , \ D_\nu \in \mathcal{D}^{\#} \ ,$$

where all functions and folpde's are real.

Now we write

$$(2.15) \qquad D_\nu = b_\nu^j \partial_{x^j} + b_\nu^0 \quad , \quad B = ((b_\nu^j(x)))_{j=0\ldots,n,\nu=1,\ldots,N} \ ,$$

where B will be considered an $(n+1)\times N$-matrix , i.e., j indicates the rows and ν the columns. It may be assumed that D_ν, $\nu=1,\ldots,M$, have $b_\nu^0=0$, while D_ν, $\nu=M+1,\ldots,N$, are of order zero. Now let $\sum'$ denote the sum (2.14) taken only over ν,λ with at least one $>M$. While $\sum'$ formally is of first order, it really is of order zero, due to $b_\lambda^0 D_\nu^0 + (b_\lambda^0 D_\nu^0)^* = (D_\nu^0 + D_\nu^{0*}) b_\nu^0 + [b_\lambda^0, D_\nu^0]$, where the right hand side is sum of products of two zero-order terms in $\mathcal{D}^{\#}$. We thus may repeat the above proceedure once more on $\sum'$, to arrive at a new

matrix $((a_{\nu\lambda}))$ with $a_{\nu\lambda}=0$ as $\nu\leq M$, $\lambda>M$ or $\nu>M$, $\lambda\leq M$. Comparing coefficients in (2.14) we then find that, with $H^{\sim}$ of (2.5) we have

$$H^{\sim -1} = BAB^{*} \quad , \tag{2.16}$$

Then (2.16) implies $H^{\sim}=(H^{\sim}B)A(H^{\sim}B)^{*}$ which translates into

$$\{D,F\} = \sum_{\nu,\lambda=1}^{N} a_{\nu\lambda}\{D,D_{\nu}\}\{D_{\lambda},F\} \text{ , for all } D,F \in \mathcal{D}^{\#} \text{ .} \tag{2.17}$$

This completes the proof of thm.2.1.

Next we will look at condition (m_5) . This condition is a consequence of (l_1), if (a_1) holds. On the other hand, under (a_1') only, prop.2.3, below, shows that (m_5) follows for the algebra C of the 'truncated classes' $A^{\#}_{U,0}$, $\mathcal{D}^{\#}_{U,0}$ of VIII,3. (We, of course, then must work with a potential q going to ∞ at ∂U , as was used in VIII,prop.3.3.)

Proposition 2.3. Suppose $\mathcal{D}^{\#}$ contains a sequence of functions (i.e. zero order expressions) χ_j , $j=1,2,\ldots,$ such that $0\leq\chi_j\leq 1$, $\chi_j = 1$ in $\mathcal{Q}_j$, where $\cup\mathcal{Q}_j = \Omega$, while

$$\lim_{j\to\infty}(\inf\{q(x): x \in \Omega\setminus\mathcal{Q}_j\}) = \infty \text{ .} \tag{2.18}$$

Then condition (m_5) holds.

Proof. Let $\Lambda_j = \chi_j\Lambda \in C$. From V,(1.7) it follows that

$$\|\Lambda_j - \Lambda\| = \|(1-\chi_j)\Lambda\| \leq \sup\{(1-\chi_j)/\sqrt{q}\} \to 0 \text{ , as } j\to\infty \text{ .} \tag{2.19}$$

Hence $\Lambda \in C$, q.e.d.

Proposition 2.4. Conditions (l_5),(m_5) and (w) imply cdn. (m_4) .

Proof. First, in the special case of the minimal classes, we have a compact commutator algebra, for which (m_4) was proven in V,3 (since a and D have compact support, V,lemma 3.3 and V,lemma 3.6 apply). Similarly, in the general case, we again use a resolvent integral of the form (and with the contour of) V,(3.6). Since we assume (m_5) we have $\Lambda^2 \in C$, hence the resolvent $R(\lambda) = (\Lambda^2-\lambda)^{-1}$ is in C for all $\lambda \in \Gamma\setminus\{0\}$, with the path Γ used in (4.6). (If C is non-unital, we at least get $\Lambda^{\varepsilon}R(\lambda)\in C$, which is enough). We get

$$[a,R(\lambda)]=\Lambda^2R(\lambda)[H,a]R(\lambda)\Lambda^2 \text{ , } [D,R(\lambda)]=\Lambda^2R(\lambda)[H,D]R(\lambda)\Lambda^2 \text{ .} \tag{2.20}$$

Using (l_5) we get $[H,a] \in \mathcal{D}!$, so that $[a,R]\Lambda^{-1-\varepsilon}$, $\Lambda^{-1-\varepsilon}[a,R] \in \mathcal{C}$.

Similarly, $[D,R]\Lambda^{-\varepsilon}$, $\Lambda^{-\varepsilon}[D,R] \in \mathcal{C}$, due to (l_5) again. Moreover, the integrands are norm continuous (even analytic). Using V,(3.9) we get estimates like those used in V,lemma 3.3 or 3.6, so that the integrals converge in norm. This completes the proof.

3. Higher order expressions within reach of $\mathcal{C}$ or $\mathcal{C}_s$.

In this section we first assume cdn's (s), (l_j) j=1,2,5, and either (m_5'), or else the minimal classes $A^{\#}_m$, $\mathcal{D}^{\#}_m$. Later on we shall replace (l_5) by a weaker condition. Let $\mathcal{C} = \mathcal{C}(A!,\mathcal{D}!) \subset L(H)$ and $\mathcal{C}^0_\infty$ be the comparison algebra and the finitely generated algebra introduced in sec.1, repectively. Our first use of (l_5) will be to identify larger classes of expressions L <u>within</u> <u>reach</u> of $\mathcal{C}$ or $\mathcal{C}^0_\infty$, as specified in V,def.6.1.

By (l_1) the class $\mathcal{D}!$ is a Lie-algebra, under the commutator product, and a two-sided $A!$-module. Let L^∞ denote the corresponding universal enveloping algebra, defined as the algebra of all differential expressions of arbitrary finite order, of the form

$$(3.1) \qquad L = \textstyle\sum_{j=1}^{M} a_j \, \Pi_{l=1}^{N^j} D_{jl} \ , \ a_j \in A! \ , \ D_{jl} \in \mathcal{D}! \ .$$

Clearly this defines an algebra under $A!$, since $\mathcal{D}!$ is a two-sided $A!$-module. We may set $a_j=1$, except for $N^j = 0$.

By L_m we denote the class of all operators in L_∞ of the form (3.1), with $N^j \le m$, for all j=1,...,M . Since $1 \in \mathcal{D}!$, by definition (or else $\mathcal{D}! = \mathcal{D}^{\#}_m$) , we may assume exactly N_j=m terms in each product of (3.1) , except for m=0 where $L_0 = A!$. Also we get $L_1 = \mathcal{D}!$, while, for m=0,1,2,...., L_m is a linear space of m-th order differential expressions over Ω, containing its formal adjoints, and all m-th order C^∞-expressions with compact support.

Note that (l_4) amounts to the condition that $H \in L_2$. Also (l_5) means that $[a,H] \in L_1$, $[D,H] \in L_2$, for all $a \in A!$, $D \in \mathcal{D}!$.

Let $A!_0$ and $\mathcal{D}!_0$ denote the ideals of $A!$ and $\mathcal{D}!$ generated by the commutators $[a,D] \in A!$, and $[D,F] \in \mathcal{D}!$, respectively. Let $L_{\infty,0} \subset L_\infty$, $L_{N,0} \subset L_N$ denote the subsets of all L of the form (3.1), where at least one factor of each product is in $\mathcal{D}!_0$. Here it is no loss of generality to assume that always the first (always the last) factor is in $\mathcal{D}!_0$.

To be more specific, we set $L_{0,0} = A!_0$. A set of generators of $L_{1,0} = \mathcal{D}!_0$ is given by $\{a[D,F] : a \in A!,\ D,F \in \mathcal{D}!\}$, or also by

$\{[D,F]a : a \in A!, D,F \in \mathcal{D}!\}$, or by $\{a[D,F]b : a,b \in A!, D,F \in \mathcal{D}!\}$.
For m=2,3,...., a set of generators is given by
$\{D_1 \dots D_{N-1}[D,F]: D_j,D,F \in \mathcal{D}!\}$, or else $\{[D,F]D_1 \dots D_{N-1}: D_j,D,F \in \mathcal{D}!\}$.
In particular, $L_{\infty,0}$ again is an algebra invariant under multiplication by $A!$, and adjoint invariant. For the minimal classes it follows easily that $A!_0 = A! = A^{\#}_m$, $\mathcal{D}!_0 = \mathcal{D}! = \mathcal{D}^{\#}_m$, $L_{m,0} = L_m$. Therefore, since a general pair always contains the minimal classes, we find that also a general $L_{m,0}$ contains all C^{∞}-expressions of order m with compact support.

The following modification of condition (l_5) will be useful:
Condition (l_5'): We have $[H,a] \in L_{1,0}$, $[H,D] \in L_{2,0}$ for all $a \in A!$, $D \in \mathcal{D}!$.

Note that (l_4) not only implies (l_5), but even (l_5'). Also, the minimal classes satisfy (l_5'), not only (l_5). Some of the results below need the stronger condition (l_5') instead of (l_5).

Proposition 3.1. The class $L_{\infty,0}$ is a two-sided ideal of L_∞. Moreover, we have LL', $L'L \in L_{m+m',0}$ for $L \in L_{m,0}$, $L' \in L_{m'}$, and

(3.2) $$[L,L'] \in L_{m+m'-1,0} \text{ for } L \in L_m,\ L' \in L_{m'}.$$

The proof is a calculation, and is left to the reader.

For every differential expression $L \in L_N$ consider the two linear operators

(3.3) $$(L_0 \Lambda^N)^{**},\ (\Lambda^N L_0)^{**},$$

formed as closures of products of unbounded operators over H, with the minimal operator L_0 of L. The closures are well defined since Λ^N is bounded, and L has a well defined formal adjoint.

Proposition 3.2. The operators (3.3) are in $L(H)$, and, moreover, they even are in the comparison algebra $C(A!,\mathcal{D}!)$. In fact, they even are in the finitely generated algebra C^0_∞ of sec.1. In particular, using the terminology of V,def.6.1, every expressions $L \in L_N$ is within reach of C and C^0_∞, as an expression of order N.

Remark: Please observe that C and C^0_∞ have the generators (1.10), each algebra with its own way of generation. In particular, the ideal of compact operators appears on its own, and is not needed among the generators.
Proof. Note that the set of generators (1.10) contains its

adjoints, and that also L_N is adjoint invariant. Thus we have $(\Lambda^N L_0)^{**} = (L_0^* \Lambda^N)^* \in C_\infty^0$ if we can show that $(L_0^* \Lambda^N)^{**} \in C_\infty^0$, and it is sufficient to consider the first kind of operator $(L^0 \Lambda^N)^{**}$. Accordingly we essentially may repeat the proof of VI,prop.3.1, noting that cdn.(1_5) just supplies the essential property.

It is clear that the assertion holds for $N \leq 1$, since $a = (a\Lambda^0)^{**}$, and $D\Lambda = (D_0 \Lambda^1)^{**}$ just are the generators of C. For a product $DF \in L_2$ we get, as in the proof of V,lemma 3.2,

$$DF\Lambda^2 u = D\Lambda^2 Fu + D\Lambda^2 [H,F]\Lambda^2 u \text{ , for all } u \in \text{im } H_0 \text{ .} \tag{3.4}$$

Using (1_5) we may substitute $[H,F]$ by a finite sum of products $D'F'$, for which the second term (3.3) assumes the form $D\Lambda^2 D'F'\Lambda^2 u = (D\Lambda)(D'^*\Lambda)^*(F'\Lambda^2)u$. All these terms involve operators of C_∞^0 only. Since im H_0 is dense, we thus get $((DF)_0\Lambda^2)^{**} \in C_\infty^0$, proving the assertion for N=2. (The first term, of course, gives $(D\Lambda)(D^*\Lambda)^* \in C_\infty^0$.)

Assume now the statement proven for L_N, $N \geq 2$. Consider a product LM, $L \in L_2$, $M \in L_{N-1} \subset L_N$. For $u \in \Lambda^{-N-1} C_0^\infty$ we get

$$LM\Lambda^{N+1} u = L\Lambda^2 M\Lambda^{N-1} u + L\Lambda^2 [H,M]\Lambda^{N+1} u \text{ ,} \tag{3.5}$$

since $\Lambda^{N \pm 1} u \in C_0^\infty$. Using (1_5) and prop.3.1 we get $[H,M] \in L_{N,0} \subset L_{N+1,0}$, so that the second term represents an operator in C_∞^0. (It may be written as $(L_0\Lambda^2)^{**}([H,M]_0\Lambda^{N+1})^{**}u$.) The space $\Lambda^{-N-1}C_0^\infty$ is dense in H in view of cdn.(s). Accordingly (3.5) implies that $LM\Lambda^{N+1} \in C_\infty^0$. Since the general expression of L_{N+1} is a sum of expressions LM of the above type, the prop. follows.

In order to simplify notation we will, for the present section, use the notation L for the realization Z of V,def.6.2, corresponding to an expression $L \in L_N$. In other words, the expression L and the unbounded (differential-) operator $L: H_N \to H_0 = H$ will be denoted by the same symbol L . (We recall that $H_N = \text{dom } \Lambda^{-N} = \text{im } \Lambda^N$. For $u \in H_N$ we write $= \Lambda^N v$, and the define

$$Lu = L\Lambda^N v = (L_0 \Lambda^N)^{**} v \text{ ,} \tag{3.6}$$

which is well defined, by prop.3.2.) Note that this notation already was in use for $D \in D!$, where we used to write $D\Lambda$, interpreting dom $D = H_1 = \text{im } \Lambda$. Also we observe that $L \subset L_0^{**}$. The notation is ambiguous, due to the fact, that an expression L of order N also has every order R>N . However, all these various operators, denoted by L , have the same closure L_0^{**} .

<u>Proposition 3.3</u>. For $L \in L_{N,0}$ we have $\Lambda L \Lambda^N$, $(\Lambda^N L)^{**}\Lambda \in E$.
<u>Proof</u>. Again it is sufficient to focus on the first kind of operator, using that $(\Lambda^N L)^{**}\Lambda = (\Lambda L^* \Lambda^N)^*$. Looking first at $A_m^\# = A!$, $D_m^\# = D!$, we find that then the operators listed are compact. Indeed, from prop.3.2 we conclude that $L\Lambda^N \in C_\infty^0$ is a finite sum of products of terms (1.10), with compactly supported a and D. But then Λa, $\Lambda D \Lambda^j$, $\Lambda^{j+1} D$ all are compact, by III,thm.3.7, so that $\Lambda L \Lambda^N$ is compact as well. Since $K(H) \subset E$, the statement follows.

In the general case we always have (m_5). For N=0 we must look at $A = \Lambda[a,D]$. Note that $[D\Lambda,a] = (D\Lambda)(\Lambda^{-1}[\Lambda,a])+[D,a]\Lambda \in E$. By prop.2.4 we have (m_4) valid as well. Thus VII,rem.3.4 implies that the first term at right is in E , hence $[D,a]\Lambda \in E$ and $\Lambda[D,a] = (\Lambda[a,D])^* \in E$ as well. Note that we have (m_4), again by prop.2.4. To discuss the case of N=1, let $L=a[D,F]$, $a \in A!$, $D,F \in D!$. Now we get, with proper interpretation of $[D,\Lambda]$,

$$(3.7) \qquad [D\Lambda,F\Lambda] = [D,\Lambda]F\Lambda + \Lambda[D,F]\Lambda + [\Lambda,F]D\Lambda \in E .$$

We already noted that $[D,\Lambda] \in E$, $[\Lambda,F] \in E$, by VII,rem.3.4, since we have (m_4) and (m_5). Therefore (3.7) implies $\Lambda[D,F]\Lambda \in E$. But we get

$$(3.8) \qquad \Lambda a[D,F]\Lambda = ([\Lambda,a]\Lambda^{-1})(\Lambda[D,F]\Lambda) + a\Lambda[D,F]\Lambda ,$$

where all terms at right are in E . Therefore the statement holds for N=1 .

For N=2 the typical term of L is of the form $F[D,E]$. We get

$$(3.9) \qquad \Lambda F[D,E]\Lambda^2 u = (\Lambda F)(\Lambda[D,E]\Lambda)u + (\Lambda F)[\Lambda,[D,E]]\Lambda u , \quad u \in H_2 ,$$

where the right hand side operator is in E , by the above.

Finally, for general $N \geq 2$, the typical term in $L_{N+1,0}$ may be written as LM , with $L \in L_{2,0}$, $M \in L_{N-1}$. Then we may write (3.5) again, multiplied by Λ from the left. Again we get $\Lambda L \Lambda^2 M \Lambda^{N-1} \in E$, $\Lambda L \Lambda^2 [H,M]\Lambda^{N+1} \in E$, completing the proof.

In the following we introduce more general algebras of differential expressions of arbitrary order, within reach of an algebra C or C_∞^0. The purpose is to replace cdn's (l_5), (l_5') by weaker conditions.

Let $P_\infty = \cup_{N=0}^\infty P_N$ be a graded algebra of differential expressions, invariant under formal adjoints, and a 2-sided $L_\infty(A!,D!)$-module. In details, we assume the expressions in P_N of order $\leq N$, that $P_N \subset P_M$ as $N<M$, (C^∞-coefficients, of course), and that

(3.10) $\quad P_N P_M \subset P_{N+M},\ P_N L_M \subset P_{N+M},\ L_N P_M \subset P_{M+N},\ N,M=0,1,2,\ldots$.

In addition let the following conditions (p_j), $j=1,2$, hold:
Condition (p_1): $[H,L] \in P_{N+1}$ for all $L \in P_N$, $N=0,1,2,\ldots$,
Condition (p_2): Every $L \in P_N$ is within reach of $L(H)$.

Notice that $(p_1),(p_2)$ imply that all $L \in P_N$ are H-compatible, in the sense of V,def.6.4.

We shall say that an algebra P_∞ with above properties is within reach of an algebra $A \subset L(H)$ if every $P \in P_N$, as an expression of order N, is within reach of A. Also, if P_∞ is within reach of $C_\infty(A!,D!)$ for some classes $A!$, $D!$, with $C_\infty \subset O(0)$ of sec.1, then we say that P_∞ is within general reach of $A!$, $D!$. For P_∞ within reach of $C_s(A!,D!)$, if we even have $\Lambda L \Lambda^N \in E_s(A!,D!)$, for all $L \in P_N$, $N=0,1,\ldots$, we shall say that P_∞ is of principal symbol type (in C_s), - the latter only for $s \in \mathbb{R}$.

For P_∞ with above properties define the classes $Q_N = P_N + L_N$, $N=0,1,\ldots$. Then $Q_\infty = \cup_{N=0}^\infty Q_N$ clearly again is a graded algebra of differential expressions and P_∞ is an ideal of Q_∞.

We will ask whether Q_∞ again is within reach of $C(A!,D!)$, given that P_∞ is within reach of $C(A!,D!)$. Cor.3.4 below gives an affirmative answer, using the following amended condition:
Condition $(1_5)(P_\infty)$: We have $[H,a] \in Q_1$, $[H,D] \in Q_2$, for $a \in A!$, $D \in D!$.

In case of an algebra P_∞, satisfying (p_j), $j=1,2$, and the above, and of principal symbol type, we also formulate a relative cdn.$(1_5')$, as follows (setting $Q_{N,0} = P_N + L_{N,0}$, $Q_\infty = \cup Q_N$):
Condition $(1_5')(P_\infty)$: We have $[H,a] \in Q_{1,0}$, $[H,D] \in Q_{2,0}$, for $a \in A!$, $D \in D!$.

Notice that $(1_5)(P_\infty)$ implies that $Q_\infty = \cup Q_N$ satisfies (p_j), $j=1,2$, by the corollary, below. In fact, this even is true without the condition that P_∞ is within reach of some comparison algebra, by the same arguments.

In the corollary, below, the special case of the minimal classes $A^\#_m$, $D^\#_m$ is excluded. (Since $A^\#_m$, $D^\#_m$ satisfy (1_5), the corollary would be useless.)

Corollary 3.4. Under the above assumptions on P_∞, if cdn.$(1_5)(P_\infty)$ holds (but not necessarily cdn.(1_5)), then the graded algebra Q_∞

is within reach of $C(A!,\mathcal{D}!)$. Moreover, if in addition $\mathcal{P}_\infty$ is of principal symbol type (in $C_0 = C$), then the algebra $\mathcal{Q}_{\infty,0} = \mathcal{P}_\infty + \mathcal{L}_{\infty,0}$ also is of principal symbol type (in C).

Proof. Since $P \in \mathcal{P}_N$ already is within reach of C we only must show this for $L \in \mathcal{L}_N$ again. For the latter one may repeat literally all arguments of the proofs of prop.3.2 and 3.3, using the ideal property of $\mathcal{P}_\infty$. In (3.4), for example, the commutator now is a sum P_2+L_2, where L_2 is treated as before, while $P_2 \in \mathcal{P}_2$ gives the term $(D\Lambda^2)P_2\Lambda^2 \in C$. Similarly for (3.5), also in view of prop.3.3.

4. Symbol calculus in H_s.

In this section we keep all assumptions of sec's 1 and 3 valid. Particularly we assume cdn's (s), (l_j), j=1,2, and (l_5) (or (l_5')), and either (m_5) for a triple $\{\Omega, d\mu, H\}$, and classes $A! = A^\#$, $\mathcal{D}! = \mathcal{D}^\# + \{\alpha 1\}$, or we use the minimal classes $A! = A^\#_m$, $\mathcal{D}! = \mathcal{D}^\#_m$. Or else, we assume $(l_5)(\mathcal{P}_\infty)$ (or $(l_5')(\mathcal{P}_\infty)$), instead of (l_5), (or (l_5')), relative to some algebra $\mathcal{P}_\infty$ within comparison reach of $C(A!,\mathcal{D}!)$, satisfying (p_j), j=1,2, and (3.10) (and of principal symbol type (in C)). In the latter case we exclude the minimal classes, and always assume (m_5').

In the interest of a unified notation, let us always at least formally work with an algebra of expressions $\mathcal{P}_\infty$, where we set $\mathcal{P}_N = \{0\}$ in the case where no preconceived such class is given. Clearly this defines the <u>trivial algebra</u> $\mathcal{P}_\infty$, satisfying all assumptions. It is within general reach of every $A!$, $\mathcal{D}!$, is of principal symbol type, gives an algebra $\mathcal{P}_\infty$, and an $\mathcal{L}_\infty$-module, satisfies cdn's (p_j), j=1,2, and $(l_5)(\mathcal{P}_\infty)$, $(l_5')(\mathcal{P}_\infty)$, for the trivial algebra $\mathcal{P}_\infty$, coincide with the ordinary conditions (l_5) and (l_5'). We appoint that for the minimal classes $A^\#_m$, $\mathcal{D}^\#_m$ only the trivial algebra $\mathcal{P}_\infty$ is to be used. We then always define $\mathcal{Q}_N = \mathcal{P}_N + \mathcal{L}_N$, $\mathcal{Q}_{N,0} = \mathcal{P}_N + \mathcal{L}_{N,0}$, so that $\mathcal{Q}_N = \mathcal{L}_N$, $\mathcal{Q}_{N,0} = \mathcal{L}_{N,0}$, in case of the trivial algebra $\mathcal{P}_\infty$.

Let us recall the <u>order</u> concept introduced in I,6 for general HS-chains. Here we use the chain $\{H_s : s \in [-\infty,+\infty]\}$ of Sobolev spaces, of course. We shall adopt an attitude similar as in theory of pseudodifferential operators to not distinguish in notation between the various extensions in $L(H_s,H_{s-m})$ of a given operator $A \in O(m)$. If $A|C_0^\infty$ is given by a differential expression L then we even use the same notation L for all above operators as well as the realization Z of V,6 . This will bring us away from the unbounded operator notions of ch.1, and will be practical in view of the fact, that there is a unique such extension, for every s, and a unique closed realization, all in view of cdn.(s).

Focusing on the algebra $\mathcal{Q}_\infty = \mathcal{P}_\infty + \mathcal{L}_\infty$ of expressions, we find that every $L \in \mathcal{Q}_N$ is within reach of $\mathcal{C}$,and even is H-compatible, in the sense of V,6, by cor.3.4 and the remarks preceeding it. Thus we find that L defines an operator in $\mathcal{O}(N)$, by V,cor.6.10.

Proposition 4.1. For $A=L\Lambda^N$, $B=(\Lambda^N L)^{**}$, with $L\in\mathcal{Q}_N$, we get

$$(4.1) \qquad \Lambda^s A\Lambda^{-s}-A = C\Lambda=\Lambda C'=E \ , \ \Lambda^s B\Lambda^{-s}-B = C''\Lambda=\Lambda C'''=E' \ , \ s\in\mathbb{R} \ ,$$

where $E, E', C, C', C'', C''' \in \mathcal{C}$ may depend on s . Moreover, if we require $(1_5')(\mathcal{P}_\infty)$ instead of $(1_5)(\mathcal{P}_\infty)$, we even get $E,E'\in \mathcal{E}$.

Proof. For a given s we apply V,thm.6.8, with $M > |s|$. Also we use that $L_j=(\mathrm{ad}\ H)^j L \in\mathcal{Q}_{N+j}$ (or $\in\mathcal{Q}_{N+j,0}$, in case of $(1_5'))$, so that

$$(4.2) \qquad L_j\Lambda^{N+j}\in \mathcal{C} \ , \text{ and } L_j\Lambda^{N+j+1}=\Lambda L_j\Lambda^{N+j}+ [L_j\Lambda^{N+j},\Lambda] \in \mathcal{E},$$

by cor.3.4, correspondingly, depending on the assumption made. (For the minimal classes we must use $\chi\Lambda$ for Λ , in the commutator with suitable χ.) Similarly the term $\Lambda L_{N+1}\Lambda^{N+1}$ in V,(6.12) is in $\mathcal{E}$, or of the form $C\Lambda$, $C\in\mathcal{C}$, respectively, hence also the reminder gives a term of the proper form for V,(6.13). (For the minimal classes $\mathcal{A}^\#_m$, $\mathcal{D}^\#_m$ the resolvent $R(\lambda)$ only is in $\mathcal{C} + \{\alpha 1\}$, but $\mathcal{C}$ is an ideal of $\mathcal{C}+\{\alpha 1\}$ so that the integrand of V,(6.12) still is in $\mathcal{C}$ or $\mathcal{E}$,resp. This gives the first (4.1) while the second follows by taking adjoints, q.e.d.

For thm.4.2 below we recall the natural isometry $\Lambda^r:H_s\to H_{s+r}$ (Actually, Λ^r stands for the extension $(\Lambda^r)_s$ of Λ^r to H_s, but the subscript 's" can safely be dropped). Note that

$$(4.3) \qquad \mathcal{L}(H_{s+r}) = \Lambda^r \mathcal{L}(H_s)\Lambda^{-r} \ , \ \mathcal{K}(H_{s+r}) = \Lambda^r \mathcal{K}(H_r)\Lambda^{-r} \ ,$$

For the remainder of this section we assume, in addition to the conditions mentioned earlier, that $\mathcal{P}_\infty$ is within general reach of the classes $\mathcal{A}!$, $\mathcal{D}!$, and of principal symbol type in $\mathcal{C}_s$, for $s\in\mathbb{R}$.

Theorem 4.2. The generators a, $D\Lambda^j$, $\Lambda^j D$, $a\in\mathcal{A}!$, $D\in\mathcal{D}!$ of (1.10) are in $\mathcal{L}(H_s)$, for every $s\in\mathbb{R}$, (and in $\mathcal{O}(0)$), so that all comparison algebras C_s (including C_∞) are well defined norm closed subalgebras of $L(H_s)$ (or $O(0)$). We have $L\Lambda^N$, $\Lambda^N L \in C_s$, $-\infty<s<\infty$, for all $L\in\mathcal{Q}_N$. Every C_s , $s\in\mathbb{R}$, contains its H_s-adjoints, i.e., C_s is a C^*-subalgebra of $L(H_s)$. For $s,t\in\mathbb{R}$, and every A in the algebra finitely generated by $C_\infty{}^0$ and $\{L\Lambda^N, \Lambda^N L: L\in\mathcal{Q}_N, N=1,..\}$ we have, (with the commutator ideal E_t of C_t)

$$(4.4)\quad \Lambda^{-s}A\Lambda^s - A \in C_t, \text{ and } '\in E_t' \text{ in case of cdn. } (1_5')(P_\infty),$$

Also,

$$(4.5)\quad C_{s+r} = \Lambda^s C_r \Lambda^{-s} , \; E_{s+r} = \Lambda^s E_r \Lambda^{-s} , \text{ for all } s,r \in \mathbb{R} .$$

In particular, the class $\mathcal{Q}_\infty$ is within general reach of $A!$, $D!$, and $\mathcal{Q}_{\infty,0}$ is of principal symbol type in every $C_s(A!,D!)$.

<u>Proof</u>. Prop.4.1 implies that a , $D\Lambda$, ΛD , $L\Lambda^N$, $\Lambda^N L \in O(0)$, hence $\in L(H_s)$, justifying the definition of C_s, also for $s=\infty$. To show $L\Lambda^N$, $\Lambda^N L\in C_s$ one repeats the argument of prop.3.2, as in cor.3.4, using that P_∞ is within reach of C_s, $-\infty<s\leq\infty$. Inspecting (3.4) and (3.5) one indeed confirms $A\in C_s$. Similarly, in case of $(1_5')(P_\infty)$, one confirms that $\mathcal{Q}_{\infty,0}$ is of principal symbol type (in C_s), $s\in\mathbb{R}$.

For an operator $A\in O(0)$ the H_s-adjoint is given by $\Lambda^{2s}A^*\Lambda^{-2s}$, with the H-adjoint A^*, since $(u,Av)_s= (\Lambda^{2s}A^*\Lambda^{-2s}u,v)$, for $u,v\in H_\infty$. We get $\Lambda^{2s}A^*\Lambda^{-2s} \in C_s$, by (4.4) ,with s,t replaced by -2s and s. Thus the adjoint invariance of C_s follows from (4.4).

Next we note that (4.4) for t=0 implies

$$(4.6)\quad \Lambda^{-s}C_\infty{}^0\Lambda^s \subset C , \text{ for all } s \in \mathbb{R} .$$

However, if A_j converges to A in $L(H)$, then $\Lambda^{-s}A_j\Lambda^s \in L(H_s)$ converges to $\Lambda^{-s}A\Lambda^s\in L(H_s)$, and vice versa. Therefore we may take closures in (4.6) and get

$$(4.7)\quad C_s \subset \Lambda^s C_0 \Lambda^{-s} .$$

This conclusion may be repeated with $C_0=C$ replaced by C_s, and Λ^s by Λ^{-s}, using (4.4) for t=s.Then we get $C=C_0\subset \Lambda^{-s}C_s\Lambda^s$, or, equality in (4.7). Thus we have the first relation (4.5) , for r=0, and the second follows since, of course, the commutator ideal is invariant under a conjugation with an isometry. Similarly for general r.

Accordingly, (4.5) and the adjoint invariance both depend on proving (only the first part of) (4.4).

However, (4.4), for $t=0$,and $s \in \mathbb{R}$, is a consequence of (4.1) , as far as the generators are concerned, and it then follows trivially for all $A \in C_\infty^{\;0}$. Similarly for A finitely generated by $L\Lambda^N$, $\Lambda^N L$, $L \in \mathcal{Q}_N$. For general t we cannot use prop.4.1, but must go back to V,thm.6.8 instead, showing that all terms at right of V,(6.11) are in C_t. Indeed, we get $L_j \in \mathcal{Q}_{N+j}$, hence $L_j\Lambda^{N+2j} \in C_t$, as already shown above, using that P_∞ is within reach of C_∞ . Also the integrand of $S_{m,s}$ is in C_t again. In fact, we have $\Lambda L_{M+1}\Lambda^{M+j} \in C_t$. In case of (m_5') the functions S^1, $S^2_{M,s}$ also take values in C_t , since $\Lambda \in C_t$,and S^j are functions of Λ . Therefore the integrand is in C_t . On the other hand, in case of the minimal classes, the resolvent is in $C_t+\{\alpha 1\}$, hence the S^j are in that algebra. But C_t is an ideal of $C_t+\{\alpha 1\}$, so that again the integrand is in C_t. Thus we only must prove convergence of the integral in $L(H_t)$.

For that we may look at the $L(H)$-convergence of the integral V,(6.12) with $L_{M+1}\Lambda^{M+N+1}$ replaced by $\Lambda^t(L_{M+1}\Lambda^{M+N+1})\Lambda^{-t} = V_t$, since Λ^t and Λ^{-t} commute with $S^j(\lambda)$. But we get $V_t \in L(H)$, by another application of the expansion V,(6.11). Accordingly the integral V,(6.12) with V_t indeed converges in $L(H)$, by V,thm.6.8 ,and we get $S_{M,-s} \in C_t$, as stated in (4.4).

Finally, for the second part of (4.4), assume that $(l_5')(P_\infty)$ holds. We may assume (4.5) true now, which only depends on the first (4.4). Thus, in order to show that the right hand side of V,(6.11) , (for $-s$), is in E_t, we must prove that $W_t = \Lambda^{-t}L_j\Lambda^{N+2j}\Lambda^t \in E$, as $j>1$, and $\Lambda^{-t}S_{M,-s}\Lambda^t \in E$, where we know that $L_j \in \mathcal{Q}_{N+j,0}$. From cor.3.4 we get $L_j\Lambda^{N+2j} \in E$. Again we use V,(6.11) on L_j , and for $-t$ instead of s, to express W_t by $L_{j+k}\Lambda^{N+2j+2k}$ and another remainder. The first terms are in E , as posted. The remainder is an $L(H)$-convergent integral with integrand in E . Thus $W_t \in E$, and $L_j\Lambda^{N+2j} \in E_t$. The Λ-powers may be taken into the remainder $S_{M,-s}$, and will give an integral V,(6.12) with $\Lambda L_{M+1}\Lambda^{M+N+1}$ replaced by V_t above. We just showed that $V_t \in E$, (we have $V_t = W_t$, for $j=M+1$). Thus also $S_{m,-s} \in E_t$, and we get (4.4) for $A = L\Lambda^N$. Similarly for the other generators $B = \Lambda^N L$. The same follows trivially for $A \in C_\infty^{\;0}$, or for A finitely generated by $L\Lambda^N$, $\Lambda^N L$, with $L \in \mathcal{Q}_N$, and the proof is complete.

In thm.4.3, below, we now summarize the consequences of thm.

4.2. These facts will be pursued in closer detail in X,4, where it will be seen that the Fredholm theory of an H_s-comparison algebra is similar, or even the same, as that in the corresponding H-comparison algebra. The assumptions are the same as for thm.4.2.

Theorem 4.3. The quotient algebras C_s/E_s , $s \in \mathbb{R}$, all are isometrically isomorphic, and also the quotient algebras $E_s/K(H_s)$, $s\in\mathbb{R}$, all are isometrically isomorphic. In each case a *-isomorphism $C_s/E_s \to C_t/E_t$ and a *-isomorphism $E_s/K(H_s) \to E_t/K(H_t)$ both are induced by the natural isometry $A \to \Lambda^r A \Lambda^{-r}$, $r=t-s$.

Accordingly, the maximal ideal spaces $\mathbb{M}_s$ and $\mathbb{M}_t$ of C_s/E_s and C_t/E_t are homeomorphic under the associate dual map of the above isometric isomorphism, and, if $E/K(H)$ is a function algebra $C(\mathbb{E},K(h))$, as in the examples of VII,1,2, and 4, then also

$$E_s/K(H_s) \simeq C(\mathbb{E}_s,K(h)) , \tag{4.8}$$

where all spaces $\mathbb{E}_s$ are homeomorphic to $\mathbb{E} = \mathbb{E}_0$.

Moreover, under condition $(1_5')(P_\infty)$, if the homeomorphisms used to identify all the spaces $\mathbb{M}_s$ with $\mathbb{M} = \mathbb{M}_0$, then the operators in the algebra finitely generated by C^0_∞ and $L\Lambda^N$, $\Lambda^N L$, for $L\in\mathcal{Q}_N$, $N=0,1,\ldots$, have their C_s-symbol independent of s .

The proof of theorem 4.3 is evident. We will continue this discussion in X,4. Note that a dependence of the C_s-symbol of a (finitely generated) operator A on s is possible, if only $(1_5)(P_\infty)$, not $(1_5')$ holds. We shall not discuss such problems here.

Finally let us shortly look at the (Frechet-)algebra C_∞ , already mentioned. Note that the algebra $C_\infty{}^0 \subset \mathcal{O}(0)$ also has a closure C_∞ in the natural topology of $\mathcal{O}(0)$, i.e., the locally convex topology of all operator norms

$$\|A\|_s = \sup\{\|Au\|_s : \|u\|_s \le 1\} , \; s \in \mathbb{R} . \tag{4.9}$$

By the Calderon interpolation theorem this is a Frechet topology; the class $\{\|A\|_j : j\in\mathbb{Z}\}$ defines the same topology (cf.I, thm.6.3) . If we assume $(1_5')(P_\infty)$ then a C_∞-symbol $\sigma: C_\infty \to C(\mathbb{M})$, with the common space $\mathbb{M}=\mathbb{M}_s$ of thm.4.3 still may be defined as the restriction to C_∞ of the C_s-symbol (which proves independent of s) (cf.X,5).

In the special case of a certain comparison algebra for Laplace-$\mathbb{R}^n$ it was shown by E.Schrohe and the author [CSch] , that the symbol of this Frechet-algebra still maps onto C(𝕄). Moreover, the same is true for the (unique) Frechet comparison algebra of a compact manifold Ω (cf. Schrohe [Sch_1]). (In both cases we have $E_s = K(H_s)$, so that the symbol of the second kind is trivial.)

For a more detailed discussion of the abstract elements of the Frechet comparison algebra C_∞, and the H_s-comparison algebras C_s cf. X,4.

5. Local properties of the Sobolev spaces H_s.

In sec's 5 and 6 we will look somewhat more closely at the spaces H_s of sec.1, which, so far, were little more than the domains of the powers of the self-adjoint operator $\Lambda = H^{-1/2}$. Here we will discuss the proofs of VI,thm.1.6, and thm. 1.2, thm.1.3, concerning local properties which were postponed so far. In sec.6 we will focus on equivalence of $\|u\|_k$, for an integer $k>0$, with a weighted L^2-norm of the 'k-jet' of u .

First let the assumptions of thm.1.2 hold: We are given

$$(5.1) \qquad \{\Omega, d\mu, H\} \quad , \quad \{\Omega, d\nu, K\}$$

on the same manifold Ω, coinciding outside some compact set. Only the minimal sets $A^\#_m$ and $\mathcal{D}^\#_m$ are needed in this discussion. We assume condition (s) for the first triple. The sets $A^\#_m$, $\mathcal{D}^\#_m$, with the first triple, then qualify for all results of III,3,V,3, V,6 and will give us a minimal algebra of expressions- explicitly known as the set of all $C^\infty(\Omega)$-differential expressions of compact support, with L_N consisting of all such expressions of order $\le N$.

Although the spaces $O(m)$, for the first triple (5.1), are established, we will not use them, or the notation of sec.4, until thm.1.2 is established, for reason of clarity. However, once we have thm.1.2, we will have a unique H_∞ and H_s , and it follows that all classes $O(m)$ for H and K coincide. Thus the notation of sec.4 then will be clear and convenient again.

Note that $\gamma = \{d\mu/d\nu\}^{1/2}$ defines a non-vanishing $C^\infty(\Omega)$-function, with $\gamma=1$ outside a compact set. We have $d\mu = \gamma^2 d\nu$, and a change of dependent variable $u=\gamma v$ will take the second triple onto the form

$$(5.2) \qquad \{\Omega, d\mu, K^\backprime\} \ , \ K^\backprime = \gamma^{-1} K \gamma \ ,$$

which has the same measure and inner product as the first (5.1).

Proposition 5.1. The second triple also satisfies condition (s).
Proof. Clearly (s) holds for $\{\Omega, d\nu, K\}$ if and only if it holds for $\{\Omega, d\mu, K^{\sim}\}$. Thus it is sufficient to prove (s) for $K^{\sim}$, i.e. one may assume that $d\mu = d\nu$.

Note that

$$(5.3) \qquad P_j = K^j H^{-j} = (H^j + (K^j - H^j))H^{-j} = 1 + L^j \Lambda^{2j} \in J ,$$

with $\Lambda = H^{-1/2}$, where $L^j = K^j - H^j$ has compact support, hence $L^j \in L_{2j}$. This operator is Fredholm, since we have

$$(5.4) \qquad \sigma_{P_j} = ((k^{jl}(x)\xi_j\xi_l)/h^{jl}(x)\xi_j\xi_l))^j ,$$

by VI,thm.2.2, since $C=J$ has $E=K(H)$, while its symbol space $\mathbb{M}$ is the 1-point compactification of the cosphere bundle. Moreover, since σ_{P_j} is always positive, the Fredholm index must be zero, since the symbol can be connected by a continuous non-vanishing family of symbols to the identity function (cf. X,thm.1.9(A)). (Notice the symbol is =1 for all x ouside a compact set.) Also,

$$(5.5) \qquad (H^{-j}u, P_j u) = (H^{-j}u, K^j H^{-j}u) \geq \|\Lambda^{2j}u\|^2$$

follows first for $u \in H^j C_0^\infty$, then also for all $u \in H$, by continuous extension. Thus $P_j u = 0$ implies $u=0$. Since P_j is Fredholm and has index zero, it must be invertible.

Now, to show that the minimal operator K_0^j is essentially self-adjoint, assume that $f \in H$, $(f, K^j u) = 0$ for all $u \in C_0^\infty(\Omega)$. We may write this as $0 = (f, P_j H^j u) = (P_j^* f, H^j u)$, $u \in C_0^\infty$. But since H_0^j is essentially self-adjoint, we have im H_0^j dense in H, so that $P_j^* f = 0$. Since P_j is invertible in $L(H)$, we conclude that $f=0$. Therefore im K_0^j is dense in H. Since $K_0^j \geq 1$ we conclude that K_0^j is essentially self-adjoint, by I, thm.3.1, q.e.d.

Denoting the Sobolev norms of the triples (5.1) by $\|.\|_{s,H}$ and $\|.\|_{s,K}$, for a moment, we note that

$$(5.6) \qquad \|u\|_{s,H} = \|H^{-s/2}u\| , \quad \|u\|_{s,K} = \|\gamma^{-1}K^{-s/2}u\| , \text{ as } u \in C_0^\infty(\Omega) .$$

where $\|.\|$ is the norm of $L^2(\Omega, d\mu)$. For a proof of thm.1.2 it is sufficient to show that both norms (5.6) are equivalent, for every $s \in \mathbb{R}$.

Proposition 5.2. For a given $s \in \mathbb{R}$ the two norms (5.6) are equivalent if and only if the operator $(H^{-s/2}K^{s/2})^{**}$ and its inverse are bounded (in $L(H)$).

Indeed, C_0^∞ is dense in both Sobolev spaces, by lemma 1.1, (since we have (s) for H and K), we get equivalence if and only if

$$(5.7) \qquad c_1\|u\|_{s,H} \le \|u\|_{s,K} \le c_2\|u\|_{s,H} \text{ for all } u \in C_0^\infty .$$

Setting $H^{-s/2}u=v$, and using that $H^{-s/2}C_0^\infty$ is dense in H we conclude that (5.7) holds if and only if

$$(5.8) \qquad c_1\|v\| \le \|\gamma^{-1}(K^{-s/2}H^{s/2})^{**}v\| \le c_2\|v\| \text{ , for all } v \in H ,$$

which means that $(H^{-s/2}K^{s/2})^{**}$ is bounded and has a bounded inverse since γ is bounded and bounded below, and by symmetry.

Next we apply I,lemma 5.1 :The operators H and $K^{\backprime}$ are self-adjoint with respect to the same inner product, and we have $K^{\backprime j}H^{-j} \in J \subset L(H)$, while the inverse H^jK^{-j} exists in $J \subset L(H)$ as well. Hence we get $K^{\backprime s/2}H^{-s/2}$ bounded for s>0, and $H^{s/2}K^{\backprime -s/2}$ bounded as well. Taking the adjoint of these operators as well, we conclude that $(K^{\backprime s/2}H^{-s/2})^{**} = \gamma^{-1}(K^{s/2}\gamma H^{-s/2})^{**} \in L(H)$, $s\in\mathbb{R}$, and that these operators have bounded inverses.

Finally we apply V, thm.3.8 for $A=\gamma$, $\Lambda^s=K^{-s/2}$, noting that the remainder is of the form $CK^{-s/2}$, with $C \in K(H)$. Since $\gamma^{-1}\in L(H)$, we get $\gamma^{-1}K^{s/2}\gamma = K^{s/2} + CK^{s/2}$, $C \in K(H) = E$. Or,

$$(5.9) \quad (1+C)K^{s/2}H^{-s/2}=\gamma^{-1}K^{s/2}\gamma H^{-s/2}\in L(H),\ H^{s/2}K^{-s/2}(1+C') \in L(H),$$

with compact operators C,C' . Clearly 1+C , 1+C' are Fredholm, hence $P_sK^{s/2}H^{-s/2}$ and $H^{s/2}K^{-s/2}Q_s$, with orthogonal projections P_s, Q_s onto spaces of finite codimension are bounded. Set $J_s = K^{s/2}H^{-s/2}$. This operator and its inverse are preclosed. If J_s is not bounded, there exists a sequence $u_j\in H$ with $u_j\to 0$, $\|J_su_j\|=1$. We must have $P_sJ_su_j\to 0$, hence $v_j=(1-P_s)J_su_j$ is bounded in a finite dimensional space. It may be assumed convergent,by compactness of closed balls in $\mathrm{im}(1-P_s)$. Thus $u_j\to 0$ $J_su_j\to v$, $\|v\|=1$. Since J_s is preclosed, this gives a contradiction. Similarly for J_s^{-1}. This completes the proof of thm.1.2.

We also have made progress in the general proof of VI,thm. 1.6, insofar as the boundedness of the operators V_t , V_t^{-1} was shown for all $t\in\mathbb{R}$. To confirm the statements about the cosets $U_t= V_t+K(H)$, we must extend VI,prop.1.8. As mentioned above, the notation of sec.4 now is in order: all operators in $\mathcal{O}(m)$ are considered as operators $H_\infty \to H_\infty$, but also are identified with their continuous extensions to any H_s, as maps $H_s \to H_{s-m}$. Specifically we have H^s, $K^s \in \mathcal{O}(2s)$.

Proposition 5.3. For $0 \le s,t,\sigma,\tau \in \mathbb{R}$ we have

$$(5.10) \qquad H^\sigma[H^{-s},K^{-t}]H^\tau \in K(H) \ , \text{ whenever } \sigma+\tau < s+t+1/2 \ .$$

Proof. With $R(\lambda)=(H+\lambda)^{-1}$, $S(\mu)=(K+\mu)^{-1}$ as in the proof of VI,prop. 1.8 we set up a formula like VI,(1.17) again, but now we will use identity VI, (6.8) to express H^{-s} and K^{-t} by integrals over higher powers of the resolvents. Instead of the commutator formula VI,(1.18) we need formula (5.11), below.

Proposition 5.4. Under cdn.(s) we have

$$(5.11) \qquad [R^m(\lambda),S^l(\mu)] = R^m(\lambda)S^l(\mu)L(\lambda,\mu)S^l(\mu)R^m(\lambda) \ ,$$

for all $\lambda,\mu > 0$, and all $l,m=1,2,\dots$, where $L(\lambda,\mu)=L_{l,m}(\lambda,\mu)$ denotes the polynomial (with $r_{pq}=2(m+l-p-q)+3$)

$$(5.12) \quad L(\lambda,\mu) =[(H+\lambda)^m,(K+\mu)^l] =\sum_{p=1}^m \sum_{q=1}^l L_{pq}\lambda^{p-1}\mu^{q-1}, \ L_{pq}\in L_{r_{pq},0} \ ,$$

where we again use the classes $A! = A_m^{\#}$, $\mathcal{D}! = \mathcal{D}_m^{\#}$.

The proof is a calculation, noting that $R(\lambda)$, $S(\mu) \in \mathcal{O}(-2)$. One has $L_{pq} = c_{pq}[H^{m-p+1},K^{l-q+1}]$, using the binomial theorem, and the fact that $[H^p,K^q]=0$, as $p=0$ or $q=0$. Also, a straight-forward extension of VI,prop.1.9 is required. We indeed get $L_{pq}\in L_{r_{pq},0}$.

Now we prove prop.5.3. Using VI,(1.6) for H^{-s} and K^{-t} write

$$(5.13) \qquad H^\sigma[H^{-s},K^{-t}]H^\tau = \sum_{p=1}^m \sum_{q=1}^l \int_0^\infty d\lambda \int_0^\infty d\mu I_{pq}(\lambda,\mu) \ ,$$

with

$$(5.14) \quad I_{pq}=c_{pq}\lambda^{m+p-s-2}\mu^{l+q-t-2}H^\sigma R^m(\lambda)S^l(\mu)L_{pq}S^l(\mu)R^m(\lambda)H^\tau, \ c_{pq}\in\mathbb{R}.$$

For given $s,t,\sigma,\tau >0$, if l and m are chosen sufficiently large, then the integrand will be norm continuous and $O((\lambda\mu)^{-1-\varepsilon})$, $\varepsilon>0$, so that the improper Riemann integrals exist in $L(H)$. Also we get $I_{pq}(\lambda,\mu) \in K(H)$, so that (5.13) is a compact operator.

Indeed, the existence of the integral at $\lambda=0$ or $\mu=0$ just requires $s<m$, $t<l$, since $p,q\ge 1$. Write

$$I_{pq}=c_{pq}(\lambda\mu)^{-1-\varepsilon}(\lambda^{\alpha}R^{m}H^{\sigma}K^{\beta-\sigma}S^{l}\mu^{\gamma})(K^{\sigma-\beta}L_{pq}K^{\tau-\beta})(\mu^{\gamma}S^{l}K^{\beta-\tau}H^{\tau}R^{m}\lambda^{\alpha})$$

(5.15)

$$=c_{pq}J_{pq}\ ,\ J_{pq}=A.B.C\ ,$$

with the three parentheses called A, B, C , where

(5.16) $\alpha=(m+p-s-1+\varepsilon)/2$, $\gamma=(l+q-t-1+\varepsilon)/2$,

while β must be chosen such that B is bounded. From $m>s$, $l>t$, we also get $0<\alpha<m$, $0<\beta<l$, for small ε . An inequality like V,(3.9) (cf. also VIII,(2.22)) implies
$\|Au\| = O(\|H^{\sigma+\alpha-m}K^{\beta-\sigma}S^{l}\mu^{\gamma}u\|) = O(\|K^{\alpha+\beta-m}S^{l}\mu^{\gamma}u\|) = O(\|K^{\alpha+\beta+\gamma-m-l}u\|)$, so that $\|A\|=O(1)$ whenever $0\geq\alpha+\beta+\gamma-m-l=\beta-(m+l-p-q+2-2\varepsilon+s+t)/2$, or,

(5.17) $2\beta \leq m+l-p-q+2-2\varepsilon+s+t = r_{pq}/2 + 1/2 - 2\varepsilon +s+t$,

with the order r_{pq} of the expression L_{pq} . For the operator C we obtain exactly the same estimate, (i.e., $C\in L(H)$ if (5.17) holds.

On the other hand we get $B\in K(H)$ if $2\beta-\sigma-\tau>r_{pq}/2$. Accordingly, β satisfying both estimates can be found if $r_{pq}/2+\sigma+\tau<r_{pq}/2 -2\varepsilon +s+t +1/2$. In other words, we must require (5.10). Then it also follows that the integrand is norm continuous, q.e.d.

The proof of VI,thm.1.6 now only requires the use of prop. 5.3, to establish the group property and adjoint invariance of the cosets U_t , since the boundedness of V_t was shown. In VI,prop.1.7 we have demonstrated how to do this if t is small. We now may use prop.5.3 instead of VI,prop.1.8 to do this for general t. Details are left to the reader.

Next, in this section we focus on the proof of thm.1.3. We will use all the notations introduced in VIII,3. Specifically let $\Lambda_j=H_j^{-1/2}$. Since this is a matter of Sobolev spaces which require cdn.(s), with our definition, we will generally assume that both triples $\{\Omega_j,d\mu_j,H\}$ satisfy cdn.(s). On the other hand, the prop's below clearly all are local. Thus cdn.(s) may be avoided, using a method similar to the one in early VI,1.

We need a few preparations.

<u>Proposition 5.6</u>. Under the assumptions of thm.1.3 let $\phi\in C^{\infty}(U)$, $0\leq\phi\leq 1$, $\phi=0$ near ∂U , $\phi=1$ outside some neighbourhood N of ∂U . Then

(5.18) $\|\Lambda_1^{-s}\phi u\| \leq c_{s,\phi}\|\Lambda_2^{-s}\phi u\|$, for all $u \in C_0^{\infty}(\Omega)$.

Proof. It is sufficient to prove (5.18). Indeed we may write $\Lambda_j^{-s}(\phi u) = \Lambda_j^{-s+2m}(\psi H^m \phi u) = \Lambda_j^{t}\psi v$, with $v=H^m\phi u$, $H = H_1|U = H_2|U$, $t=-s+2m$, where ψ satisfies the same assumptions as ϕ , and also $\psi\phi = \phi$, and where the integer $m=1,2,\ldots$, may be choosen such that $t>0$. Then (5.18) for all v, ψ and t instead of u, ϕ and -s will imply our statement for the general case.

Let $\chi \in C^\infty(U)$, $0\le\chi\le 1$, $\chi=0$ near ∂U , $\chi\phi=\phi$, and let $\omega=1-\chi$. We get

$$\|\Lambda_1{}^t\phi u\| \le \|\chi\Lambda_1{}^t\phi u\| + \|\omega\Lambda_1{}^t\phi u\| = J_1+J_2 , \tag{5.19}$$

where we apply V,prop.6.7 to the term J_2 . (Here we work with the operator $A = \phi\Lambda^0$ involving the H-compatible expression ϕ of order zero.) The terms J_j in (5.19) will be treated separately.

Proposition.5.7. With ϕ, ω, satisfying the above assumptions, there exists a constant $\gamma = \gamma_{r,t,\phi,\omega}$ such that

$$\|\omega\Lambda_1^t\phi u\|_1 \le \gamma\|\Lambda_2^r\phi u\|_2 , \text{ for all } u \in C_0^\infty(\Omega_1) , \tag{5.20}$$

where $r\in\mathbb{R}$ is arbitrary.

Proof. It is enough to prove (5.20) valid for every sufficiently large integer $r=2N$. Thus let $N>t>0$. Again, we may replace ϕu at right of (5.20) by u . For if that is true for all such ϕ,ω ,then we use it for $\phi^\sim$ instead of ϕ with $\phi\phi^\sim = 1$,and replace u by ϕu in the inequality achieved. Also we may replace u by $H_2^{2N}u$, for the equivalent relation (we used cdn.(s))

$$\|\omega\Lambda_1^t\phi H_2^{2N}u\|_1 \le \gamma\|u\|_2 , \quad u \in C_0^\infty(\Omega_2) . \tag{5.21}$$

Again in (5.21) we may replace u by ψu with some function ψ with $\psi\phi = 1$, and then get (5.22), below, equivalent to (5.20).

$$\|\omega\Lambda_1^t\phi H_1^{2N}\psi u\|_1 \le \gamma\|u\|^2 , \quad u \in C_0^\infty(\Omega_2) . \tag{5.22}$$

But this is an immediate consequence of V,(6.9) and (6.10). For we may assume $\psi\phi = 0$, and also, $((\text{ad } H_1)^l\omega)\phi = 0$. Therefore, if the expansion V,(6.9) is applied to $A=\omega$, with $N=0$, $s=t$, and the adjoint is right-multiplied by ϕ, then the remainder is the only non-vanishing term of the expansion. By choosing M large the reminder term can be made L^2-bounded, even compensating the power H_1^{2N} . Hence we get $\|\omega\Lambda_1^t H_1^{2N}\psi u\|_1 \le \gamma\|\psi u\|_1 = \gamma\|\psi u\|_2$, q.e.d. (Note that V,(6.9)-(6.10) are valid since ω is H-compatible.)

Proposition 5.8. For every $\rho,r,t>0$ we have

$$(5.23)\qquad \Lambda_j^{-\rho}(\chi\Lambda_1^t\chi - \chi\Lambda_2^t\chi)\Lambda_k^{-r} \in K(H)\ ,\ j,k=1,2\ .$$

Proof. This is a matter of the following identity, where we write $R_j(\lambda)=(\Lambda_j{}^2-\lambda)^{-1}$. We have, with $M_j=(\mathrm{ad}\,H_k)^j\chi$, for $N,N'=0,1,\ldots$,

$$(5.24)\qquad \begin{aligned}&\chi\Lambda_1{}^2R_1(\lambda)\chi - \chi\Lambda_2{}^2R_2(\lambda)\chi = \\ &(-1)^N\lambda^{N+N'}\Lambda_j{}^{2N}R_j{}^N(\lambda)(M_N\Lambda_1{}^2R_1(\lambda)M_{N'} - M_N\Lambda_2{}^2R_2(\lambda)M_{N'})\Lambda_1{}^{2N'}R_1{}^{N'}(\lambda)\end{aligned}$$

For $N=1$, $N'=0$ relation (5.24) coincides with VII,(3.11). For more general N,N' it follows by induction, using that $\Lambda_j{}^2R_j(\lambda) = (1-\lambda H_j)^{-1} = T_j$, so that $\lambda M_{N+1} = M_N(1-\lambda H_1) - (1-\lambda H_j)M_N = M_NT_1{}^{-1} - T_j{}^{-1}M_N$, i.e.,

$$(5.25)\qquad \lambda\Lambda_j{}^2R_j(\lambda)M_{N+1}\Lambda_1{}^2R_1(\lambda) = \Lambda_j{}^2R_j(\lambda)M_N - M_N\Lambda_1{}^2R_1(\lambda).$$

Relation (5.25) serves for an induction proof of (5.24). On the other hand, (5.23) follows from (5.24) with the same resolvent integral procedure used over and over, q.e.d.

Finally, to complete the proof of prop.5.6, we write

$$(5.26)\qquad \chi\Lambda_1^t\phi u = \chi\Lambda_2^t\phi u + (\chi\Lambda_1^t\chi - \chi\Lambda_2^t\chi)\phi u\ ,$$

where the first term is $O(\|\Lambda_2^t\phi u\|_2)$, and the second term as well, due to (5.23), q.e.d.

We now may approach the proof of thm.1.3. The second statement is easily reduced to the first. If ∂U is smooth, then the diffeomorphism identifying the two subdomains U_j of Ω_j , to constitute the open subdomain U may be extended beyond ∂U such that Ω_1 and Ω_2 contain the commom subdomain W , with compact boundary, while $U\cup\partial U \subset W$. Perhaps then the triples are different over $W\setminus U$. However, we then may use thm.1.2 to remedy this, at least in a subdomain of W , still satisfying the conditions. Then (1.6) holds for U , as a subset of W.

To prove (1.6), let $V \subset U$ be an open subset, with $V^{\mathrm{clos}}\subset U$. Since ∂U is compact, there is a compact neighbourhood N of ∂U , relative to $U\cup\partial U$ such that $U\setminus N\cap V=\emptyset$. Then we may select a function ϕ meeting the conditions of prop.5.7, with $\phi=1$ outside N , so particularly $\phi=1$ on a neighbourhood of V^{clos}. By V, prop.6.7, applied to the 0-order operator $\phi \in C^\infty(\Omega_j)$ with $[H,\phi] \in \mathcal{D}^{\#}_m$, we find that $\Lambda^{-s}\phi\Lambda^s$ is a bounded operator of $L(H_j)$, hence $\phi \in L(H^j_s)$.

Now let $u \in H^1_s$. There exists a sequence $u_k \in C_0^\infty(\Omega_1)$ such that $u_k \to u$ in H^1_s. We have $\phi u \in H^1_s$ as well, and $\phi u_k \to \phi u$ in H^1_s. However, we get $u|V = (\phi u)|V$, since ϕu may be interpreted as the product of the distribution u and the C^∞-function ϕ, and since $\phi=1$ near V. Also, the distribution ϕu has support in U since supp $\phi \subset U$. Accordingly, we may interpret ϕu as a distribution on Ω_2 as well, and also we get $\phi u_k \to \phi u$ in $\mathcal{D}'(\Omega_2)$, i.e., in weak convergence of distributions over Ω_2. Also, by (5.18), the sequence ϕu_k is a Cauchy sequence in H^2_s, hence converges in H^2_s to some $v \in H^2_s$. Since this also implies distribution convergence (on Ω_2), we find that $v=\phi u \in H^2_s$. This shows that every restriction $u|V$ of some $u \in H^1_s$ may also be obtained as a restriction to V of $v=\phi u \in H^2_s$. Since the assumptions are symmetric, this implies the statement of thm.1.3.

6. Sobolev norms of integral order.

In this section, we look at Sobolev norms $\|u\|_k$ for the case of an integer $k>0$. Clearly, with the inner product $(.,.)$ of H,

$$(6.1) \qquad \|u\|_k^2 = (\Lambda^{-k}u,\Lambda^{-k}u) = (u,H^k u)\ ,\quad u \in C_0^\infty(\Omega)\ ,$$

expressing $\|u\|_k$ in terms of derivatives of u. In $\mathbb{R}^n$, where H has constant coefficients, one may integrate by parts for

$$(6.2) \qquad \|u\|_k^2 = \sum_{|\alpha|\le k} \gamma_\alpha \|u^{(\alpha)}\|^2\ ,$$

with positive (multinomial coefficients) γ_α (cf.[C_1],III,(2.11)).

In order to derive a similar formula under more general conditions, let us assume a triple $\{\Omega,d\sigma,q-\Delta\}$ in Sturm-Liouville normal form, i.e., $d\mu = dS = \sqrt{h}dx$, $h= (\det((h^{jk})))^{-1}$. (From III,prop.6.1 we know that this is no loss of generality, but it should be kept in mind that cdn (q_3), below, refers to this form.) We assume $\{\Omega^\wedge,d\mu^\wedge,H^\wedge\}=\{\Omega,d\mu,H\}$, and $q\ge1$ on Ω. Moreover, we will use the Riemannian co-variant derivatives induced by the metric tensor h^{jk} (cf.app.B.).

Under these general assumptions we impose the following.

Condition (q_2): The Riemannian curvature tensor R (cf.(B.10)) is bounded on Ω , and its co-variant derivatives of all orders also are bounded over Ω .

Condition (q_3) : The potential q satisfies the estimates $\nabla^k q = O(q^{1+k/2})$, $k=1,2,\dots$, $x \in \Omega$, where $\nabla^k q$ denotes the k-th order co-variant derivative (tensor) of q.

Our problem formally suggests introduction of the norms

$$\langle u\rangle_N = \{|q^{N/2}u|^2+|q^{(N-1)/2}\nabla u|^2+\dots+|\sqrt{q}\nabla^{N-1}u|^2+|\nabla^N u|^2\}^{1/2}, \tag{6.3}$$

$$<u>_N^2 = \int_\Omega \langle u\rangle_N^2 d\mu = \|q^{N/2}u\|^2 + \dots + \|\nabla^N u\|^2 ,$$

with the formal covariant tensor ∇^k of the k-th covariant derivative, and with $|t|$, $\|t\|$, for $t=(t_{i_1\dots i_k})$, defined by

$$|t|^2= h^{i_1 j_1}\dots h^{i_k j_k}\bar{t}_{i_1\dots i_k}t_{j_1\dots j_k} , \quad \|t\|^2= \int_\Omega |t|^2 d\mu . \tag{6.4}$$

Theorem 6.1. Under cdn.s (q_2), (q_3) the Sobolev norm $\|.\|_N$, on the dense space $C_0^\infty \subset H_N$, is equivalent to the norm $<.>_N$. That is,

$$c_1<u>_N \le \|u\|_N \le c_2<u>_N , \text{ for all } u \in C_0^\infty(\Omega) , \tag{6.5}$$

where c_1 , c_2 are positive constants independent of u .

Proof. The second estimate (6.5) essentially is formal: For N=2m, an even number, one writes (6.1) in the form $\|u\|_N^2= \|H^m u\|^2$.
In order to simplify the formalism introduce, for a moment, the $(n+1)\times(n+1)$-matrix $((g^{\nu\mu}))_{\nu,\mu=0,\dots n}$, $g^{jk}=h^{jk}$, $j,k\neq 0$, $g^{00}=q$ $g^{0j}=g^{j0}=0$, $j\neq 0$, and its inverse $((g_{\nu\mu}))$. Use summation convention from 0 to n (over greek indices only, one up, one down). Also define $a^{jk}=-h^{jk}$, $j,k\neq 0$, $a^{00}=q$, $a^{0j}=a^{j0}=0$, $j\neq 0$, and, with formal tensor notation, (greek indices from 0 to n), $[u]_k$ the 'tensor' with components $u_{\nu_1\dots\nu_k} = \nabla_{\nu_1}\dots\nabla_{\nu_k}u$, with $\nabla_0 u=u$, $\nabla_\nu u=u_{|\nu}$, as $\nu \neq 0$.

Recall that the metric tensor h^{jk} acts like a constant; it has all its covariant derivatives zero. Suppose, for a moment, that also the co-variant derivatives of q vanish (i.e.,q=const.). Then we may write

$$H^m u = a^{\mu_1\nu_1}\dots a^{\mu_m\nu_m} u_{\mu_1\nu_1\dots\mu_m\nu_m} . \tag{6.6}$$

This is just the contraction of the tensors $(a)\times(a)\times\dots\times(a)$, with m factors $(a)=(a_{\nu\mu})$, and $[u]_N$. Thus it is a matter of Schwarz' inequality that

$$(6.7)\qquad |H^m u|^2 \le (g_{\mu\nu}g_{\sigma\tau}\bar{a}^{\mu\sigma}a^{\nu\tau})^m (g^{\mu_1\nu_1}\dots g^{\mu_N\nu_N}\bar{u}_{\mu_1\dots\mu_N}u_{\nu_1\dots\nu_N}) .$$

The first term at right is a constant. Integrating over Ω we get the second estimate (6.5). The second term at right of (6.7) may be written explicitly as

$$(6.8)\qquad \begin{aligned} &|\nabla^N u|^2 + \binom{N}{1}q|\nabla^{N-1}u|^2 + \dots + \binom{N}{j}q^j|\nabla^{N-j}u|^2 + \dots + q^N|u|^2 , \\ &\le c\langle u\rangle_N^2 . \end{aligned}$$

However, in general the derivatives of q will not vanish. Thus (6.7) must be replaced by a more general representation, containing additional terms reflecting the interchange of q with co-variant differentiation. The point then is, that the estimates of cdn.(q_3) give

$$(6.9)\qquad H^m u = a^{\nu_1\dots\nu_N} u_{\nu_1\dots\nu_N} ,$$

where the 'tensor' $(a^{\nu_1\dots\nu_N})$ satisfies the same estimates as $(a)\times(a)\times\dots\times(a)$, above, so that (6.7) still follows. (If a derivative 'lands' on q instead of u , then we will get one more power $\sqrt{q}$, but one less derivative on u, precisely as required).

This proves the second relation (6.5) , for an even N . The proof for an odd N is similar. One will set N=2m+1 ,and work with $H^m u$ as before. Then, however, $\|u\|_N^2 = (H^m u, H^m u)_1$, so that one more derivative must be worked under the coefficients of H^m . There is no new point, however. Details are left to the reader.

For the first inequlity (6.5) we require the prop.,below.

Proposition 6.2. Under cdn's (q_2) , (q_3) we have

$$(6.10)\qquad H(u_{|i_1\dots i_N}) = (Hu)_{|i_1\dots i_N} + O(\langle u\rangle_N) .$$

Proof. This is a matter of pulling out the covariant derivatives from the expression H . Clearly the interchange of $\Delta = h^{jk}\nabla_j\nabla_k$ with $\nabla_{i_1}\dots\nabla_{i_N} = \nabla^{\sim}$ will give additional terms involving the curvature tensor R , and its derivatives up to order N-1 . We get

$$(6.11)\qquad \Delta\nabla^{\sim} - \nabla^{\sim}\Delta = \sum_{p=1}^{N} {}^{p}t_{i_1\dots i_N}^{\;j_1\dots j_p}\, u_{|j_1\dots j_p} ,$$

with bounded tensors p_t , in view of (q_2) .

On the other hand, Leibniz' formula gives

$$\nabla^{\sim}(qu) = q\nabla^{\sim}u + \sum\gamma(\nabla'q)(\nabla''u) , \tag{6.12}$$

where ∇', ∇'' are covariant differentiations for complementary subsets $(i_{j_1}\dots i_{j_r})$, $(i_{k_1}\dots i_{k_s})$ of $(i_1\dots i_N)$, with $j_1\dots j_r$ and $k_1\dots k_s$ in increasing order each, and where the sum is taken over all such splittings of $(i_1\dots i_N)$, with certain individual constants γ for each term. Again (q_3) must be used; it will give $q\nabla^{\sim}u = \nabla^{\sim}(qu) + O(\langle u\rangle_{N-1})$. From (6.12) we get the corresponding for Δu , where it must be used that $q\geq 1$. Q.E.D.

For the first inequality (6.5) we introduce $\langle u,v\rangle_N$, the form obtained by polarizing the hermitian form (6.8), integrated over Ω (i.e., $\langle u\rangle_N$ is equivalent to $(\langle u,u\rangle_N)^{1/2}$; only the positive binomial coefficients distinguish (6.3) and (6.8)). An integration by parts gives

$$\begin{aligned}\langle u,v\rangle_{N+1} &= g^{\mu_1\nu_1}\dots g^{\mu_N\nu_N}\,\bar{u}_{\mu_1\dots\mu_N}\,(H(v_{\mu_1\dots\mu_N})) = \\ &= \langle u,Hv\rangle_N + O(\langle u\rangle_N\langle v\rangle_N) = \langle u,Hv\rangle + O(\|u\|_N\|v\|_N) ,\end{aligned} \tag{6.13}$$

by induction. Iterating (6.13) , we get

$$\begin{aligned}\langle u,u\rangle_{N+1} &= \langle Hu,Hu\rangle_{N-1} + O(\|u\|_N^2 + \|u\|_{N-1}\|Hu\|_{N-1}) \\ &= \langle Hu,Hu\rangle_{N-1} + O(\|u\|_{N+1}^2) .\end{aligned} \tag{6.14}$$

Continuing on we either get $\langle u,u\rangle_{N+1} = \|H^m u\|^2 + O(\|u\|_{N+1}^2)$, if $N+1=2m$ is even. Or else, the first term at right takes the form $(H^m u,H^m u)_1$, if $N+1 = 2m+1$ is odd. In either case (6.5) follows, q.e.d.

Note that thm.6.1 is applicable to both of our typical cases, that of a regular boundary, with a singular q , and the case of a complete manifold with q=1 , or q slowly increasing. For example, the expression $H = x^{-2}-d^2/dx^2$, in the interval $(0,\infty)$ has a q satisfying (q_3) , as easily checked.

7. Examples for higher order theory. The secondary symbol.

In this section we will consider examples of classes $A!$, $\mathcal{D}!$ satisfying the general conditions imposed. Clearly we must insist on cdn.(s), for each example considered. As we know, the conditions (1_j) in particular give us the possibility of looking at operators $A=L\Lambda^N$ as members of a comparison algebra. It is easily checked that many of the classes $A^{\#}$, $\mathcal{D}^{\#}$ singled out so far do not give an algebra or Lie-algebra. For example, regarding $A^{\#}_c$, $\mathcal{D}^{\#}_c$, $A^{\#}_C$, $\mathcal{D}^{\#}_C$ of V,4 and VII,4 one finds that $A^{\#}_c$, $A^{\#}_C$ are algebras, but $\mathcal{D}^{\#}_c$, $\mathcal{D}^{\#}_C$ are not in general Lie-algebras.

On the other hand, some of the classes used for $\mathbb{R}^n$ and $\mathbb{B}^{jk}$ have this property. For example, let us look at $[C_1]$,III,(9.1), and define $A! = A$, with the class A defined there. (In detail, $A = A!$ is the class of all bounded $C^{\infty}(\mathbb{R}^n)$-functions having their partial derivatives of all orders vanishing at infinity.) Let $\mathcal{D}!$ denote the class of all folpde's with coefficients in A , with the standard coordinates of $\mathbb{R}^n$, of course. Then it is easily seen that this pair $A!$, $\mathcal{D}!$ has all the properties (1_j) . In particular $\mathcal{D}!$ is a Lie-algebra. Also, the universal algebra L_{∞} is completely known, in this case. L_N is just the algebra of all differential expressions of order $\leq$ N, with coefficients in A , and $L_{\infty}=\cup L_N$.

On a general manifold Ω we always have complete control of the minimal classes $A^{\#}_m$, $\mathcal{D}^{\#}_m$, and their enveloping algebra, which consists of all expressions with compact support. The generated algebra will be the minimal algebra J_0 , or, in $L(H_s)$, the corresponding minimal algebra $J_s \subset L(H_s)$. Now, thm.6.1 will imply a simple generalization of V,(1.8) to higher order expressions, allowing us a better control of operators $A=L\Lambda^N$ in our algebras.

Lemma 7.1 Assume H in SL-normal form, with (s),(q_2),(q_3). Let an $\leq$N-th order expression L over Ω be written in the form

$$(7.1)\qquad L = {}_Na\nabla^N + {}_{N-1}a\nabla^{N-1} + \dots + {}_1a\nabla + {}_0a \ ,$$

with (locally L^{∞}-) symmetric contravariant tensor fields ${}_ja = ({}_ja^{i_1\dots i_j})$ defined over Ω , where ${}_ja\nabla^j$ denotes the j-fold contraction with the formal tensor ∇ : ${}_ja\nabla^j={}_ja^{i_1\dots i_j}\nabla_{i_1}\dots\nabla_{i_j}$.

For such expressions let a norm $\|L\|_H$ be defined by

(7.2) $$\|L\|_H = \sum_{j=0}^{N} \|q^{-(N-j)/2} |{}_j a| \|_{L^\infty(\Omega)} ,$$

with the local norm $|{}_j a|(x)$ of the tensor ${}_j a$. There exists a constant c, independent of u and the choice of tensors ${}_j a$, such that

(7.3) $$\|L\Lambda^N u\| \leq c\|L\|_H \|u\| , \text{ for all } u \in \Lambda^{-N} C_0^\infty(\Omega) \subset H .$$

Proof. Clearly (7.3) is equivalent to

(7.4) $$\|Lv\| \leq c\|L\|_H \|v\|_N , \text{ for all } v \in C_0^\infty(\Omega) .$$

On the other hand (7.4) simply is a matter of Schwarz' inequality in view of (6.5), q.e.d.

Theorem 7.2. Assume H in SL-normal form with (s),(q_2),(q_3). Then the minimal comparison algebra J_0 contains every operator $A=L\Lambda^N$, with L of the form (7.1), with continuous tensors ${}_j a$, satisfying

(7.5) $$|{}_j a| = o_\Omega(q^{(N-j)/2}) , \ j=0,1,\dots,N .$$

Moreover, let $R_\infty = \cup R_N$, where R_N denotes the class of all expressions of the form (7.1) with $C^\infty(\Omega)$-coefficients satisfying

(7.6) $$\nabla^k({}_j a) = o_\Omega((\sqrt{q})^{N-j}) , \ j=0,1,\dots,N , \ k=0,1,\dots .$$

Then the collection $R_\infty = \cup_{N=0}^\infty R_N$ is an adjoint invariant graded algebra satisfying (p_1),(p_2), and (3.10) for the minimal algebra J_0. Also, R_∞ is within reach and of principal symbol type, for J_0. Furthermore we have $A=L\Lambda^N \in O(0)$,and $\Lambda A \in K(H_s)$, for $L \in R^N$, $s \in \mathbb{R}$.

Proof. We know that every A corresponding to C^∞-tensors ${}_j a$ with compact supports is in J_0 (cf. VI,prop.3.1), since such L has compact support. In fact, such A even is contained in the finitely generated algebra C_∞^0 corresponding to the minimal classes, hence $A \in J_s$, $s \in \mathbb{R}$. For a general continuous ${}_j a$ satisfying (7.5) there exists an approximation under the norm $\|L\|_H$ by compactly supported tensors, and the corresponding sequence of operators converges to A in $L(H)$, so that $A \in J_0$. Next a calculation confirms that R_∞ is invariant under formal adjoints. In view of V,cor.6.10 it is clear that the remaining statements of thm.7.2 depend on the verification of cdn.(p_1) for the class R_∞ defined. (Note that (p_2) is a consequence of the H-boundedness of A, already shown.)

In other words, we must verify that $[H,L] \in R_{N+1}$ for $L \in R_N$. This is shown by a calculation which we shortly indicate: Let

$L_j = a.\nabla^j$, with a j-tensor a. Using all rules of covariant differentiation we get

$$(7.7) \qquad [H,L_j] = [H,a].\nabla^j + a.[H,\nabla^j] ,$$

where the second commutator already was considered in prop.6.2. Using (q_j) , j=2,3 , one finds that the same arguments give $a.[H,\nabla^j] \in R_{N+1}$, using (7.6) for $a={}_j a$. On the other hand,

$$(7.8) \qquad [H,a] = [\Delta,a] = 2(a^{i_1\ldots i_j})^{|k} \nabla_k + (\Delta a) ,$$

with the contravariant derivative '$|k$' (i.e., covariant, raised with the metric tensor). Again it is evident, looking at the estimates (7.6), that $[H,a].\nabla^j \in R_{N+1}$

It is clear that $R_\infty = \cup R_N$ of thm.7.2 is within reach of J_0 . For J_0 the classes L_N and $L_{N,0}$ coincide. Every $A=L\Lambda^N$, for $L\in R_N$ was seen to be a limit in norm of a sequence $L_k\Lambda^N$, with compactly supported L_k, i.e., $L_k \in L_{N,0}$. Thus $\Lambda L\Lambda^N = \lim_{k\to\infty}\Lambda L_k \Lambda^N \in K(H)=E$, q.e.d.

It is clear that R_∞ also is within reach (and of principal symbol type) with respect to any arbitrary comparison algebra $C(A!,D!)$ with the property

$$(***) \qquad A!R_N \,,\; R_N A! \subset R_N \,,\; D!R_N \,,\; R_N D! \subset R_N ,$$

(which implies (3.10)), since $J_0 \subset C$ and $K(H) \subset E$ are always true. This fact proves useful for an $A!$, $D!$ for which $(1_5)(R_\infty)$ or even $(1_5')$ can be confirmed, as in ex.7.3, below.

On the other hand, while thm.7.2 did not state that R_∞ is within general reach (or princial symbol type) for the minimal classes, it did assert $L\Lambda^N \in O(0)$, $\Lambda L\Lambda^N \in K(H_s)$, for $L\in R_N$. Therefore, while thm.4.2 and thm.4.3 do not apply in their present form, they will apply if the operators $L\Lambda^N$, $\Lambda^N L$, for all $L\in R_\infty$, are added to the generators (1.10) of C_s, the other conditions granted, as the reader is invited to check.

Example 7.3. For a triple in SL-normal form satisfying (q_j), j=2,3, consider the following classes $A!$, $D!$.

$A!$ is the class of all bounded $C^\infty(\Omega)$-functions with all (Riemannian) co-variant derivatives $o_\Omega(1)$.

$D!$ is the class of all $C^\infty(\Omega)$-folpde's $D=b\nabla + p$ with $b=O(1)$ and $p= O(\sqrt{q})$, and

$$(7.9) \qquad \nabla^k b = o_\Omega(1) \,,\; \nabla^k p = o_\Omega(\sqrt{q}) .$$

A calculation confirms that, for these classes, we get (***)

above, as well as cdn's (1_1), (1_2), and also

(7.10) $[H,a] \in R_1$, $[H,D] \in R_2$, for all $a \in A!$, $D \in \mathcal{D}!$.

In other words, $A!$ and $\mathcal{D}!$ also satisfy $(1_5{}')(R_\infty)$. Accordingly, cor.3.4 applies : All operators in $\mathcal{Q}_\infty = R_\infty + L_\infty$ are within reach of $C(A!,\mathcal{D}!) \subset L(H)$. We also obtain (slightly enlarged) comparison algebras in $L(H_s)$, and in $O(0)$, containing $L\Lambda^N$ and Λ^N, for $L \in \mathcal{Q}_N$.

Also all expressions $L \in \mathcal{Q}_{N,0} = R_N + L_{N,0}$ give operators $A = L\Lambda^N$ in $K(H_s)$, $s \in \mathbb{R}$.

We observe that the class $\mathcal{D}!$ used here is less restricted in the first derivatives of the coefficients, than the class $\mathcal{D}^\#_c$ of V,4, since $\varepsilon<1$ is required in V,(3.16). However, we have no curvature restriction in sec.5, instead. Also the first derivatives of functions in $A!$ are stronger restricted as those of $A^\#_c$.

Following the example of $\mathbb{R}^n$ let us look at the above pair $A!$, $\mathcal{D}!$, in the special case of the Laplace triple of a complete Riemannian manifold Ω. The corresponding comparison algebra was discussed by the author in [CS],IV,11, under cdn.(1_4) , (which poses a rather severe restriction), and, almost in the present generality, by McOwen in his thesis $[M_0]$. In fact, our discussion here uses some ideas of McOwen in $[M_0]$, in a generalized form.

Example 7.4. We are given the Laplace triple $\{\Omega, ds, 1-\Delta\}$ of a complete Riemannian manifold Ω , with $A!$, $\mathcal{D}!$ as in example 7.3.

This triple is in SL-normal form, with q=1 satisfying (q_3). cdn.(s) follows from IV,thm.1.8. We have the additional requirement (q_2), i.e., the Riemann curvature tensor and all its derivatives are bounded. $A!$ and $\mathcal{D}!$ consist of all bounded functions (folpdes) with covariant derivatives of coefficients 0 at ∞ .

Notice that $A! \subset A^\#_c$, $\mathcal{D}! \subset \mathcal{D}^\#_c$, with the classes of VII,4. Recall that the algebra $C = C(A!,\mathcal{D}!)$ has compact commutators, hence gives the simplest ideal chain. Also, cdn's (1_1), (1_2) hold trivially, for $A!$, $\mathcal{D}!$ (we assume $\Omega^\wedge = \Omega$, of course).

We still observe, that although the algebras C and C_s of the present example are well defined on a general space Ω with bounded curvature tensor (having bounded covariant derivatives), we get

(7.11) $\lim_{x\to\infty} b^l R^i{}_{ljk} = \lim_{x\to\infty}(b^i{}_{|jk} - b^i{}_{|kj}) = 0,$

for every tensor (b^j) occuring as principal part of some $D \in \mathcal{D}!$, as a consequence of app.B. This shows that the curvature is restricted at ∞, if folpdes $D\in\mathcal{D}!$ exist which are not in $\mathcal{R}_1$. For example, on a manifold Ω having a chart $\{x\in\mathbb{R}^n : |x|\geq 1\}$ near ∞ if there exist n vectorfields with linearly independent limits at ∞ , occurring as principal parts of folpde's in $\mathcal{D}!$, then we must have $\lim_{x\to\infty} R = 0$, R being the curvature tensor.

Note that even on $\mathbb{R}^n$ (or $\mathbb{G}^{jk}$) the comparison algebra $\mathcal{C}(\mathcal{A}!,\mathcal{D}!)$ is inconveniently large, for some discussions. Its symbol space $\mathbb{M}$ exhibits some features of the Stone-Cech compactification. This is why we often prefer to work with the smaller algebra $\mathcal{S}$ of [CHe_1], corresponding to the functions and folpde's having directional limits at infinity. Similarly, working on a noncompact Riemannian space one might tend to restrict the geometrical structure of the space at infinity, and in addition the generating classes.

From example 7.4 we conclude that compact commutator algebras which are 'free at infinity', in the sense that their symbol space contains unrestricted cotangent spaces also over ∞ , only can be achieved on spaces Ω which are (asymptotically) flat at infinity. On the other hand the few examples of non-compact commutator algebras with a 2-link ideal chain we have discussed in VIII,1, 2, 4, and in [C_1],V, may be offered for examination, in view of the following points.

(a) The longer ideal chain may reflect a more complicated structure of the compactification $M_{A^\#}$ at ∞, as in the first example of VIII,1. Or, (b) it may result from a nonvanishing curvature tensor at ∞, as in VIII,2. Or, (c) it may reflect the necessity of boundary conditions (at ∞) as in [C_1],V .

In every case examined, existence of a finite set of generators of $\mathcal{D}!$ (mod $\mathcal{A}!$) gives a simpler theory.

In case of example 7.4 - Laplace comparison algebra of the bounded functions and folpdes with covariant derivatives vanishing at ∞, on a complete Riemannian manifold, we now want to supply a criterion, allowing the construction of the secondary $A = L\Lambda^N$, for $L\in\mathcal{L}_N$.

Above we have discussed validity of (s),(l_1),(l_2),(l_5)($\mathcal{R}_\infty$) It is found that (m_1'), (m_4), (m_5), (m_7) hold trivially, but we now <u>require</u> <u>the</u> <u>separation</u> <u>condition</u> (m_3). Then VII,thm.3.6 is applicable and gives the secondary symbol of $D\Lambda$, for $D\in\mathcal{D}!$. On the other hand, for $L\in\mathcal{L}_N$, the operator $A = L\Lambda^N$ first must be writ-

as a sum of products of such $D\Lambda$, (mod E), before its secondary symbol can be obtained. For the principal symbol we know, of course, that only the principal part of L matters, and that $\sigma_A|\mathbb{W}$ equals the principal part, restricted to $S^*\Omega$.

Theorem 7.5. Let all above assumptions be satisfied.

(a) The operator $A = L\Lambda^N$, for a product

(7.12) $$L = D_1 D_2 \cdot \ldots \cdot D_N \ , \ D_j \in \mathcal{D}! \ , \ j=1,\ldots,N \ ,$$

is contained in C , and its symbol is given as the restriction to $\mathbb{M}$ of the product

(7.13) $$\tau_{D_1\Lambda}\,\tau_{D_2\Lambda}\cdot\ldots\cdot\tau_{D_N\Lambda} \quad .$$

(b) Let $x_0 \in \partial M_{A!}$ have a neighbourhood N in $M_{A!}$ with the following property: There exists a cut-off function χ with the properties of cdn.$(m_3)_{x_0}$, such that, in local coordinates any-

where at $N\cap\Omega$ we have

(7.14) $$L = \sum_{j=0}^{N} {}^{j}a^{l_1\ldots l_j}\nabla_{l_1}\ldots\nabla_{l_j} \quad , \ {}^{j}a^{l_1\ldots l_j} \in C^\infty(U) \ ,$$

where (with ${}^{k}_{r}b$, c_k depending on j, of course)

(7.15) $$\chi({}^{j}a) = \sum c_k ({}^{k}_{1}b)\times\ldots\times({}^{k}_{j}b) \quad \text{for all } j=0,\ldots,N,$$

with finite sums, and with bounded tensors $({}^{k}_{r}b)$ of one variable index, defined over all Ω such that $c_k \in A!$, $({}^{k}_{r}b^{l}\partial_{x^l}) \in \mathcal{D}!$.

Then we have $A = \chi L\Lambda^N \in C$. Moreover, the secondary symbol of A over x_0 is given explicitly as the restriction to the (possibly void) set $\iota^{-1}(x_0)\subset\partial\mathbb{P}^*\Omega$ of the continuous extension of the function (7.16) (which is contained in $CB(T^*\Omega)$) :

(7.16) $$\sum_{j=0}^{N}(\chi(x)({}^{j}a^{l_1\ldots l_j}(x))\xi_{l_1}\ldots\xi_{l_j})/(h^{km}(x)\xi_k\xi_m+1)^{N/2}.$$

(c) For any $L\in L_{N,0}$, and, more generally, for $L\in R_N$, we get $A = L\Lambda^N \in C_\infty$, and the secondary symbol of A is zero. Moreover, we have $A \in K(H_s)$, for all $s\in\mathbb{R}$, whenever $L \in R^N$ is of order $<N$, so that its entire symbol vanishes.

Proof. Assertion (a) is trivial for N=1 ,and follows by induction: Write

(7.17) $$D_1\ldots D_N D\Lambda^{N+1} = (D_1\ldots D_N\Lambda^N)((D\Lambda) + (\Lambda^{-N}(D\Lambda)\Lambda^N - (D\Lambda))) \ ,$$

where the first term's symbol can be written in the form (7.12), by induction, while the last term is compact, by cor.3.4.

The key to (b) is found in (c): Any commutator [D,F] has coefficients vanishing at infinity, by V,(2.13), where the derivatives may be written as co-variant derivatives. Hence the formal symbol vanishes there too. Accordingly, using (a), it is clear that the secondary symbol is zero for any $L\Lambda^N$, where $L\in L_N$, since for such L there is at least one commutator in each of the N-fold products. It thus follows that, for the construction of a local decomposition of $L\Lambda^N$ into N-fold products the operators $D\Lambda$ may be treated as if they were commuting. Hence the polynomial decomposition corresponding to (7.15) gives the proper symbol. Also, since $L\Lambda^N$, for $L\in R_N$, has vanishing secondary symbol, it is sufficient to calculate modulo R_N . This completes the proof.

Remark 7.6. Notice that thm.7.5 will be valid for the C_s-symbol as well, if the operators $L\Lambda^N$, $\Lambda^N L$, for $L\in R_\infty$, are added to the generators (1.10). Also we observe that McOwen's result has (m_3) replaced by another condition - that the dimension δ_x of VII,2 be constant on connected components of $\partial M_{A!}$.

A local version of thm.7.5 is valid too, requiring only $(m_3)_{x_0}$, at some $x_0\in \partial M_{A!}$. Also, one of the other representations in VII,2,3,4 may be applicable, if $(m_3)_{x_0}$ fails.

Example 7.7. Consider the comparison algebra C of VIII,4 again. Here we have a Laplace triple on a complete Riemannian manifold again, but the classes $A!$ and $D!$ of example 7.4 appear as too restrictive. The algebra C of VIII,4 corresponds to a different choice of $A!$ and $D!$, which we discuss in X,5. These new classes contain functions (folpde's) with covariant derivatives not tending to zero at infinity (only the t-derivatives must vanish). On the other hand, there exists a limit function (or folde), at each end, which is not in general true for example 7.4.

Example 7.8. For the one-dimensional example of Sohrab, discussed in VII,thm.4.8, we trivially have cdn.(q_2) satisfied, since R=0 , and then must impose (q_3) , which simplifies to

$$(7.18) \qquad q^{(k)} = d^k q/dx^k = O(q^{1+k/2}) \ , \ k=1,2,\ldots,$$

assuming SL-normal form, of course.

Covariant derivatives are ordinary derivatives, in this case. Accordingly, L_N consists of all N-th order expressions

(7.19) $L=\sum_{j=0}^{N} a_j(x)\partial_x^j$, with $a_j=O(q^{(N-j)/2})$, $a_j^{(k)}=o(q^{(N-j)/2})$, $k>0$

Control of the secondary symbol is easy, in this case. We get

(7.20) $$\sigma_{L\Lambda^N}(\pm\infty,\xi) = \lim_{x\to\pm\infty}(\sum_{j=0}^{N} a_j(x)\xi^j)/(\xi^2+q(x))^{N/2} ,$$

for all $L\in L_N$.

<u>Example 7.9</u>. Consider again the triple of V,thm.4.1 (and VII,4), now considered with cdn's (q_2), (q_3), and without a sub-extending triple. We require cdn.(s), hence VII,(4.7) valid for every positive γ , not for γ_1 only, and then use the classes $A!$, $D!$ of example 7.3.

Here we only notice that the conditions imposed are consistent. For example, consider the 1-dimensional example

(7.21) $$\Omega = (0,\infty) , d\mu = dx , H = -d^2/dx^2 + 1+x^{-3} .$$

A calculation shows that (q_3) holds as well as (s) (IV,thm.1.5.). Accordingly we will get a special case of example 7.3. The discussion of classes of differential operators within reach of the generated comparison algebra should follow that of example 7.3. Also, of course, we will have to characterize the secondary symbol space of this algebra. This will not be investigated here.

CHAPTER 10. FREDHOLM THEORY IN COMPARISON ALGEBRAS.

In this final chapter we are going to discuss the main application of theory of comparison algebras to the derivation of Fredholm properties, and essential spectrum properties for differential operators within reach.

Note that the general concepts were developed in [C_1], and they were demonstrated there in relation to elliptic differential operators on $\mathbb{R}^n$, or on the half space $\mathbb{R}^{n+1}_+$. There is no change at all when we apply these abstract concepts to general manifolds. A more detailed introductory discussion is given in sec.1, below.

In particular we choose to discuss this abstract theory after our full scale investigation of the symbol spaces of a comparison algebra, above, to direct the focus of attention accordingly. Also , many proofs will be omitted.

Note that we shortly touched the concepts of Fredholm operator, Fredholm inverse (and Green inverse) in VI,3, while results on compact manifolds were discussed, requiring this terminology.

In sec.1 we discuss the general facts around Fredholm operator, Fredholm index, and C^*-algebras with symbol. In sec.2 we discuss some facts concerning the Fredholm properties of differential operators within reach of a comparison algebra, and about essential spectra. In sec.3 we discuss comparison algebras with compact operator valued symbol, and operators acting on crosssection of vector bundles over a noncompact manifold.

In section 4 we discuss abstract Fredholm theory for a C^*-algebra with a two-link ideal chain. In sec.5 we turn to the Frechet algebra of operators of order 0, bounded over all Sobolev spaces, reviewing some abstract general facts, discussed in [C_1] .

1. Fredholm theory in $C \subset L(H)$.

In this section we will continue the abstract discussions of ch.I, looking at the concept of Fredholm operator in general, bounded or unbounded, and, especially, at the theory of Fredholm operator and index in a comparison algebra $C \subset L(H)$, $H=L^2(\Omega)$, and for differential operators within reach. This material has been discussed by the author and his associates in [C_1], [BC_1], [BC_2], [CM_1], [CM_2], [CHe_1], [CS]. Some slightly different approaches are given by Coburn [Cb_j], Douglas [Dg_j], Barnes et al.[BMSW] . Accordingly we will omit some detailed proofs, referring to [C_1], where our views were presented in close detail.

In these notes we have focused on the Hilbert spaces H_s , $s \in \mathbb{R}$, and the Frechet space H_∞ only. Accordingly we restrict our review to Hilbert spaces only, except in sec.4, where the Frechet algebra C_∞ and differential operators within its reach are studied In Hilbert space many features become simpler, and discussions are less costly. However, most features extend to bounded or unbounded operators in Banach spaces as well (cf. [C_1],App.AI) .

On the other hand, a general Fredholm theory in Frechet spaces seems to have pathological features. For example, the linear group of all invertible operators of $L(X)$, for a Frechet space X , no longer needs to be open in $L(X)$. Therefore it perhaps is even more interesting that the Fredholm theory of C_∞ still remains intact. It looks like that of a C^*-algebra (cf.sec.4).

In fact, C_∞ is a ψ^*-subalgebra of $L(H)$, in the sense of Gramsch [G_1]. As a subalgebra of $L(H)$ it contains all its inverses in the larger algebra, just as this is known for C^*-algebras Moreover, in [CSch], [Sch_1] it was seen that, at least for some important special cases, the symbol homomorphism of C_∞ still is a surjection onto $C(\mathbb{M})$ (This property is known to be generally false for comparison algebras in L^p-spaces or other Banach spaces.)

Perhaps, in the interest of a more systematic approach, the present discussion should have been in order at the begin of ch.5. However, we wanted to establish the Banach algebra structure of a comparison algebra as the focus of our attention. Thus we were pushing aside the important motivation of this theory from theory of regular and singular elliptic boundary problems in partial differential equations. The latter will proceed now.

Returning to abstract theory of (unbounded closed) linear operators in Hilbert space of I,1 we recall that a closed operator $A: \text{dom } A \to H$, in an abstract (separable) Hilbert space H, is called a <u>Fredholm operator</u> if (i) im A is a closed subspace of H, and (ii) we have

$$\text{(1.1)} \qquad \dim \ker A < \infty \text{ , and } \operatorname{codim} \operatorname{im} A = \dim H/(\operatorname{im} A) < \infty \text{ .}$$

It is not hard to see that (ii) implies (i), but it is useful to keep (i) in mind, as an important property of Fredholm operators. In particular, for operators with closed range (with im A closed) the second condition (1.1) may be stated by requiring that $\dim(\operatorname{im} A)^{\perp} < \infty$. For a Fredholm operator one defines the <u>Fredholm index</u> by setting

$$\text{(1.2)} \qquad \operatorname{ind} A = \dim \ker A - \dim(H/\operatorname{im} A) = \dim \ker A - \dim \ker A^*.$$

The concept of Fredholm operator is nontrivial only for infinite dimensional spaces. This may be the reason for the fact that they were systematically studied only in the 1950-s, although the concept is quite simple. Special classes of Fredholm operators(integral operators and differential operators) were studied in the late 19-th and early 20-th century (cf. Fredholm [Fh_1] and F.Noether[Nt_1], who showed that Fredholm operators with nonvanishing index occur in the theory of the 2-dimensional oblique derivative boundary problem).

We summarize the principal properties of a Fredholm operator in thm.1.1, below. We will express the statement that A is Fredholm by stating that 'A is Φ ', or that ' A has property Φ '.

<u>Theorem 1.1.</u>(a) Property Φ and ind A are invariant under small bounded and arbitrary compact perturbations; (b) Property Φ is adjoint invariant, and $\operatorname{ind} A^* = -\operatorname{ind} A$; (c) The product AB of closed Fredholm operators A and B is always a well defined closed operator; we have $(AB)^* = B^*A^*$, and

$$\text{(1.3)} \qquad \operatorname{ind}(AB) = \operatorname{ind} A + \operatorname{ind} B \text{ ;}$$

(d) A closed operator is Φ if and only if either $0 \in \operatorname{Rs}(A)$ or 0 is an isolated point of Sp(A) , and an eigenvalue of finite multiplicity for A and A^* .

<u>Remark 1.2.</u> Point (a) of thm.1.1 may be expressed as follows.

(a): For a closed A with preperty Φ , and $B \in L(H)$, $C \in K(H)$ we have $A+C \in \Phi$ as well as $A+\varepsilon B \in \Phi$, for small ε , and

$$\text{(1.4)} \qquad \text{ind } A = \text{ind } (A+C) = \text{ind } (A+\varepsilon B) .$$

The same statement holds for unbounded operators B and C which are 'A-bounded' and 'A-compact', respectively, i.e.,

$$\text{(1.5)} \qquad B|(\text{dom } A) \in L(\text{dom } A, H) \quad , \; C|(\text{dom } A) \in K(\text{dom } A, H) \; ,$$

where dom A is regarded as a Hilbert space of its own under the graph norm I,(2.8) .

For a proof of thm.1.1 cf.[C_1] (for Banach spaces) , [CLa] for Hilbert spaces, with generalized perturbations, also [Nb_j] for generalized perturbations in Banach spaces. For a discussion of bounded and unbounded Fredholm operators between different Banach spaces cf.[C_1] ,AI .

We also will consider Fredholm operators between different (separable) Hilbert spaces. However, since any two separable Hilbert spaces H_1 , H_2 are isometrically isomorphic (with an isomorphism U defined by mapping $\phi_j \to \psi_j$ for any fixed pair $\{\phi_j\}$, $\{\psi_j\}$ of orthonormal bases of H_1 and H_2) we simply define $A \in Q(H_1, H_2)$ to be Fredholm if AU^{-1} is a closed Φ-operator in H_2 , and define $= \text{ind}(AU^{-1})$. This reduces our discussion to the case of Φ-operators in $Q(H)$, $H=H_2$.

The key to our present theory of Fredholm theory of comparison algebras is the following fact stated for bounded Fredholm operators. Let $\Phi(H)$ denote the class of bounded Φ-operators.

Theorem 1.3. (a) An operator $A \in L(H)$ is Fredholm if and only if there exists an operator $B \in L(H)$ such that

$$\text{(1.6)} \qquad AB = 1 + E \; , \; BA = 1+F \; ,$$

where E and F are operators of finite rank (with finite dimensional image) .

(b) An operator $A \in L(H)$ is Fredholm if and only if its coset $A^\vee = A + K(H) \bmod K(H)$ is an invertible element of the C^*-algebra $L(H)/K(H)$.

(c) Let $[A^\vee]$ be the homotopy class of $A^\vee$, in the group $GL(L(H)/K(H))$ of invertible elements of $L(H)/K(H)$. These homotopy classes form a group $H\Phi$ under operator multiplication. There exists a group homomorphism $\upsilon : H\Phi \to \mathbb{Z}$ such that

(1.7) $\quad$ $\operatorname{ind} A = \upsilon[A^\vee]$, for all $A \in \Phi(H)$.

Remark 1.4. An operator $B \in L(H)$ satisfying (1.6) is called a Fredholm inverse (or Φ-inverse) of A . The homomorphism υ of (1.7) is called the index homomorphism (of the C^*-algebra $L(H)$). Clearly a Fredholm operator will have many Fredholm inverses. Among these we have focused on the special Fredholm inverse in $[C_1]$,AI and, with respect to an HS-chain, the distinguished Fredholm inverse (VI,3).

For a proof of thm.1.3 cf.$[C_1]$,AI , for Banach spaces.

Now let us recall the following property of C^*-algebras, not normally shared by other Banach algebras:

Proposition 1.6. Let A be a *-invariant sub-algebra of the C^*-algebra B , where both A and B have (the same) unit. Then, if $A \in A \subset B$ is invertible in B , its inverse A^{-1} is contained in A .

For a proof cf. $[C_1]$, AII , or $[R_1]$.

Combining thm.1.4 and prop.1.6 we now get a result useful for Fredholm theory in comparison algebras.

Theorem 1.7. Let A be a C^*-subalgebra of $L(H)$, containing the identity operator and the compact ideal $K(H)$. Then an operator $A \in A$ is Fredholm if and only if its coset $A^\vee = A+K(H)$ is an invertible element of the quotient algebra $A^\vee = A/K(H)$.

Indeed, we then have $A/K(H)$ a C^*-subalgebra of $L(H)/K(H)$, both with the unit $1+K(H)$, so that prop.1.6 applies.

Remark 1.8. Under the assumptions of cor.1.7 we also may replace the homomorphism υ by a homomorphism from the group of homotopy classes of $GL(A/K(H))$ to $\mathbb{Z}$ (cf. $[BC_1]$).

Now, if $A = C$ is a comparison algebra, as studied in chapters V - IX, then the algebra $C/K(H)$ has a very simple structure at least in the following two cases.

Case (A): The commutator ideal E is $K(H)$ (Such comparison algebras were investigated in V,4 and VII,2).

Case (B): The commutator ideal E is not $K(H)$, but has the property that $E/K(H)$ is an algebra of compact-operator-valued functions over a locally compact space (such as the two examples discussed in VIII,1 and VIII,2.)

Note that other cases are possible: Not every comparison is either in case (A) or (B) . In particular there may be ideal chains longer than the two-link $C \supset E \supset K(H)$, with other proper ideals fitting between E an $K(H)$.

First consider case (A) . Then $C/K(H) \simeq C(\mathbb{M})$, with the symbol space $\mathbb{M}$. A continuous complex-valued function $\sigma_A(m)$ over the compact Hausdorff space $\mathbb{M}$ is invertible if and only if it never vanishes. Therefore we have the following result.

<u>Theorem 1.9(A).</u> Under cdn.s $(a_0),(a_1),(d_0)$ only let the comparison algebra C have compact commutators. Let $\mathbb{M}$ denote the symbol space of C (cf.VII,1), and σ_A the symbol of $A \in C$. Then an operator $A \in C$ is Fredholm if and only if its symbol $\sigma_A(m)$, $m \in \mathbb{M}$, does never vanish. There exists a group homomorphism υ from the group $\mathbb{H}^1(\mathbb{M})$ of homotopy classes $[\sigma]$ of non-vanishing continuous complex-valued functions σ over $\mathbb{M}$ such that

$$(1.8) \qquad \text{ind } A = \upsilon([\sigma_A]) \text{ for all } A \in C \cap \Phi(H) .$$

The proof of thm.1.9(A) is an almost trivial consequence of our above discussion. For a more detailed derivation cf. [BC_1] .

For a partial discussion of case (B) we refer to sec.4 below. Two inversions, modulo the two ideals E and K are needed, in this case. We also refer to [CPo] , [CDg_1] and [CMe_1] , where we treat a useful generalization for the special 2-link-algebras of VIII,1,2,4 , and for more general such algebras. Instead of the two inversions we there have only one symbol, assuming singular-integral-operator-values at certain points of the symbol space.

2. Fredholm properties of operators within reach of C.

Note that thm.1.9(A) gives a very simple necessary and sufficient criterion for the Fredholm property and index of operators in the algebra C , provided that we know details about the symbol space $\mathbb{M}$ of the algebra, and also know how to explicitly obtain the symbol of an operator $A \in C$. That precisely was the intent of our effort, in the preceeding sections.

The present thm.1.9(A) only represents the simplest non-trivial case of a chain of similar results. Later we will discuss a similar result for case (B) as well. Also some cases involving longer ideal chains will not be discussed. Also we will have to deal with the case of a more general matrix $A = ((A_{jk}))$ of operators $A_{jk} \in C$, where even infinite dimensional matrices will be admitted. Also, this will lead us into comparison algebras acting on $L^2(\Omega)$-crossections of a vector bundle on Ω .

Let us use the present example to explain the principle of comparison: Often one will not be interested in property Φ (or the index) for operators in C, but rather, for an unbounded operator of the form

$$(2.1) \qquad L = AQ^{-1} \text{ , with } A, Q \in C \text{ .}$$

Example 2.1. Let $Q = \Lambda = H^{-1/2}$, with the expression H of some triple $\{\Omega, d\mu, H\}$, and let $A = D\Lambda$, with some $D \in \mathcal{D}$. Assuming cdn.(m_5), it then follows that Λ and A both are in C . However, the operator $\Lambda^{-1} = H^{1/2}$ is an unbounded closed Fredholm operator of H , with index 0 , since it is self-adjoint, positive definite, and ≥ 1 , (thm.1.1(d)). Suppose now that A is Fredholm. For example, assume that C has compact commutator, and that $\sigma_A(m) \neq 0$ for all $m \in \mathbb{M}$. Then apply thm.1.9(A)). Then $A\Lambda^{-1}$ is an unbounded Fredholm operator, and $\text{ind}(A\Lambda^{-1}) = \text{ind } A$, by thm.1.1(c),(d) , since $\text{ind } \Lambda^{-1} = 0$. However, we find that

$$(2.2) \quad Du = b^j u_{|x^j} + pu = A\Lambda^{-1}u, \text{ for all } u \in \text{dom}(A\Lambda^{-1}) = \text{dom } \Lambda^{-1} = H_1,$$

with the first Sobolev space H_1. In other words, $Z = A\Lambda^{-1}$ is a realization of the first order differential operator D, with domain H_1, in the sense of V,1 . Thus,if σ_A does never vanish on $\mathbb{M}$, then we have found a realization Z of the differential operator D which is Fredholm. In fact, since $\Lambda^{-1}: H_1 \to H$ acts as an isometry, (cf.V,1) one concludes at once that the realization D constructed (with domain H_1) is Φ if and only if A is Φ , i.e., if and only if $\sigma_A(m) \neq 0$, $m \in \mathbb{M}$.

The above conclusion is not very far reaching, since there are much simpler methods available, to solve the single first order partial differential equation Du=f, with or without boundary conditions. (In fact, looking at the symbol $\sigma_A = b^j(x)\xi_j$ over the wave front space $\mathbb{W} \subset \mathbb{M}$, it is clear at once that the condition $\sigma_A \neq 0$ on $\mathbb{W}$ cannot be satisfied, unless n=1.)

This changes, however, if we either consider a matrix $A = ((A_{jk}))_{j,k=1,\dots,m}$, of operators $A_{jl} \in C$ or an operator linked to a higher order expression $L \in L_N$, as in VIII,3. We look at matrices of operators in C in sec.3, below, although this is the simpler problem. Here we next look at an $L\in L_N$.

Example 2.2. Consider the case $Q=\Lambda^N$, for some integer N=1,2,..., with a differential expression $L\in L_N$. Here we must ask for cdns. (s),(m_5),(l_j), j=1,2, (apart from (a_0),(a_1),(d_0)), and $(l_5)(P_\infty)$, for a suitable class P_∞, within reach of $A!$, $D!$, so that IX,prop. 3.2 or cor.3.4 hold. (Note that (s) may be replaced by (s_k), for sufficiently large k, as easily seen.) Also, cdn.(m_1) or a cdn. (m_1)^, for suitable q^, may be useful. Since compactness of commutators is required, we assume a triple as in early V,4, with a large constant γ_1 of V,(5.3) to get cdn.(s_k), not only (w), and cdn.(q_1), with classes $A!\subset A^\#_C$, $D!\subset D^\#_C$. Or else, we assume the case of IX,example 7.3, with cdn's (q_2), (q_3). In any such case we get

$$(2.3)\qquad A = LQ = L\Lambda^N \in C \ .$$

Thus the conclusion of example 2.2 may be repeated. We have $\operatorname{im}\Lambda^N = \operatorname{dom}\Lambda^{-N} = H_N$, with the Sobolev space of order N of IX,1. Formula (2.4) defines a closed realization of the expression L . (It was already discussed in V, def.6.2 , and called Z there. Here we find it more convenient to use the same symbol for the expression L and its realization.)

$$(2.4)\qquad L = A\Lambda^{-N} \ , \ \operatorname{dom} L = H_N = \operatorname{im}\Lambda^N \ .$$

This realization L is a closed Φ-operator, if and only if

$$(2.5)\qquad \sigma_A(m)\neq 0 \text{ for all } m\in\mathbb{M} \ .$$

We again are interested in checking on (2.5) explicitly. In that respect note that formula VI,(3.9) indeed gives the symbol on the wave front space. We get

$$(2.6)\qquad \sigma_A(x,\xi)=\lim_{\rho\to\infty}(\rho^{-N}a(x,\rho\xi))=a_N(x,\xi), \text{ as } |\xi|=h^{jk}\xi_j\xi_k=1,$$

where L=a(x,D), in local coordinates. In other words, $\sigma_A|\mathbb{W}$ is just the principal symbol of the expression L, at the unit co-sphere, in the conventional sense, and the condition (2.5), restricted to $\mathbb{W}^{\text{clos}}$, amounts to the statement that

$$(2.7)\qquad |a_N(x,\xi)| \geq \eta >0 \text{ for all } (x,\xi) \in \Sigma^*\Omega = \mathbb{W} \ ,$$

i.e., that the expression L is uniformly elliptic on Ω, with respect to the metric tensor h_{jk} .

It is important to observe that we are not yet in control of the full symbol, except in a case where the secondary symbol space is empty. However, we have the results of VII,2, 3, 4, and IX,7 to investigate the secondary symbol. For example, under (q_1) of VII,4, and (m_1'), (m_3), (m_7), or the corresponding '^'-conditions, we conclude from VII,thm.4.7 that there is no secondary symbol space over $\partial\Omega$, although there may well be points of the secondary symbol space over $\partial M_{A^\#}\setminus\partial\Omega=\partial M_A^{\infty\#}$, as our examples of V,4, VII,2, VII,4, and IX,7 show.

Even if the space $\mathbf{M}_s$ is completely known, as in case of VII, thm.4.8, for example, the secondary symbol of $A=L\Lambda^N$, for a higher order expression L is known only if a representation of A as a sum of products $(D_1\Lambda)(D_2\Lambda)\dots(D_k\Lambda)$ is explicitly given. For the factors $D\Lambda$ of such a product we may invoke VII,thm.2.2, thm.2.3, thm.2.6, etc. Again, under (q_2), (q_3), we may refer to IX,thm.7.5, given its assumptions.

In some cases, however, a complete knowledge of the symbol is not required, while significant statements still can be made.

Note that the differential operator L of example 2.2, regardless whether Fredholm or not, always is a restriction of the closure of the minimal operator L_0 , since $(\Lambda^{-N}|C_0^\infty)^{\text{clos}} = \Lambda^{-N}$, due to cdn.(s). Assuming the expression L to be formally self-adjoint, the operator L will be hermitian, and a restriction of the closure L_0^{**} of L_0 . In fact, we now have the following.

<u>Theorem 2.3</u>. Assume cdn's (a_0), (a_1), (d_0), (s), (l_1), (l_2), (m_5), $(l_5)(P_\infty)$ with respect to a suitable class P_∞ within reach of C , and of principal symbol type, and assume that C has compact commutator. Also assume that $\sigma_Q = \sigma_{\Lambda^N} \neq 0$ on all of $\mathbf{M}_s$. Then for every self-adjoint uniformly elliptic expression $L\in L^N$ (i.e., L satisfies (2.7)) the realization L of (2.4) is a self-adjoint operator.

<u>Remark 2.4</u>. The condition '$\sigma_{\Lambda^N}\neq 0$ (or $\sigma_\Lambda\neq 0$) on $\mathbf{M}_s$ ' follows from VII,cor.4.3, if q=1 for large x of $\Omega\cup\partial\Omega$, since then $D=D_0=1$ may be taken in VII,thm.3.6.

<u>Proof of thm.2.3</u>. Let us define $B = \Lambda^{N/2}L\Lambda^{N/2}$, where, for a

moment we use the convention of IX,4 (i.e., regard all operators and expressions as operators on H_∞). It follows that (the H-closure of) B defines a bounded self-adjoint operator of H. Moreover, from IX,prop.4.1 we conclude that

$$(2.8) \qquad A - B = A - \Lambda^{N/2} A \Lambda^{-N/2} = \Lambda C, \text{ with } C \in \mathcal{C} ,$$

so that also $B \in \mathcal{C}$, since $A \in \mathcal{C}$. In fact, we get $\sigma_A = \sigma_B$ on $\mathbb{W}$, since $\sigma_\Lambda = 0$ on $\mathbb{W}$.

Since B is a self-adjoint operator, its symbol must be real-valued. Since we assumed L uniformly elliptic, it follows that $\sigma_B = \sigma_A$ is bounded away from zero on $\mathbb{W}$. On the other hand we assumed $\sigma_\Lambda \neq 0$ on $\mathbb{M}_s = \mathbb{M} \setminus (\mathbb{W}^{clos})$. Since Λ also is self-adjoint, σ_Λ is real-valued as well. Hence we conclude that

$$(2.9) \qquad \sigma_{(B+i\varepsilon\Lambda^N)}(m) \neq 0 \text{ for all } m \in \mathbb{M} ,$$

since $\sigma_B \neq 0$ in a neighbourhood N of $\mathbb{W}^{clos}$, while $\mathbb{M} \setminus N \subset \mathbb{M}_s$ is compact, so σ_{Λ^N} bounded away from 0 on $\mathbb{M} \setminus N$, and the two terms cannot cancel each other, one being real, the other purely imaginary. Conclusion: $B + i\varepsilon\Lambda^N$ is Fredholm, hence $L + i\varepsilon = \Lambda^{-N/2}(B + i\varepsilon\Lambda^N)\Lambda^{-N/2}$ also is Fredholm, whenever $\varepsilon \neq 0$, as follows from thm.1.1(c). From the same thm. it also follows that

$$(2.10) \qquad (L+i\varepsilon)^* = \Lambda^{-N/2}(B+i\varepsilon\Lambda^N)^*\Lambda^{-N/2} = \Lambda^{-N/2}(B-i\varepsilon\Lambda^N)\Lambda^{-N/2} = L - i\varepsilon ,$$

in the precise sense of adjoint of unbounded operator, using that both operators $\Lambda^{-N/2}$ and $(B+i\varepsilon\Lambda^N)$ are Fredholm. Hence it follows that $L^* = L$, or that L is self-adjoint, q.e.d.

For a general, not necessarily self-adjoint, but uniformly elliptic expression we can derive knowledge on the essential spectrum of the realization (2.4) from the knowledge of the secondary symbol. Here the essential spectrum of an unbounded closed operator $A \in \mathcal{Q}(H)$ is defined as the collection of all $\lambda \in \mathbb{C}$ such that $A - \lambda$ is not a Fredholm operator. From thm.2.1 it follows that the essential spectrum consists precisely of all points $\lambda \in \mathrm{Sp}(A)$ which are not isolated point-eigenvalues (of A or A^*) of finite multiplicity. (Note that other definitions of the concept 'essential spectrum' are in common use (cf. Kato [K_2], for example)

Theorem 2.5. Assume cdn's (a_0), (a_1), (d_0), (s), (l_1), (l_2),

(m_5), $(l_5)(P_\infty)$, as above, and that the algebra C has compact commutators. Let $\sigma_Q=\sigma_{\Lambda^N}\neq 0$ on all of $\mathbb{M}_s$. Let $L\in L_N$ be a uniformly elliptic expression (satisfying (2.7)), and assume that the set

$$(2.11) \qquad \{\sigma_A(m)/\sigma_{\Lambda^N}(m) : m\in\mathbb{M}_s\}$$

leaves out at least one complex number. Then the realization of L defined by (2.4) is a closed operator (and the closure of the minimal operator L_0). Moreover, the essential spectrum of this realization is given by the set (2.11).

Proof. In essence we repeat the arguments of the proof of thm. 2.3 in the following simpler form. First, the realization L again is a restriction of L_0^{**}. Next, we write

$$(2.12) \qquad L-\lambda = (A-\lambda\Lambda^N)\Lambda^{-N} ,$$

where $A_\lambda=A-\lambda\Lambda^N \in C$, and $\sigma_{A_\lambda}=\sigma_A \neq 0$ on $\mathbb{M}_p=\mathbb{W}^{clos}$, due to $\sigma_\Lambda=0$ on $\mathbb{M}_p$, and (2.7). There exists at least one $\lambda_0\in\mathbb{C}$, not contained in the set (2.11). We conclude that $\sigma_{A_{\lambda_0}}=\sigma_A-\lambda_0\sigma_{\Lambda^N}\neq 0$ on all of $\mathbb{M}$, so that A_{λ_0} is Fredholm. Accordingly $L-\lambda_0$, hence L is closed. Also $L-\lambda_0$ is Fredholm, by thm.1.1(c). The argument may be repeated for any $\lambda\in\mathbb{C}$ not assumed by the set (2.11), showing that $L-\lambda$ is Fredholm for every such λ. Hence the essential spectrum is a subset of the set (2.11). On the other hand, if λ is contained in the set (2.11), then we know that $A-\lambda\Lambda^N$ is not Fredholm. But the operator $\Lambda^{-N}:H_N=\mathrm{dom}\,\Lambda^N \to H$ is an isometry between the Sobolev spaces H_N and H, as we know. Relation (2.12) may be interpreted in two different ways, either the product is a product of two unbounded operators, or a composition of $A_\lambda:H\to H$ and $\Lambda^{-N}:H^N\to H$. Both interpretations describe exactly the same map, (even though not topologically). With the second interpretation we find that $A_\lambda\Lambda^{-N}$ is not a bounded Fredholm operator $H_N\to H$, hence either codim im $=\infty$, or dim ker $=\infty$, or both. But the bounded operator $H_N\to H$ and the unbounded operator of H with domain dom L $= H_N$ have exactly the same image and kernel. Therefore L cannot be Fredholm as well, q.e.d.

The above examples were designed as a demonstration, showing how a (singular) elliptic problem may be handled with the 'comparison techique'. We selected the simplest nontrivial case for this demonstration. Results like thm.2.3 and thm.2.6 have straight generalizations to the case of a (finite or infinite) matrix of differential expressions, as discussed in sec.3, and to expressions formally acting on the type of vector bundles of sec.3 as well. We shall not discuss details.

Perhaps we should also point to a possible use of resolvent integrals for realizations of the form (2.4). A criterion like thm.2.5 gives very precise control on the Fredholm domains of the analytic operator family $L_\lambda = L-\lambda$: They are just the connected components of the complement of the (closed) set of (2.11). By a result of Gramsch [G_2] it is sufficient to know existence of the resolvent in only one point of a Fredholm domain, to conclude existence of the resolvent as a meromorphic function defined in the entire domain, with poles clustering at the boundary only. Such knowledge then allows definition of functions of L as resolvent integrals.

Let us finally note that one needs not necessarily choose the operator Q of (2.3) as a power of Λ. Clearly, Q^{-1} should be Fredholm, and perhaps should have other properties. It could be self-adjoint (and positive definite), for example. Note that the condition $\sigma_Q \neq 0$ on $\mathbb{M}_s$, required in thm's 2.3 and 2.5, does not hold for the problems of Sohrab (VII,thm.4.8).There we have $\sigma_\Lambda = 0$ on the entire symbol space, including $\mathbb{M}_s$.

Our technique for those thm's depended on the fact, that the principal symbol space coincides with the set where $\sigma_Q=0$. In general one could redesign the results, replacing the set $\mathbb{M}_p$ by ker σ_Q .

Note also that we have been using the comparison technique in [C_1],V,14, with an operator Q (called T there) of a different kind. Each T used there represents an elliptic boundary condition (of Lopatinskij-Shapiro type), and we were solving a regular (not a singular) boundary-value problem.

3. Systems of operators, and operators acting on vector bundles.

We continue our review of Fredholm theory in comparison algebras, and now discuss operators with compact operator-valued symbol, as in $[C_1]$,V,8 , and $[BC_2]$. Recall that such algebras already were used in VIII,1,2,4. In particular it was found that the commutator ideal E ,if not compact, often has a natural compact-operator-valued symbol. Here, however, we work from the start in a Hilbert space H^∞ which is a topological tensor product

$$(3.1) \qquad H^\infty = h\hat{\otimes}H \ , \ H = L^2(\Omega) \ ,$$

where h is a separable Hilbert space. (We normally assume $h=\ell^2 = L^2(\mathbb{Z}^+)$, $\mathbb{Z}^+ = \{1,2,3,....\}$). In $[BC_2]$ the space H also is an abstract Hilbert space. Here we consider $L(H)$ as the home of our comparison algebra C, acting on the complex-valued functions of $H = L^2(\Omega)$. We now want to consider the C^*-algebra

$$(3.2) \qquad C^\infty = k\hat{\otimes}C \ , \text{ where } k = K(h) \ .$$

Note that an operator $K \in k$ is an infinite matrix $K = ((k_{jk}))$, acting on a sequence of $h=\ell^2$ by matrix multiplication. All matrices with only finitely many non-zero entries are necessarily compact, and every compact matrix is a limit in norm convergence of a sequence of such matrices of finite rank. Correspondingly, it can be shown that every operator of $k\hat{\otimes}L(H)$ is an infinite matrix

$$(3.3) \qquad A = ((A_{jk}))_{j,k=1,2,...} \ , \text{ where } A_{jk} \in L(H) \ .$$

Vice versa, a given such matrix (3.3) is in $k\hat{\otimes}L(H)$ if and only if it is a limit, in norm convergence of $L(H^\infty)$, of a sequence of matrices (3.3) each having only finitely many non-zero entries A_{jk} . We have $K(H^\infty) = k\hat{\otimes}K(H) \subset k\hat{\otimes}L(H) \subset L(H^\infty)$, all with proper inclusions, and $K(H^\infty)$ is a two-sided ideal of $k\hat{\otimes}L(H)$. We get $A \in C^\infty$ if and only if all entries $A_{jk} \in C$, $j,k=1,2,....$, and, again, A is limit of a norm convergent sequence of such matrices with only finitely many non-zero entries.
For such operator $A \in C^\infty$ we define a symbol σ by setting

$$(3.4) \qquad \sigma_A(m) = ((\sigma_{A_{jk}}(m)))_{j,k=1,2,...} \ .$$

Now it follows easily that $\sigma_A : \mathbb{M} \to k$ is a continuous function in norm topology of $k = K(h)$. We have the result, below, where

$E = K(H)$ is assumed for the remainder of sec.3, although an extension to the 2-link-case would meet no obstacles.

Theorem 3.1. Let $(a_0),(a_1),(d_1)$ hold, and let $C \subset L(H)$ be a comparison algebra with compact commutator .

(a) Then an operator $B = 1+A$, with $A \in C^\infty$ is Fredholm if and only if the function $1+\sigma_A(m)$ is invertible in $L(h)$, for every $m \in \mathbb{M}$. (b) There exists a homomorphism υ from the group $H\Phi k(\mathbb{M})$ of homotopy classes $[\sigma]$ of continuous maps $\sigma\colon \mathbb{M} \to GL(1+k)$, under operator multiplication as group operation, into $\mathbb{Z}$, such that

$$(3.5) \qquad \text{ind } A = \upsilon[\sigma_A] \ , \text{ for all } A \in C^\infty \cap \Phi(H^\infty) \ .$$

For detailed proofs we refer to $[BC_2]$, thm.7 and thm.10. There we also discuss the precise definition of the topological tensor products $h\hat{\otimes}H$,and $P\hat{\otimes}Q$, for C^*-subalgebras $P \subset L(h)$, and $Q \subset L(H)$.

The most important applications of thm.3.1 aim at (finite dimensional) matrices of operators with entries in a comparison algebra C , and, more generally, at operators acting on crossections of a (finite dimensional) vector bundle over Ω . Note that, in case of a compact manifold Ω , the homomorphism υ of thm.3.1 essentially coincides with the relation between topological and analytical index expressed by the Atiah-Singer-Bott index theorem.

Finding υ explicitly, i.e., obtaining an explicit index formula for a comparison algebra C^∞ on a noncompact mainfold, will be a task of algebraic topology. We are not equipped to discuss this problem here, but, perhaps should make a few comments.

Remark 3.2. A homotopy $A\colon[0,1]\times\mathbb{M} \to GL(1+k)$, in detail,

$$(3.6) \qquad A(t,m)=a(t,m) + K(t,m) \ , \ 0 \le t \le 1 \ , \ m \in \mathbb{M} \ ,$$

always can be deformed into a homotopy $U\colon[0,1] \to U(1+k)$, where $U(1+k)$ denotes the group of unitary operators in the algebra $1+k \subset L(h)$. This follows, because a polar decomposition $A = BU$ is possible within the algebra $1+k$, where $B=(AA^*)^{1/2}$, $U=A(AA^*)^{-1/2}$. Here U is unitary, while the invertible self-adjoint positive definite operator B may be deformed into the identity, by the deformation $t \to tB+(1-t) \in GL(1+k)$.

Remark 3.3. The problem of discussing the index homomorphism υ may be reduced to a more conventional form in homotopy theory by inducing the Bott group $U(\infty)$, defined as follows. For N=1,2,..., let $h_N \subset h = l^2$ be the subspace of all sequences $u=(u_1,u_2,\dots)$ with $u_{N+1}=u_{N+2}=\dots=0$, and let U(N) denote the subgroup of $U(1+k)$ consisting of all unitary matrices equal to 1 in $h^\perp$. Regarding U(N) as a topological group with (Euclidean) topology one defines $U(\infty)$ as inductive limit of the sequence U(N) , with inductive limit topology. Concretely, we thus have

$$(3.7) \qquad U(\infty) = \cup_{j=1}^{\infty} U(N) .$$

A subset $M \subset U(\infty)$ is closed if and only if $M \cap U(N)$ is closed, for all N=1,2,... . A compact subset of $U(\infty)$ can have a non-void intersection with at most a finite number of U(N) .

Note that a continuous map $A: X \to U(1+k)$, for any compact space X can be written uniquely in the form

$$(3.8) \qquad A(x) = a(x)(1+k(x)),\ x\in X\ ,\ \text{with } a\in C(X,S^1)\ ,\ k\in C(X,K)\ ,$$

with the unit circle $S^1 = \{c\in\mathbb{C}: |c|=1\}$ in $\mathbb{C}$. On the other hand, every continuous map $F:X\to U(\infty)$ is continuous from X to $U(1+k)$, and vice versa, a continuous map $G:X \to U(1+k)$ of the form (3.8), with $a(x) \equiv 1$, can be continuously deformed into a map $F:X\to U(\infty)$. (Essentially, to construct F from G one uses the fact that $U(\infty)$ is dense in the class of all unitary operators on h of the form $1+K$, $K\in k$, in norm topology of $L(h)$, and the polar decomposition of rem.3.2.)

This reduces the discussion of the homotopy classes of thm. 3.1 to that of homotopy classes of maps $\mathbb{M} \to U_N$, with the unitary group U_N of large dimension N.

Note that the theory of N×N-matrices $A=((A_{jk}))_{j,k=1,\dots,N}$ of operators $A_{jk}\in C$ may be treated as a special case of thm.3.1. Indeed, for such a matrix we define an operator $A^\sim \in C^\infty$ by

$$(3.9)\ A^\sim=((A^\sim_{jk}))_{j,k=1,2,\dots},\ A^\sim_{jk}=A_{jk},\ \text{for } j,k\le N,\ =\delta_{jk}\ \text{otherwise.}$$

Clearly $A^\sim$ will be a Fredholm operator of $L(H^\infty)$ if and only if A is a Fredholm operator of $C^N = L(\mathbb{C}^N)\otimes C$, and A, $A^\sim$ then will have the same index. The symbol of A will of course be defined as

$$(3.10) \qquad \sigma_A(m) = \sigma_{A^\sim}(m)|h_N\ ,$$

noting that $\sigma_A(m)$ is reduced by the decomposition $h=h_N \oplus h_N^\perp$.

It is just as easy, of course, and, in fact, simpler, to

redo the above theory with h replaced by the finite dimensional space $h_N = \mathbb{C}^N$.

In the following let $C^m = L(\mathbb{C}^m) \otimes C$, where C denotes any comparison algebra. Let us come back to the comparison principle of (2.1), and again set $Q = \Lambda = ((\Lambda\delta_{jk}))$, and $A = D\Lambda$, with

$$(3.11) \qquad D = ((D_{jk}))_{j,k=1,\dots,m} \, , \; D_{jk} \in \mathcal{D}^{\#} \, .$$

Again we get a realization $D = A\Lambda^{-1}$ of the ($m\times m$-matrix of) expression(s) D . That realization always is a restriction of the closure of the minimal operator D_0 , regardless of any condition (w), or (s_k) . In the results below we skip looking at the Fredholm index. An abstract index formula always follows in a similar way as in thm.1.9(A), thm.3.1,etc. Obtaining a concrete index formula should be easiest, in the present cases, but will not be a subject of our concern.

Theorem 3.4. Assume cdn's (a_0), (a_1), (d_0), and that C has compact commutator. The above realization D is a closed (unbounded) Fredholm operator of $L(H^m)$, with $H^m = \mathbb{C}^m \otimes H$, if and only if

$$(3.12) \qquad \text{the } m\times m\text{-matrix } \sigma_A(m) = ((\sigma_{D_{jl}\Lambda}(m))) \text{ is invertible for } m \in \mathbb{M} \, .$$

Note that no cdn.(w) or (s) is required in thm.3.4. The proof is an immediate consequence of thm.3.1.

Let us try to discuss the result. Again, we first consider condition (3.12) restricted to the wave front space $\mathbb{W} \subset \mathbb{M}$. There it just means that the first order $m\times m$-system $D = ((D_{jk}))$ is uniformly elliptic, with respect to the tensor h_{jk} . That is, if

$$(3.13) \qquad D = -iB^j(x)\partial_{x^j} + P(x)$$

locally, with $m\times m$-matrix-valued functions $B^j(x)$, $P(x)$, then the function $D(x,\xi) = B^j(x)\xi_j$ (globally defined as a function in $C^\infty(T^*\Omega, L(\mathbb{C}^m))$) is invertible for every $(x,\xi) \neq 0$, and satisfies

$$(3.14) \qquad \|D^{-1}(x,\xi)\| \le \chi < \infty \, , \text{ as } x\in\Omega, \; \xi\in\mathbb{R}^n, \; h^{jk}(x)\xi_j\xi_k = 1 \, ,$$

with the matrix inverse $D^{-1}(x,\xi)$, at x,ξ .

For a compact manifold Ω this is the only condition, since then $\mathbb{M} = \mathbb{W}$. On the other hand, if Ω is non-compact then there still may be comparison algebras with $\mathbb{M}_s = \emptyset$. For example, this is true under the assumptions of VII,thm.4.7, if $\Omega\cup\partial\Omega$ is compact.

On the other hand, our examples of $[C_1]$, and those of Sohrab show that we may have $\mathbb{M}_s \neq \emptyset$. Then we can apply thm.3.4 only if $\mathbb{M}_s$ and $\sigma_A|\mathbb{M}_s$ are accessible.

Note that thm.3.4 at least gives a sufficient criterion for the Fredholm property, if we have the assumptions of VII,thm.3.6 , (or thm.3.6^, for a suitable $\{\Omega\hat{},d\mu\hat{},H\hat{}\}$ (c> $\{\Omega,d\mu,H\}$). Then the symbol is a restriction of the formal symbol of VII,(3.25). For a compact commutator algebra $\mathcal{C}$ the operator D of (3.11) then is Fredholm if (though not necessarily only if) the formal symbol $\tau_{D\Lambda}$ is invertible at each point of $\partial\mathbb{P}^*\Omega$.

Finally we will use our above theory to also consider operators acting on crossections of a vector bundle over Ω. We have avoided vector bundles, so far, because our C^*-algebra approach to symbol and symbol space is developed best for complex-valued symbols, while its extension to vector bundles is a simple abstract matter, to be discussed now.

Let V be a vector bundle over Ω with fiber an inner product space V_x of dimension m, with norm $|u|_x$, $u \in V_x$. We only consider (C^∞-)vector bundles of the form $V=V_{A^\#}|\Omega = \theta^{-1}(\Omega)$, where $V_{A^\#}$ is a (continuous) vector bundle over the compactification $M_{A^\#}$ of Ω induced by $A^\#$ (cf.VII,1), with bundle projection θ . We assume the euclidean metric $|.|_x$ of the fiber defined for $x\in M_{A^\#}$, continuous in x , and C^∞ over Ω.

For the above we assume given a comparison triple on Ω , and generating classes $A^\#$, $\mathcal{D}^\#$ satisfying $(a_0),(a_1),(d_0)$, and then will introduce operators acting on $H_V = L^2(\Omega,V,d\mu)$, the space of crossections u of V satisfying $\|u\|^2 = \int_\Omega |u(x)|_x^2 d\mu < \infty$.

Let $M_{A^\#}=U_1\cup\ldots\cup U_N$ be a finite atlas of the (compact) bundle $V_{A^\#}$, and $1=\psi_1+\ldots+\psi_N$ a subordinate partition of unity, $\psi_j|\Omega\in C^\infty(\Omega)$

Under the above assumptions we will construct an isomorphic imbedding of the Hilbert space H_V into our above space $H^R=h_R\otimes H\subset H^\infty$ with orthogonal projection $P_V:H^\infty\to H_V$ in $\mathcal{C}_R \subset \mathcal{C}^\infty$, $\mathcal{C} = \mathcal{C}(A^\#,\mathcal{D}^\#)$. For two such vector bundles V and W one then may consider operators of the form $A\hat{}=P_W A\tilde{} P_V$, where $A\tilde{}\in \mathcal{C}^\infty$. We get $A\hat{}\in \mathcal{C}^\infty$; the restriction $A^\wedge|H_V = A$ is a map in $L(H_V,H_W)$, and the set of such maps will be denoted by C_{VW}. Fredholm results for $A\in C_{VW}$ will be obtained.

Let $e_p^l=(\delta_{l\lambda})_{\lambda=1,\dots,p}$, $l=1,\dots,p$, denote the standard base of $\mathbb{C}^p$. Define a linear map Θ_j from the space of local crossections over U_j into the $\mathbb{C}^N\otimes\mathbb{C}^m$-valued functions over U_j by $u(x)\to e_N^j\otimes u(x)$, in the bundle coordinates of the chart U_j. Here $\psi_j(x)u(x)\to e_N^j\otimes\psi_j(x)u(x)$, where both sides may be extended 0 outside U_j to obtain global crossections of V and the product bundle $\Omega\times(\mathbb{C}^N\otimes\mathbb{C}^m)=Z$, respectively. A (global) crossection u of V may be written in the form $u=\sum\psi_j u$, and a map $\Theta:T_V\to T_Z$, between the global crossections of f V and Z is defined by $\Theta u=\sum\Theta_j(\psi_j u)$, (where each $\psi_j u$ is mapped in the coordinates of U_j).

The above construction defines a bundle homomorphism $V\to Z$, and an isomorphism: For an $x\in U_k$, in the bundle coordinates of U_k, we get $\Theta u(x)=\sum_{j=1}^N e_N^j\otimes R_{jk}(\psi_j u)$, with the transition-matrices R_{jk}. Thus $\Theta u(x)=0$ if and only if $\psi_j u=0$, $j=1,\dots,N$, i.e., $u=0$.

In particular, if u_{jl} denotes the local crossection in U_j with $u_{jl}(x)=e_m^l$ in the coordinates of U_j, then $\psi_j u_{jl}=v_{jl}$ defines a global crossection, and $f_k(x)=\sum_{j=1}^N e_N^j\otimes v_{jl}(x)$, $x\in\Omega$, defines a global m-frame over $M_{A^\#}$, C^∞ over Ω. Write $V(x)=\operatorname{span}\{f_1,\dots,f_m\}$.

We may identify the space $\mathbb{C}^N\otimes\mathbb{C}^m$ with the space h_R, $R=mN$, of rem.3.3, writing $e_N^j\otimes e_m^k=e_R^{N(j-1)+k}$. Then the map Θ identifies the space H_V with the closed subspace of all measurable functions over Ω with value $u(x)\in V(x)$, such that $\|u\|_V$ of (3.15) is finite:

$$\|u\|_V=\int_\Omega |u(x)|_x^2 d\mu\ ,\quad u\in H_V\ ,\quad |u|_x^2=\{\sum m_{jk}(x)\bar{u}_j u_k\}^{1/2}, \tag{3.15}$$

with a positive definite matrix $((m_{jk}(x)))$, $m_{jk}\in C_{A^\#}$. Moreover, the global frame f_k is $C_{A^\#}$ as well.

It follows that the orthogonal projection $P_V(x):h\to V(x)$ (with the inner product of h) has matrix coefficients in $C_{A^\#}$, and hence is a (multiplication) operator in $C^R\subset C^\infty$.

For two vector bundles V,W, satisfying the above assumptions consider the class $C_{V,W}$ above. It is natural to introduce a symbol σ_A (mod $E_{V,W}=P_W E^\infty|V$, where $E^\infty=k\hat{\otimes}E$) by setting

$$(3.16)\qquad \sigma_A(m) = P_W(\iota(m))\sigma_{A^\sim}(m)|V(\iota(m)) \ , \ m \in \mathbb{M} \ .$$

Clearly σ_A may be regarded as a homomorphism $V_{\mathbb{M}} \to W_{\mathbb{M}}$, with the liftings $V_{\mathbb{M}}$, $W_{\mathbb{M}}$ of V and W onto $\mathbb{M}$ (by means of the map ι of VII,1) Again consider only the case of $E=K(H)$

Theorem 3.5. An operator $A\in C_{VW}$ is Fredholm in $L(H_V,H_W)$ if and only if its symbol $\sigma_A \in \hom(V_{\mathbb{M}},W_{\mathbb{M}})$ is an isomorphism onto $W_{\mathbb{M}}$.

Proof. Let m,m' be the fiber dimensions of V,W. Consider the case $m\le m'$; in the other case we may take adjoints. The above global frame $f_1,\ldots f_m$ for V(x) (and g_j for W(x)) allows us to define a global isomorphism $V(x)\to W(x)$ by assigning $f_j\to g_j$. Using the corresponding isomorphism $H_V\to H_W$ we may get restricted to the case $V(x)\subset W(x)$. For m=m' get V=W and note that $A^\Delta=1-P_V+AP_V\in C^\infty$ is ϕ iff A is ϕ, reducing thm.3.5 to thm.3.1. For m<m' σ_A cannot be surjective, and A cannot be ϕ : The projection P_0 onto span$\{g_{m'}\}$ is noncompact, but $\sigma_{P_0AP_V}=0$, hence $P_0AP_V\in K(H^\infty)$, a contradiction, if A is ϕ . Q.E.D.

It is easy to derive an abstract index formula of the form (3.5) again. Details are left to the reader.

Also, most of the results of this section can be generalized to the case of a multi-link ideal chain, with no or very little change in proof.

4. Discussion of algebras with a two-link ideal chain.

In this section we want to look at case (B) of sec.1. That is, in detail, we assume that we have

$$(4.1)\qquad C \supset E \supset K(H) \ ,$$

where the commutator ideal E of the comparison algebra C properly contains $K(H)$, and where

$$(4.2)\qquad E/K(H) = C(\mathbb{M}_E,K(h)) \ ,$$

with a locally compact space $\mathbb{M}_E$, called the symbol space of E , and the compact ideal $K(h)$ of a (separable) Hilbert space h .

We of course also get

(4.3) $A/E = C(\mathbb{M})$,

with another (compact) space $\mathbb{M} = \mathbb{M}_C$, called the symbol space of C , since by design C/E is a commutative C^*-algebra with unit, so that the Gel'fand-Naimark theorem applies.

We have discussed two examples of the present form in VIII, 1, 2, and a third more complicated example in VIII,4. Another example, corresponding to the regular elliptic boundary-value problem of a half-space, was discussed in $[C_1]$,V .

Again it should be mentioned, that we know examples with longer ideal chains, i.e., $E/K(H)$ is not a function algebra, but

(4.4) $E \supset F \supset G \supset \ldots \supset K(H)$,

with each successive quotient of the form $C(\mathbb{X},K(h))$. (cf.Dynin $[Dy_1]$). For most examples the space h has a concrete meaning; it occurs in some tensor product decomposition $H=h\hat{\otimes}H^{\sim}$ of $H=L^2(\Omega,d\mu)$.

On the abstract side of this discussion, Kaplanski $[Kp_1]$ has introduced a class of C^*-algebras, called GCR-algebras, having a similar ideal chain, perhaps of infinite length (cf.Dixmier $[Dx_1]$,p.87).(There the type of algebra is called 'algebre post-liminaire'.) Sakai $[Sk_1]$ has shown that a C^*-algebra is 'GCR' if and only if it is of type 1, in the sense that every representation is of type 1).

It is easy to formulate an abstract result extending thm. 4.1, below, to longer chains. However, we will not discuss this, since all our examples are of type (A) or (B).

Presently, for a two-link-chain, we have the following result.

Theorem 4.1.(a) An operator $A \in C$ is invertible modulo E if and only if its (C-)symbol σ_A, a certain continuous complex-valued function over $\mathbb{M} = \mathbb{M}_C$ does never vanish; (b) If an E-inverse P of $A \in C$ exist, so that we have

(4.5) $AP = 1+E$, $PA = 1+F$, with $E,F \in E$,

then 1+E and 1+F will be Fredholm if and only if the operator-valued functions $1+\tau_E$, $1+\tau_F$ are invertible at all points of $\mathbb{M}_E$, and the inverses are bounded on $\mathbb{M}_E$, where τ_E , τ_F denote the E-symbols of E and F, respectively; (c) If the E-inverse P exists, and G, H are $K(H)$-inverses of 1+E and 1+F, then PG and HP are $K(H)$-inverses of A, and A is Fredholm; (d) Vice versa, if $A \in C$ is Fredholm, then there exists a $K(H)$-inverse P of A, which also is an E-inverse, so that (4.5) holds with $E,F \in K(H) \subset E$. We then have $\sigma_A \neq 0$, for all $m \in \mathbb{M}$, and $\tau_E = \tau_F = 0$ on $\mathbb{M}_c$, so that $1+\tau_E = 1+\tau_F = 1$ are invertible.

The proof is a standard application of theory of commutative C^*-algebras.

Let us practice once more the comparison technique, by applying thm.4.1 to the example of VIII,4, involving a complete Riemannian manifold with cylindrical ends. First, in that respect, recall that the secondary symbol space of that algebra C is void, by VII,thm.4.2. Second, we note that we have cdn's (s),(m_1'),(m_6), (m_j), (l_j), j=1,2,3,4,5, and the assumption of VII,thm.3.6, all true. Accordingly we have the result, below.

Theorem 4.2. Let C be the comparison algebra of VIII,4. Then we have $A = L\Lambda^N \in C$, for every $L \in L_N$. Moreover, the operator $A=L\Lambda^N$ admits an E-inverse if and only if L is uniformly elliptic on Ω with respect to the tensor h_{jk} .

Indeed, IX,prop.3.2 is valid, and the **algebra** C has no secondary symbol space, so that the theorem is evident.

As a consequence of thm.4.2 we must look at the E-symbol of $E = 1-AP$ and $F=1-PA$, for a suitable E-inverse P of A, in order to decide whether or not A is Fredholm. Clearly, again, the realization $L=A\Lambda^{-N}$ of (2.4) of the expression L will be a closed Fredholm operator if and only if A is Fredholm.

First let us investigate the class L_N (of IX,3) in the present case. The description of the classes $A^\#$, $D^\#$, is found in VIII,4(i.e, they are determined by VIII,(2.2) , at each end of Ω).

Note that the classes $A^\#$, $D^\#$ of VIII,4 were 'minimal', in the sense that we were interested only in supplying enough generators for the algebra we had in mind, and to satisfy our basic conditions. Presently we should enlarge these classes (without changing the comparison algebra C), in order to have 'large' classes L_N of differential expressions.

Clearly we may admit into $A^\#$ any function $a \in C^\infty(\Omega) \cap C(M_{A^\#})$, without enlarging the algebra. Since $D^\#$ is a left $A^\#$-module, and $D^\#$ a Lie-algebra we must watch about properties of the derivatives in order to keep our cdn's (m_j),(l_j) preserved. A more careful investigation shows that we can reach the following classes $L_N^\# \supset L_N$

of N-th order differential expressions.

Definition 4.3. The class $L_N^{\#}$ consists precisely of all expressions of order N with C^{∞}-coefficients, defined on Ω , such that at each end, in the coordinates (x,t) of VIII, (2.1), L has the form

(4.6) $L=\sum_{j=0}^{N} L_j(t)\partial_t^j$, $\lim_{t\to\infty} L_j(t)=L_j(\infty)$, $\lim_{t\to\infty}(\partial_t^k L_j)(t)=0$, $k>0$.

Remark 4.4. Recall, that, at each end, we have a 'chart'

(4.7) $$U = \{(x,t):\ x\in B_j\ ,\ 0<t<\infty\} \quad .$$

In local coordinates $(x_1,\dots,x_{n-1})$ of the compact manifold B_j formula (4.6) means that $L=\sum L_j(t)\partial_t^j$, where

(4.8) $$L_j(t) = \sum_{|\alpha|\le N-j} a_{j,\alpha}(x,t)D_x^{\alpha}\ ,\ \lim_{t\to\infty} a_{j,\alpha}(x,t)=a_{j,\alpha}(x,\infty),$$
$$\lim_{t\to\infty}\partial_t^k a_{j,\alpha}(x,t) = 0\ ,\ k>0\ .$$

A quick calculation shows that the classes $L_N^{\#}$ are generated in the sense of IX,3 by the function algebra $A!=L_0^{\#}$ and Lie-algebra $D! = L_1^{\#}$, while the Stone-Weierstrass theorem may be used to show that $A^{\#}$ and $D^{\#}\Lambda$ are dense in $A!$ and $D!\Lambda$, respectively, in norm convergence of $L(H)$, so that $C(A^{\#},D^{\#}) = C(A!,D!)$.

Let us go somewhat more into the detail of construction of an E-inverse P , for an operator $A=L\Lambda^N$, since it is evident, that thm.4.1 is of little use for Fredholm theory, unless P is known, since the symbols $\sigma_{(1-AP)}$, $\sigma_{(1-PA)}$ otherwise cannot be obtained.

Let $L\in L_N^{\#}$ be uniformly elliptic, so that $\sigma_A\ne 0$ on $\mathbb{M}$, for $A=L\Lambda^N$. We focus on one of the cylindrical ends $\Omega_j^0=B_j\times\mathbb{R}$, which will be called $\Omega^{\blacktriangle} = B\times\mathbb{R}$. Correspondingly we have $H^{\blacktriangle}$, $C^{\blacktriangle}$, $A!^{\blacktriangle}$, $D!^{\blacktriangle}$, $\Lambda^{\blacktriangle}$, all defined as the corresponding concept on the straight cylinder $\Omega^{\blacktriangle}$, with only 2 ends, which has the common domain $U=B\times\mathbb{R}^+$ with Ω , onto which we apply algebra surgery.

One may construct a uniformly elliptic expression $L^{\blacktriangle}\in L^{\blacktriangle\#}_N$, coinciding with L on U , and then observe that

(4.9) $$\chi L\Lambda^N\chi - \chi L^{\blacktriangle}\Lambda^{\blacktriangle N}\chi \in K(H_U)\ ,\ H_U = L^2(U)\ .$$

This shows that construction of the E-inverse P , and even the later construction of a Fredholm inverse can be accomplished separately, at each end. Once we construct a Fredholm inverse $P^{\blacktriangle} = P_j$ for $A^{\blacktriangle} = L^{\blacktriangle}\Lambda^{\blacktriangle N}$, at each end, a Fredholm inverse P is given by

$$(4.10)\qquad P = \sum_{j=0}^{M} \psi_j P_j \psi_j \ , \quad M = \text{number of ends},$$

with cut-off functions ψ_j, $j=1,\dots,M$, and a suitable $\psi_0 \in C_0^\infty(\Omega)$, making $\sum_0^M \psi_j{}^2 = 1$. Also, P_0 is just any operator in J_0 such that $\sigma_{AP}=1$ over a compact neighbourhood of supp ψ_0.

Working on the straight cylinder $B\times\mathbb{R} = \Omega^{\blacktriangle}$ now, let us drop the superscript '$\blacktriangle$' again, keeping in mind, that $L=L^{\blacktriangle}$ only is determined on the right end, $t\to\infty$, and is quite arbitrary for small negative t. According to (4.6) the limits $L_j(\infty) = \lim_{t\to\infty} L_j(t)$ exist, and we may define the 't-translation invariant' expression

$$(4.11)\qquad L_\infty = \sum L_j(\infty)\partial_t^j \ .$$

Since L is arbitrary for negative t we may assume that L_∞ also is the limit of L, as $t\to-\infty$. Then one concludes that $L_\infty \in L_N^\#$, and

$$(4.12)\qquad A - A_\infty = L\Lambda^N - L_\infty\Lambda^N \in J_0 \ .$$

The theorem, below, is self-explanatory.

Theorem 4.5. If P_∞ is an E-inverse of A_∞, then we have

$$(4.13)\quad AP_\infty = J + 1 + E \ , \ P_\infty A = J' + 1 + E' \ , \ J,J' \in J_0 \ , \ E,E' \in E \ ,$$

with the minimal comparison algebra J_0. Then, if we find $K(H)$-inverses $1+F$, $1+F'$, $F,F' \in E$, of $1+E$ and $1+E'$, respectively, we get

$$(4.14)\ A(P_\infty(1+F)) = 1+K+C,\ ((1+F')P_\infty)A = 1+K'+C',\ K,K' \in J_0,\ C,C' \in K(H).$$

Indeed, we get $(A-A_\infty)P_\infty \in J_0$, $P_\infty(A-A_\infty) \in J_0$, explaining (4.13), and then (4.14) is evident.

It is clear now that the right hand sides of the relations (4.14) are contained in the compact commutator algebra $J = 1 + J_0$, so that existence of $K(H)$- inverses for $1+K$ and $1+K'$ is a matter of checking on the symbols. In addition, a proper choice of the cut-off function χ_j in (4.10) will allows us to use any operator G, $G' \in 1+J_0$ satisfying $\sigma_G \sigma_{(1+K)} = 1$, $\sigma_{G'} \sigma_{(1+K')} = 1$ near $t=\infty$. Thus we have proven the result, below.

Theorem 4.6. Assume L of thm.4.2 uniformly elliptic. At each end let $L^{\blacktriangle} = L(\infty)$ be the translation invariant limit operator of L, and let $\Lambda^{\blacktriangle} = (1-\Delta^{\blacktriangle})^{-1/2}$ be the (translation invariant) inverse comparison operator square root. Then $A = L\Lambda^N$ is Fredholm if and only if, at each end, the translation invariant limits $A^{\blacktriangle} = L^{\blacktriangle}\Lambda^{\blacktriangle N}$ are Fredholm operators of $L(H^{\blacktriangle})$.

One will be tempted, of course, to conjugate $A^{\blacktriangle}$ with the t-Fourier transform F_t of VIII,2. Note that this takes $A^{\blacktriangle}$ onto the operator

$$(4.15) \qquad A^{\blacktriangle\wedge} = \sum_{k=0}^{N} A_k(\tau\Lambda_x)^k T^k(t),\ A_k = i^k L_k(\infty)\Lambda_x^k ,$$

with

$$(4.16) \qquad \Lambda_x = (1-\Delta_x)^{-1/2} ,\ T(\tau) = (1+\tau^2\Lambda_x{}^2)^{-1/2} ,$$

Recall also that the commutator ideal here is represented by the class $CO(\mathbb{R},k)+K(H^{\blacktriangle})$, $k=K(L^2(B))$, cf.VIII,2. A simple left multiplication of $A^{\blacktriangle\wedge}$ by a Fredholm inverse $P^{\blacktriangle}$ of the elliptic operator A_0 on B will convert (4.15) into the form

$$(4.17) \qquad P^{\blacktriangle}A^{\blacktriangle\wedge} = 1+K(\tau) ,\ K(\tau) \in CB(\mathbb{R},k) .$$

This, however, is no inversion mod E . Still the problem is reduced to an equation involving compact operators over the compact manifold B . (For more facts cf. [CPo].)

Remark 4.7. It is clear that the operator $K(\tau)$ of (4.17) is not only continuous but even analytic in the parameter τ . Using a result of Gramsch [G_2] it therefore follows that the Fredholm operator-valued analytic operator function $1+K(\tau)$ has a meromorphic inverse if only it is invertible at a single $\tau=\tau_0$. Existence of such τ_0 indeed can be established, using the Sobolev estimate of VI, thm.4.3, by a simple estimate, due to Agmon and Nirenberg, cf. [AN_1]. Then, of course, $1+K(\tau)$ must be invertible at all $\tau \in \mathbb{C}$ near the real axis, except at most countably many without finite cluster. In fact, it follows that there are at most finitely many such points in some strip $|\mathrm{Im}\ \lambda| \le \eta$, $\eta > 0$.

We will get a Fredholm operator if none of these exceptional points lies on the real axis. While in general this may or may not occur for a general expression L , it always will be true for the expression $e^{\delta t}Le^{-\delta L}$ for all but finitely many $|\delta| < \eta$, $\delta \in \mathbb{R}$.

5. Fredholm theory and comparison technique in Sobolev spaces.

First let us point to a serious restriction of applicability of the present discussion. We generally assume here cdn.(s), which is never satisfied for a comparison triple with any nonvoid regular boundary, - i.e., if any open nonvoid subset of $\partial M_{A^{\#}}$ is an n-1-dimensional manifold $\partial\Omega$ such that $d\mu$ and H extend to $\Omega\cup\partial\Omega$, satisfying all the conditions they otherwise satisfy on Ω .

Thus, in particular, while much of our L^2-theory applies to the regular elliptic boundary problem, with Lopatinskij-Shapiro type boundary conditions, this is not so for theory in Sobolev spaces.

In fact, even our definition of Sobolev spaces H_s does not supply the commonly used Sobolev spaces if (s) is violated. In that case there is a choice of self-adjoint extensions of H_0^m , for some integers m. One will get the common Sobolev space H_m only if Λ^m is replaced by the inverse square root of the extension giving the largest domain for the corresponding form $(u,H^m v)=(u,v)_m$, which then is the Sobolev space H_s . This normally does not correspond to the Friedrichs extension (which gives the smallest domain for the form). In case of a regular boundary the Friedrichs extension corresponds to the Dirichlet boundary condition, while the Sobolev space corresponds to the Neumann condition. In particular also, the Λ_s to be substituted for Λ^s in IX,(1.1) no longer will be the power of a fixed operator Λ .

Thus we assume cdn.(s), for the remainder of sec.5. Also we require the same conditions as in IX,4, i.e., (1_j), j=1,2, (m_5), and $(1_5')(P_\infty)$, for a suitable P_∞ within general reach of $A^{\#}$, $D^{\#}$, of principal symbol type. (We always have (a_j), j=0,1, and (d_0), of course.)

Under these assumptions IX,thm.4.3 is valid in its strong form. A comparison algebra C_s is defined in H_s, $s\in\mathbb{R}$, from the generators IX,(1.10). Any two such comparison algebras C_s , C_t are isometrically isomorphic, under the natural isomorphism $\Lambda^r : H_s \to H_t$, r=t-s. For $A\in C_s$ we have $\Lambda^{-r}A\Lambda^r \in C_t$, and vice versa. The commutator ideals also map onto each other, under this conjugation.

The symbol spaces are homeomorphic under the associate dual map of the isomorphism $C_s/E_s \to C_t/E_t$ induced. Also, for $A \in C^0_\infty$, and even for A finitely generated by $L\Lambda^N$, $\Lambda^N L$, $L\in Q_N=P_N+L_N$, we have $\Lambda^{-r}A\Lambda^{r}-A \in E$, so that the symbol of A in H_s coincides with the symbol of A in H_t, if the associate dual map is used to identify the two symbol spaces. With that interpretation we obtain a common symbol space $\mathbb{M}$ for all algebras C_s , and the symbol σ_A is independent of s, whenever $A\in C^0_\infty$, or A is finitely generated from $L\Lambda^N$, $\Lambda^N L$, for $L\in \mathcal{Q}_N$, and C^0_∞ .

Taking closure in the topological algebra $O(0)$ we obtain the Frechet algebra C_∞ . Since convergence $A_j\to A$ in C_∞ implies convergence $A_j\to A$ in every C_s , it follows that every $A\in C_\infty$ has its symbol independent of s . Accordingly we define a symbol for the Frechet algebra C_∞ by setting σ_A equal to the symbol of $A\in C_\infty\subset C_s$ in C_s , for any fixed s, this definition being independent of s.

We also get any two algebras $E_s/K(H_s)$, $E_t/K(H_t)$, $s,t\in\mathbb{R}$, isometrically isomorphic, under the same conjugation with Λ^r , $r=t-s$. Thus, if the algebra $E=E_0$ has a symbol of the second kind, i.e., if $E/K(H)$ is isometrically isomorphic to $C(\mathbb{E},K(h))$, as in the examples of VIII,1, 2, and 4, then the same holds true for all the algebras $E_s/K(H_s)$, $s\in\mathbb{R}$. We get $E_s/K(H_s)\approx C(\mathbb{E},K(h))$. If $\gamma:E \to K(H)$ denotes the symbol homomorphism of E , then the C_s-symbol γ^s_E of an operator $E \in C_s$ is given by

$$(5.1)\qquad \gamma^s_E = \gamma_{E_s} , \quad E_s = \Lambda^{-s}E\Lambda^s \in C ,$$

where we know from IX,(4.4), here valid in its strong form, that $E_s\in E$, so that γ_{E_s} is defined.

It now is natural to focus on the commutator ideal E^0 of the finitely generated algebra C^0_∞ . Note that E_s is the closure of E^0 in $L(H_s)$, $s \in \mathbb{R}$, and in $O(0)$,for $s=\infty$. For an operator $E\in E^0$ we again have γ^s_E defined for all $s\in\mathbb{R}$, and the same is true for $E\in E_\infty$, by continuous extension in $O(0)$. While we found the C_s-symbol of an $A\in C_\infty$ independent of s , in all cases examined, the corresponding property for the E_σ-symbols would depend on the relation

$$(5.2)\qquad \Lambda^{-r}E\Lambda^{r} - E \in K(H) \text{ for all } E\in E^0,\ r\in \mathbb{R} .$$

A check on our examples in sec.VIII,1,2,4 , and [C_1],V reveals that (5.2) normally is <u>not</u> satisfied.

As a substitute for 'γ^s_E=const.' , using V,thm.6.8, it follows that E_s is an entire analytic function of s , defined for all real and complex s, and taking values in E , for all s . All derivatives of E_s (for s) again are in E .

Then, using that the symbol homomorphism $\gamma: E/K(H)\to C(\mathbb{E},K(h))$ is an isometry, one concludes that also γ_{E_s} is an entire function of s . Indeed, we get, with $F_s=dE_s/ds$,

$$(5.3)\quad \sup\|\gamma_{E_s}-\gamma_{E_{s'}}-(s-s')\gamma_{F_{s'}}\| \le \|E_s-E_{s'}-(s-s')F_{s'}\| = o(|s-s'|).$$

showing that the complex derivative of γ_{E_s} for s exists and is a continuous function over $\mathbb{E}$, with values in $K(h)$ again.

Now, for a general operator $E \in E_\infty$, the Frechet closure of E^0 in $O(0)$, we obtain E as limit of a sequence $E_j\in E^0$, for which the sequence of symbols $\gamma_{E_{js}}$ converges uniformly in s , on compact subsets of $\mathbb{C}$. Therefore γ_{E_s} is entire again. We have proven:

Proposition 5.1. Under the assumptions of sec.5 the symbol $\gamma_E^s(e)$, for $E\in E_\infty$, $e\in\mathbb{E}$, is an entire function of s with values in $K(h)$, continuous in s and e , and $CO(\mathbb{E})$, for fixed s. Moreover, γ_E even is entire as a function $\mathbb{C}\to CO(\mathbb{E},K(h))$. For fixed $e\in\mathbb{E}$ and $E\in E^\infty$ the (entire Fredholm-operator-valued) function $1+\gamma_E^s(e)$, either has a nontrivial kernel for every (real and complex) s, or else has a meromorphic inverse, singular only at discrete points of $\mathbb{C}$.

We have the following analogue of thm.4.1:

Theorem 5.2. Under the above assumptions, an operator $A \in C_s$ has an E_s-inverse if its (primary) symbol σ_A does never vanish on the space $\mathbb{M}$. In case of $A\in C_\infty$ the primary symbol is independent of s , so that an E_s-inverse exists either for no s or for all $s \in \mathbb{R}$.

An 'E_s-inverted' operator $K=1+E$, $E\in E_s$ is Fredholm if and only if its E_s-symbol $1+\gamma_E^s$ is nonsingular for all $e\in\mathbb{E}$, and its inverse is in $1+CO(\mathbb{E},K(h))$.

A Fredholm operator of C_s must have an E_s-inverse, hence a non-vanishing primary symbol. Moreover, it must have an E_s-inverse B such that AB and BA have a non-singular E_s-symbol on all of $\mathbb{E}$.

Among the examples offered one finds that the Hilbert space h of IX,(4.8) always has a concrete meaning as a factor in a topological tensor decomposition

$$(5.4)\qquad H = H_0= h\hat{\otimes}h' ,$$

where then the E-symbol γ_E of a generator $E\in E^0$ may be easily calculated from that decomposition. Generally the operators Λ^s do not separate relative to (5.4).

For example, in the algebra of VIII,2 (with generators VIII,(2.2)), setting n'=1, we have $h = h_x$, $h' = h_t$, and

$$(5.5) \qquad \Lambda^s = (1 - \partial_t^2 - \Delta_x)^{-s/2} ,$$

with ∂_t , Δ_x , but not Λ^s acting on the factor spaces. After an F_t-conjugation, as in VIII,2, we get the operator function

$$(5.6) \quad M^s(\tau) = \Lambda_x{}^s T^s(\tau) \in K(h), \text{ with } T(\tau) = (1+\tau^2\Lambda_x^2)^{-1/2} ,$$

acting as multiplication operator on the factor h' . Here it may be observed that $M^s(\tau)$ provides an isomorphism $h \to h^s$ (although not an isometry), depending on τ , for each fixed τ , where, for a moment, h^s denotes the L^2-Sobolev space of order s of the compact manifold B , with respect to the metric on hand. Thus H_s , in this case, corresponds to the class of all h^s-valued functions $u(\tau)$ over $\mathbb{R}$ with $\int_{-\infty}^{\infty} \|M^{-s}(\tau)u(\tau)\|^2 d\tau < \infty$.

One confirms easily that, in this case the symbol γ_E , for an $E \in E^0$, assumes values in the algebra C_B^0 , the finitely generated algebra within the unique Laplace comparison algebra C_B of the compact space B . Thus $1+\gamma_E(e)$, for fixed $e \in \mathbb{E}$, is an operator of $O(0)$, relative to B , and is K-invariant under H-conjugation, in the sense of I,def.6.5, in view of IX,(4.4), and (5.6). Accordingly, I,thm.6.6 applies, and it follows that $\ker(1+\gamma_E^s)$ and $\ker(1+\gamma_E^s)$ have dimension independent of s . Accordingly, in this case, although γ_E^s is not independent of s , it still follows that $1+\gamma_E^s(e)$ is invertible for one $s=s_0$ if and only if it is invertible for all s .

To apply the last observation for construction of Fredholm inverses of an $A \in C_\infty$ we would require an E_∞-inverse B of A . We will not attempt its construction here, and leave further investigations to the reader.

Instead we now get restricted to the case $E=K(H)$. We know that then also $E_s = K(H_s)$, for $s \in \mathbb{R}$, by IX,thm.4.2.

As in VI,3 two operators A, B $\in O(\infty)$ are called <u>Green-inverses</u> of each other, if we have

$$(5.7) \qquad AB - 1 \in F_\infty , \ BA - 1 \in F_\infty ,$$

where F_∞ denotes the ideal of operators of finite rank in $O(-\infty)$, as introduced in I, problem 6.8. There we stated that the $O(0)$-closure F_∞ coincides with the class of K_∞ of operators $A \in O(0)$ with $A \in K(H_s)$ for all $s \in \mathbb{R}$.

Starting with any comparison algebra $C_\infty = C_\infty(A!, D!)$, under the assumptions of the present section (with C of compact commutator), we define the classes (subspaces of $O(s)$)

$$(5.8) \qquad \Psi C_s = \{A\Lambda^{-s} : A \in C_\infty\} \; , \; \Psi C_\infty = \cup\{\Psi C_s : s \in \mathbb{R}\} \; .$$

It then follows that ΨC_∞ is a graded algebra containing the (differential operators of the) expressions in $\mathcal{Q}_\infty$ within general reach. In details, we have

$$(5.9) \qquad \Psi C_s . \Psi C_t \subset \Psi C_{s+t} \; , \; s,t, \in \mathbb{R} \; , \; \mathcal{Q}_N \subset \Psi C_N \; , \; N=0,1,2,\ldots \; .$$

Also $\Psi C_0 \supset K_\infty \supset F_\infty$.

For an operator $K = A\Lambda^{-s} \in \Psi C_s$ the symbol σ_A of A is called the <u>symbol quotient</u> of order s.

<u>Theorem 5.3</u>. An operator $K \in \Psi C_s$ admits a Green inverse in $O(-s)$ if and only if its symbol quotient of order s does never vanish on the symbol space $\mathbb{M}$. Moreover, any such Green inverse is contained in ΨC_{-s} .

The essential point of the proof is that operators in $\Psi C_0 = C_\infty$ are K-invariant under H-conjugation, in the sense of I,def.6.5, so that I,thm.6.6 applies for all Fredholm operators. This in essence implies the statement for s=0, while the general case is easily reduced to s=0. For details cf. [C_1],IV,thm.4.1.

<u>Theorem 5.4</u>. Under the present assumptions, with a compact commutator algebra C, the algebra C_∞ is a ψ^*-subalgebra of $L(H)$.

<u>Proof</u>. Suppose $A \in C_\infty$ admits an inverse $B \in L(H)$. Then it is Fredholm in C , hence $\sigma_A \neq 0$ on $\mathbb{M}$. By thm.5.2 it then admits a Green inverse $B_0 \in C_\infty$. In fact, the proof of I,thm.6.6 shows that the distinguished Green inverse $B_0 : (\ker A)^\perp \to \operatorname{im} A$ of VI,3, with the orthogonal complement in H, is in C_∞ . But that must be B, since $\operatorname{im} A = H$, $\ker A = 0$, A being invertible, q.e.d.

Appendix A. Auxiliary results, concerning functions on manifolds.

Let the manifold Ω satisfy the general assumptions of III,1. In particular we assume the existence of a countable locally finite atlas $\{\Omega_j : j=1,2,\dots\}$, where each Ω_j^{clos} is compactly contained in some U_j, where $\{U_j\}$ is another locally finite atlas of Ω. Let $\Omega^{\sim}$ be an open subdomain of Ω (where $\Omega = \Omega^{\sim}$ is permitted.) Suppose $f(x)$ and $g(x) > 0$ are functions over $\Omega^{\sim}$ and Ω, respectively. We will use the Landau symbols in the following sense:

Write $f=O(g)$ (in $\Omega^{\sim}$) if $f(x)/g(x)$ is bounded over $\Omega^{\sim}$; write $f=o_{\Omega}(g)$ (in $\Omega^{\sim}$) if $f=O(g)$ (in $\Omega^{\sim}$) and $\lim_{x\to\infty}(f(x)/g(x))=0$ (in Ω). (That is, for $\varepsilon>0$ there exists a compact set $K \subset \Omega$ such that $|f(x)/g(x)| < \varepsilon$ for all $x \in \Omega^{\sim}\setminus K$.) We shall write $f=O(g)$, and $f=o(g)$, (without "(in $\Omega^{\sim}$)", etc.) if no confusion can arise.

Lemma A.1. Let f,g be as above, and let g be continuous over Ω. If $f = o_{\Omega}(g)$, then there exists a positive $C^{\infty}(\Omega)$-function ψ such that $f = O(\psi)$ (in $\Omega^{\sim}$), and $\psi = o_{\Omega}(\gamma)$.

Proof. Let $\phi(x) = f(x)/g(x)$, so that we have $\phi(x)$ bounded over $\Omega^{\sim}$ and $\lim_{x\to\infty}\phi(x) = 0$. Consider ϕ extended to Ω by setting $\phi(x) = 0$ outside Ω, then the limit still is zero. With our partition ω_j define $\eta_j = \sup\{|\phi(x)| : x \in \operatorname{supp}\omega_j\}$. Observe that $\eta_j>0$, and $\lim_{j\to\infty}\eta_j = 0$. For there exists a compact set $K_\varepsilon \subset \Omega$ such that $|\phi|<\varepsilon$ outside K_ε, for every $\varepsilon>0$. Only finitely many of the sets $O_j = \operatorname{supp}\omega_j$ can have points in common with K_ε. Else there exists $x_{j_k} \in \Omega_{j_k}\cap K_\varepsilon$ with a limit point $x_0 \in K_\varepsilon$. Due to localy finite coverage x_0 has a neighbourhood N contained in only finitely many Ω_j, so cannot be a limit point of the x_{j_k}. Hence, for $\varepsilon>0$ there exists N_ε such that all $\operatorname{supp}\omega_j$, $j>N_\varepsilon$ are completely outside K_ε. Accordingly, $\eta_j = \sup\{|\phi_j(x)| : x \in \operatorname{supp}\omega_j\} \le \varepsilon$.

Now we just define the function $\psi(x)=g(x)\sum_{j=1}^{\infty}\eta_j\omega_j(x)$, where the sum is locally finite, hence represents a $C^{\infty}(\Omega)$-function. Since γ is only continuous, the function ψ thus defined is not yet $C^{\infty}(\Omega)$, but is at least continuous. It also satisfies the other conditions: For any $x \in \Omega$ we get $|\phi(x)| = \sum|\phi(x)|\omega_j(x) \le \sum\eta_j\omega_j(x) = \psi(x)/g(x)$. This implies $|f(x)| = g(x)|\phi(x)| \le \psi(x)$, i.e.,

$f = O(\psi)$ (in $\Omega^{\sim}$) . Also we note that $\lim_{x\to\infty}\sum\eta_j\omega_j(x) = 0$, so that $\psi = o_\Omega(g)$. Indeed, let $K_N = \Omega_1\cup\ldots\cup\Omega_N$. Clearly K_N is compact, while $x \in \Omega\setminus K_N$ implies that $\sum_{j=1}^{\infty}\eta_j\omega_j(x) = \sum_{j=N+1}^{\infty}\eta_j\omega_j(x) \leq \sup\{\eta_j : j\geq N+1\} \to 0$, as $N \to \infty$. To fully establish the lemma we now must make a C^∞-correction of ψ which does not disturb the other conditions already established. This is accomplished in lemma A.2, below.

Lemma A.2. Let $f,g \in C(\Omega)$, $g > 0$. Then there exist positive $C^\infty(\Omega)$-functions γ , δ such that

$$\text{(A.1)}\qquad f(x) \leq \gamma(x) \leq f(x) + \delta(x)\ ,\ x \in \Omega\ ,\ \delta = o_\Omega(g)\ .$$

Proof. Let $\varepsilon_j = 2^{-j}\operatorname{Min}\{g(x) : x \in \Omega_j^{\text{clos}}\}$. Clearly $\varepsilon_j > 0$, since Ω_j^{clos} are compact, and $g(x) > 0$. Using the coordinate transform of U_j we may regard $\Omega_j\subset \Omega_j^{\text{clos}}\subset U_j$ as subsets of $\mathbb{R}^n$. Let $\omega_j \geq 0$, $\omega_j \in C_0^\infty(\Omega_j)$, $\sum_{j=1}^{\infty}\omega_j(x) = 1$ in Ω , i.e., $\{\omega_j\}$ is a partition of unity . Using regularizing techniques and the compactness of Ω_j^{clos} it is possible to find $f_j\in C_0^\infty(\Omega_j)$ such that

$$\text{(A.2)}\qquad \omega_j(x)f(x) \leq f_j(x) \leq \omega_j(x)f(x) + \varepsilon_j\ ,\ x \in \Omega_j\ .$$

For each j let $\chi_j(x) \in C_0^\infty(U_j)$, $0\leq\chi_j\leq 1$, $\chi_j(x) = 1$, as $x \in \Omega_j$. Then (A.2) implies that

$$\text{(A.3)}\qquad f(x) \leq \sum f_j(x) \leq f(x) + \sum \varepsilon_j\chi_j(x)\ ,\ x \in \Omega\ .$$

Here $\gamma(x) = \sum f_j(x)$, and $\delta(x) = \sum \varepsilon_j\chi_j(x)$ are well defined $C^\infty(\Omega)$-functions, since the sums are locally finite. Moreover,

$$\text{(A.4)}\qquad 0 < \delta(x)/g(x) \leq \sum_{x\in \operatorname{supp}(\chi_j)} 2^{-j}\ .$$

For every N=1,2,..., a compact set K_N may be found such that $\operatorname{supp}\chi_j \cap K_N^{\text{compl}} = \emptyset$, for all $j \leq N$, due to local finiteness of $\{U_j\}$. Therefore the right hand side of (A.4) is $< \sum_{j=1}^{N+1}2^{-j}=2^{1-N}$, as $x \in K^{\text{compl}}$. We get $\lim_{x\to\infty}\delta(x)/g(x) = 0$, or, $\delta = \omega_\Omega(g)$, q.e.d.

Appendix B. Covariant derivatives, and curvature.

By 'covariant derivatives' we mean the Riemannian covariant derivatives, with respect to the connection defined by the Christoffel symbols of the metric tensor h_{jk} , where

$$\text{(B.1)} \qquad ((h_{jk}))_{j,k=1,\dots,n} = ((h^{jk}))^{-1} .$$

Let b_j and c^j be covariant and contravariant tensors, (i.e., tangent or cotangent vector fields, respectively, depending smoothly on $x \in \Omega$. Then we have

$$\text{(B.2)} \qquad b_{j|k} = b_{j|x^k} - \Gamma^l_{jk} b_l \ , \ c^j{}_{|k} = c^j{}_{|x^k} + \Gamma^{jk}_m c^m,$$

where the Christoffel symbols

$$\text{(B.3)} \qquad \Gamma^i_{jk} = h^{il}\Gamma_{jk,l} \ , \ \Gamma_{jk,l} = \tfrac{1}{2}(h_{jl|x^k} + h_{lk|x^j} - h_{jk|x^l}) \ ,$$

are symmetric in j and k, and are not tensors. Both covariant derivatives are tensors with two variable indices.

Covariant derivatives of tensors with multiple indices are defined similarly. For example,

$$\text{(B.4)} \qquad a_{ij|k} = a_{ij|x^k} - a_{ih}\Gamma^h_{jk} - a_{hj}\Gamma^h_{ik} \ ,$$

and,

$$\text{(B.5)} \qquad a^{ij}{}_{|k} = a^{ij}{}_{|x^k} + a^h_j\Gamma^i_{hk} + a^{hj}\Gamma^i_{hk} \ .$$

In general,

$$\text{(B.6)} \qquad a^{r^1r^2}_{s^1s^2\dots|i} = a^{r^1r^2}_{s^1s^2\dots|x^i} + \sum_\alpha a^{r^1\dots r^{\alpha-1}jr^{\alpha+1}\dots}_{s^1\dots} \Gamma^{r^\alpha}_{ji}$$

$$- \sum_\beta a^{r^1\dots}_{s^1\dots s^{\beta-1}ls^{\beta+1}\dots} \Gamma^l_{s^\beta i}$$

Higher covariant derivatives are defined recursively. For example $a^i{}_{|jk} = (a^i{}_{|j})_{|k}$: The covariant derivatative of the tensor $a^i{}_{|j}$, of one covariant and one contravariant index. Etc. For sake of completeness one defines $a_{|j} = a_{|x^j}$. That is, the first covariant derivative of a (scalar) function a(x) is its gradient.

Covariant derivatives do not necessarily commute, even if the coefficients are smooth. We have

(B.7) $$a_{|ik} = a_{|ki} ,$$

for a scalar a . However , for a tensor a_i and b^i we get

(B.8) $$a_{i|jk} - a_{i|kj} = a_l R^l_{ijk} ,$$

and

(B.9) $$b^i{}_{|jk} - b^i{}_{|kj} = -b^l R^i{}_{ljk} ,$$

with the Riemann curvature tensor R^l_{ijk} , and R_{iljk} . Here we have

(B.10) $$R^l{}_{ijk} = \Gamma^l_{ik|x^j} - \Gamma^l_{ij|x^k} + \Gamma^m_{ik}\Gamma^l_{mj} - \Gamma^m_{ij}\Gamma^l_{mk}, \quad R_{ilkj} = h_{ir}R^r{}_{lkj} .$$

Similarly

(B.11) $$a_{ij|kl} - a_{ij|lk} = a_{ih}R^h_{jkl} + a_{hj}R^h_{ikl} ,$$

etc. It is known that the metric tensor h^{jk} , as well as h_{jk} , behave like constants under covariant differentiation, i.e.,

(B.12) $$h_{jk|l} = 0 \quad , \quad hjk_{|l} = 0 .$$

Also, covariant derivatives satisfy the usual differentiation rules. For example,

(B.13) $$(a_{jk}b^{jk})_{|l} = a_{jk|l}b^{jk} + a_{jk}b^{jk}{}_{|l} ,$$

etc. Finally we note the symmetry relations of the Riemann tensor:

(B.14) $$R_{ijkl} = -R_{jikl} = -R_{ijlk} , \quad R_{ijkl} + R_{iklj} + R_{iljk} = 0 .$$

Also the Ricci tensor will be used:

(B.15) $$R_{ij} = R^k_{ijk} ,$$

as well as the divergence formula

(B.16) $$(a_{|j})^{|j} = h^{jk}a_{|jk} = \sqrt{h}^{-1}(\sqrt{h}h^{jk}a_{|x^j})_{|x^k} .$$

(Notice that, as for all tensors, a covariant derivative may be 'raised', giving a contravariant derivative:

(B.17) $$a_j{}^{|k} = h^{jl}a_{j|l} , \text{ etc.}$$

Appendix C. Summary of the conditions (x_j) used.

In the following we give a list of the various conditions used to describe features of comparison algebras and their generating sets, in alphabetical order, with the section listed, where they were introduced.

(a_0) (V,1): $a \in A^{\#}$ implies $\bar{a} \in A^{\#}$, and $A^{\#} \supset C_0^{\infty}(\Omega)$.

(a_1) (V,1): $A^{\#}$ contains the constant functions.

(a_1') (V,1): For every $D \in \mathcal{D}^{\#}$ there exists an $a \in A^{\#}$ such that ∇a has compact support, and $aD = Da = D$.

(d_0) (V,1): $D=b^j\partial_{x^j}+p\in \mathcal{D}^{\#}$ implies $\bar{D}=\bar{b}^j\partial_{x^j}+\bar{p}\in \mathcal{D}^{\#}$, and $\mathcal{D}^{\#}$ contains all C∞-coefficient folpde's with compact support.

(l_1) (IX,1): $A^{\#}$ is an algebra under the pointwise product of functions, and $D^{\#}$ is a Lie-algebra, under the commutator product of folpde's. Also, $\mathcal{D}^{\#}$ is a two-sided $A^{\#}$-module, as under (m_2), and $A^{\#}$, interpreted as a set of zero-order expressions, is a subset of $\mathcal{D}^{\#}$. Also, for $a \in A^{\#}$, $D \in \mathcal{D}^{\#}$ we have $[a,D] \in A^{\#}$.

(l_2) (IX,1): $D\in D^{\#}$ implies that $D^{*}\in\mathcal{D}^{\#}$.

(l_3) (IX,1): Cdn.(m_3) holds, and, in addition, for every $a \in A^{\#}$ with $a\neq 0$ at a closed set $\mathcal{Q} \subset M_{A^{\#}}$ there exists a function $b \in A^{\#}$ with $ab=1$ near $\mathcal{Q}$.

(l_4) (IX,1): The expression H can be written as a finite sum of terms aDF, with $a \in A^{\#}$, $D,F \in \mathcal{D}^{\#}$.

(l_5) (IX,1): For every $a \in A^{\#}$, $D \in \mathcal{D}^{\#}$ we have $[a,H] \in \mathcal{D}^{\#}$, and $[D,H]$ is a finite sum of terms bEF, $b\in A^{\#}$, $E,F\in\mathcal{D}^{\#}$.

(l_5') (IX,3): We have $[H,a] \in L_{1,0}$, $[H,D] \in L_{2,0}$, for all $a \in A!$, $D \in \mathcal{D}!$.

$(l_5)(P_\infty)$ (IX,3): Cdn.(l_5) with L_j replaced by $\mathcal{Q}_j=L_j+P_j$, for a graded algebra $P_\infty=\cup P_N$ within reach.

$(l_5')(P_\infty)$ (IX,3): Cdn.(l_5') with $L_{j,0}$ replaced by $\mathcal{Q}_{j,0} = P_j+L_{j,0}$, with an algebra $P_\infty=\cup P_N$ within reach and of principal symbol type.

(m_1) (VII,2): For each $D \in \mathcal{D}^{\#}$ the expression

$$\{D,D\}(x) = h_{jk}(x)\bar{b}^j(x)b^k(x) + \bar{p}(x)p(x)/q(x)$$

defines a function in $A^{\#}$.

$(m_1)^\wedge$ (VII,4): Same as (m_1), but with $\{D,D\}$ replaced by

$\{D,D\}^\wedge$, formed with the coefficients of some triple $\{\Omega^\wedge, d\mu^\wedge, H^\wedge\}$ (c> $\{\Omega, d\mu, H\}$.

(m_1') (VII,3): In addition to $\{D,D\}$ of (m_1) also $\{D,D\}_0(x) = |p(x)|^2/q(x)$ defines a function in $A^\#$.

$(m_1')^\wedge$ (VII,4): Same as (m_1') , but with respect to a triple $\{\Omega^\wedge, d\mu^\wedge, H^\wedge\}$ (cf. $(m_1)^\wedge$).

$(m_1)_x$, $(m_1')_x$, $(m_1)^\wedge_x$, $(m_1')_x^\wedge$ (VII,4): The corresponding conditions, (m_1),..., stated only for the given x.

(m_2) (VII,2): $A^\#$ and $\mathcal{D}^\#$ are complex vector spaces, and $\mathcal{D}^\#$ is a left $A^\#$-module, i.e., $aD \in \mathcal{D}^\#$ for all $a \in A^\#$, $D \in \mathcal{D}^\#$.

(m_3) (VII,3): For each $x_0 \in M_{A^\#}$ and open neighbourhood N of x_0 there exists a function $a \in A^\#$ with $0 \le a \le 1$, and a=1 near x_0, and a=0 outside N, and with $D_a = h^{jk} a_{|x^j} \partial_{x^k} \in \mathcal{D}^\#$.

$(m_3)_x$ (VII,4): Condition (m_3) imposed only for $x = x_0$, not necessarily for any other $x_0 \in M_{A^\#}$.

$(m_3)^U$ (VIII,3): For every $a \in A^\#$, $D \in \mathcal{D}^\#$ with a=0 , D=0 near ∂U we have $\chi_U^j a \in A^\#$, $\chi_U^j D \in \mathcal{D}^\#$, with the characteristic function χ_U^j of $U \subset \Omega_j$.

(m_4) (VII,3): There exists an $\varepsilon > 0$ such that $\Lambda^{-1-\varepsilon}[a,\Lambda] \in C$, and $\Lambda^{-\varepsilon}[D,\Lambda] = \Lambda^{-e}(D\Lambda - (\Lambda D)^{**}) \in C$, for all $a \in A^\#$, $D \in \mathcal{D}^\#$.

(m_5) (VII,3): We have $\Lambda \in C$.

(m_5') (IX,1): $\mathcal{D}^\#$ contains the 0-order folpde 1 .

$(m_6)_x$ (VII,3): The dimension δ_x of $S_x^\# = \mathcal{D}^\# / \mathcal{D}_x^\#$, with $\mathcal{D}_x^\# = \{D \in \mathcal{D}^\# : \{D,D\}(x) = 0\}$ is maximal (i.e., $\delta_x = n+1$) .

(m_6) (VII,3): Means $(m_6)_x$ for all $x \in M_{A^\#}$.

$(m_6)^\wedge$ (VII,4): Cdn. (m_6), but with respect to $\{D,D\}^\wedge$, formed with the coefficients of some triple $\{\Omega^\wedge, d\mu^\wedge, H^\wedge\}$ (c> $\{\Omega, d\mu, H\}$.

$(m_6)_x^\wedge$: Cf. $(m_6)^\wedge$.

$(m_7)_x$ (VII,3): There exists a $D_0 \in \mathcal{D}^\#$, such that $\{D,D\}(x) \neq 0$, but $\{D,D\}_1(x) = 0$.

(m_7) (VII,3): Means cdn. $(m_7)_x$ for all $x \in M_{A^\#}$.

$(m_7)_x^\wedge$ (VII,4): Means cdn. $(m_7)_x$,but with $\{D,D\}$ replaced by $\{D,D\}^\wedge$, as n $(m_6)^\wedge$.

$(m_7)^\wedge$ (VII,4): See $(m_7)_x^\wedge$.

(p) (V,6): (Same as 'H-compatible, V,def.6.4).

(p_1) (IX,3): $[H,L] \in P_{N+1}$ for all $L \in P_N$, N=0,1,... .

(p_2) (IX,3): Every $L \in \mathcal{P}_N$ is within reach of $\mathcal{L}(\mathcal{H})$.

(q_1) (VII,4): We have q=q^ in some neighbourhood of $\partial\Omega$.

(q_2) (IX,6): The Riemann curvature tensor and all its co-riant derivatives are bounded over Ω .

(q_3) (IX,6): The potential q satisfies $\nabla^k q = O(q^{1+k/2})$, k=1,..

(s) (V,1): For all m=1,2,... the minimal operator $(H^m)_0 = H_0{}^m$ of the m-th power H^m is essentially self-adjoint.

(s_j) (V,1): (where j=1,2,...,∞) $(H^j)_0 = H_0{}^j$ is essentially self-adjoint.

(w) (V,1): The minimal operator H_0 is essentially self-adjoint.

List of symbols used

References

[AN_1] S.Agmon and L.Nirenberg, Properties of solutions of ordinary differential equations in Banach space; Commun. Pure Appl. Math. 16 (1963) 121-239.

[Al_1] W.Allegretto, On the equivalence of two types of oscillations for elliptic operators; Pac.J.Math. 55(1974) 319-328.

[Al_2] ________, Spectral estimates and oscillation of singular differential operators; Proc.Amer.Math.Soc.73(1979)51.

[Al_3] ________, Positive solutions and spectral properties of second order elliptic operators;Pac.J.Math.92(1981) 15-25.

[APS] M.Atiyah, V.Patodi, and I.Singer, Spectral assymmetry and Riemannian geometry; Math. Proc. Phil. Cambridge Phil. Soc. 77 (1975) 43-69.

[AS] M.Atiyah and I.Singer, The index of elliptic operators; I: Ann. of Math. 87 (1968) 484-530; III: Ann. of Math. 87 (1968) 546-604; IV: Ann. of Math. 92 (1970) 119-138; V: Ann of Math. 92 (1970) 139-149.

[BMSW] B.Barnes, G.Murphy, R.Smyth, T.West, Riesz- and Fredholm theory in Banach algebras; Pitman, Boston 1982.

[BDT] P.Baum, R.Douglas, M.Taylor, Cycles and relative cycles in analytic K-homology; to appear.

[B_1] R.Beals, A general calculus of pseudo-differential operators; Duke Math.J.42 (1975) 1-42.

[B_2] ________, Characterization of pseudodifferential operators and applications; Duke Math.J. 44 (1977) ,45-57 ; correction, Duke Math.J.46 (1979),215 .

[B_3] ________, On the boundedness of pseudodifferential operators; Comm.Partial Diff.Equ. 2(10) (1977) 1063-1070.

[Bz_1] I.M.Berezanski, Expansions in eigenfunctions of self-adjoint operators; AMS transl. of math. monographs, vol.17, Providence 1968.

[BW_1] B.Booss and K Wojciechowski, Desuspension of splitting elliptic symbols I,II; to appear.

[BdM_1] L.Boutet de Monvel, Boundary problems for pseudo-differential operators; Acta Math. 126 (1971) 11-51.

[BS_1] J.Bruening and R.Seeley, Regular singular asymptotics; Adv. Math. 58 (1985) 133-148.

[BS_2] ________ , The resolvent expansion for second order regular singular operators; Preprint Augsburg 1985.

[BS_3] __________ , An index theorem for first order regular singular operators; Preprint Augsburg 1985.

[BC_1] M.Breuer and H.O.Cordes,On Banach algebras with σ-symbol; J.Math.Mech. 13 (1964) 313-324 .

[BC_2]__________, part 2 ; J.Math.Mech.14 (1965) 299-314 .

[Ca] A.Calderon, Intermediate spaces and interpolation,the complex method; Studia Math.24 (1964) 113-190.

[Ca_1] T.Carleman, Sur la theorie mathematique de l'equation de Schroedinger; Arkiv f. Mat., Astr.,og Fys. 24B N11 (1934).

[Ct_1] M.Cantor, Some problems of global analysis on asymptotically simple manifolds; Compos. Math. 38 (1979) 3-35.

[CV_1] A.Calderon and R.Vaillancourt,On the boundedness of pseudodifferential operators; J.Math.Soc.Japan 23 (1971) 374-378.

[CV_2] __________, A class of bounded pseudodifferential operators; Proc.Nat.Acad.Sci.USA 69 (1972) 1185-1187.

[Che_1]J.Cheeger, Spectral geometry of spaces with cone-like singularities;Proc.Nat.Acad.Sci. USA (1979) 2103.

[CGT] ____, M.Gromov, and M.Taylor, Finite propagation speed, kernel estimates for functions of the Laplace operator, and the geometry of complete Riemannian manifolds; J. Diff. Geom. 17 (1982) 15-53.

[Ch] P.Chernoff, Essential selfadjointness of powers of generators of hyperbolic equations ; J.Functional Analysis 12 (1973) 402-414 .

[CBC] Y.Choquet-Bruhat,and D.Christadolou, Elliptic systems in $H_{s,\delta}$-spaces on manifolds, elliptic at infinity; Acta Math.146 (1981) 129-150.

[$Cb_{1,2}$] L.Coburn, The C^*-algebra generated by an isometry, I:Bull. Amer.Math.Soc.73 (1967) 722-726; II: Transactions Amer.Math.Soc.137 (1969) 211-217.

[CL_1] L.Coburn and A.Lebow, Algebraic theory of Fredfholm operators; J.Math.Mech.15,577-584.

[CM] R.Coifman and Y.Meyer, Au dela des operateurs pseudodifferentielles; Asterisque 57 (1979) 1-184.

[CC_1] P.Colella and H.O.Cordes,The C^*-algebra of the elliptic boundary problem;Rocky Mtn.J.Math. 10 (1980) 217-238 .

[Cd_1] E.A.Coddington,The spectral representation of ordinary self-adjoint differential operators;Ann.of Math.60 (1954) 192-211.

[CdL_1] _____ and N.Levinson, Theory of ordinary differential equations,McGraw Hill,New York 1955.

[C_1] H.O.Cordes,Elliptic pseudo-differential operators,an abstract theory.Springer Lecture Notes in Math.Vol.756, Berlin,Heidelberg,New York 1979.

[C_3] __________,Techniques in pseudodifferential operators;monograph,to appear.

[C_4] __________, On some C^*-algebras and Frechet-*-algebras of pseudodifferential operators; Proc.Sympos.P.Appl. Math.43 (1985) 79-104.

[CQ] __________, A matrix inequality, Proc.Amer.Math.Soc.11 (1960) 206-210.

[CC] __________, On compactness of commutators of multiplications and convolutions,and boundedness of pseudodifferential operators; J.Functional Analysis,18 (1975) 115-131.

[CD] __________, A pseudo-algebra of observables for the Dirac equation; Manuscripta Math.45 (1983) 77-105.

[CE] __________, A version of Egorov's theorem for systems of hyperbolic pseudo-differential equations;J.of Functional Analysis 48 (1982) 285-300.

[CEd] __________, Ueber die Eigenwertdichte von Sturm-Liouville Problemen am unteren Rande des Spectrums; Math. Nachr.19(1958), 64-71.

[CF] __________, A pseudodifferential Foldy-Wouthuysen transform; Comm.in Partial Differential Equations,8(13) (1983) 1475-1485.

[CFo] __________, On maximal first order partial differential operators; Amer.J.Math.82 (1960) 63-91.

[CG] __________, On geometrical optics;lecture notes,Berkeley,1982

[CI] __________, The algebra of singular integral operators in $\mathbb{R}^n$; J.Math.Mech.14 (1965) 1007-1032.

[CL] __________, On pseudodifferential operators and smoothness of special Lie-group representations;Manuscripta math. 28 (1979) 51-69 .

[CP] __________, A global parametrix for pseudodifferential operators over $\mathbb{R}^n$;Preprint SFB 72 (Bonn) (1976) (available as preprint from the author).

[CS] ________, Banach algebras,singular integral operators and partial differential equations;Lecture Notes Lund,1971.(Available from the author).

[CT] ________, An algebra of singular integral operators with two symbol homomorphisms;Bulletin AMS ,75 (1969) 37-42 .

[CWS] ________, Selfadjointness of powers of elliptic operators on noncompact manifolds;Math.Ann. 195 (1972) , 257-272 .

[CPo] ________, On the two-fold symbol chain of a C^*-algebra of singular integral operators on a polycylinder; Revista Mat.Iberoamericana 1986, to appear.

[CDg_1] ____, and S.H.Doong, The Laplace comparison algebra of a space with conical and cylindrical ends; To appear; Proceedings, Topics in pseudodifferential operators; Oberwolfach 1986; Springer LN.

[CEr] ____,and A.Erkip, The N-th order elliptic boundary problem for non-compact boundaries;Rocky Mtn.J.of Math. 10 (1980) 7-24 .

[CHe_1] ____,and E.Herman , Gelfand theory of pseudo-differential operators; American J.of Math. 90 (1968)681-717.

[CHe_2] ______________ ,Singular integral operators on a half-line; Proceedings Nat. Acad. Sci. (1966) 1668-1673.

[CLa] ____, and J.Labrousse, The invariance of the index in the metric space of closed operators;J.Math.Mech. 15 (1963) 693-720.

[CM_1] ____,and R.McOwen , The C^*-algebra of a singular elliptic problem on a non-compact Riemannian manifold; Math.Zeitschrift 153 (1977) 101-116 .

[CM_2] ______________ ,Remarks on singular elliptic theory for complete Riemannian manifolds; Pacific J.Math. 70 (1977) 133-141 .

[CMe_1]______, and S.Melo, An algebra of singular integral operators with coefficients of bounded osecillation; To appear.

[CSch]________,and E.Schrohe, On the symbol homomorphism of a certain Frechet algebra of singular integral operators; Integral equ. and operator theory (1985) 120-133.

[CW] ____,and D.Williams, An algebra of pseudo-differential operators with non-smooth symbol ;Pacific J.Math. 78 (1978) 279-290 .

[CH] R.Courant and D.Hilbert,Methods of mathematical physics,I: Interscience Pub. ltd. New York 1955; II: Interscience Pub. New York 1966.

[Dd_1] J.Dieudonne, Foundations of modern analysis,Acad.Press , New York 1964 .

[Do_1] H.Donnelly, Stability theorems for the continuous spectrum of a negatively curved manifold; Transactions AMS 264 (1981) 431-450.

[Dg_1] R.Douglas, Banach algebra techniques in operator theory; Academic Press, New York 1972.

[Dg_2] ________, Banach algebra techniques in the theory of Toeplitz operators, Expository Lec. CBMS Regional Conf. held at U. of Georgia, June 1972, AMS, Providence, 1973.

[Dg_3] _________, C^*-algebra extensions and K-homology; Princeton Univ.Press and Tokyo Univ.Press, Princeton 1980.

[DG_1] J.Duistermat and V.Guillemin, The spectrum of positive elliptic operators and periodic bicharacteristics; Inventiones Math. 29 (1975) 39-79.

[$DS_{1,2,3}$] N.Dunford and J.Schwarz, Linear operators, Vol.I,II,III, Wiley Interscience,New York 1958,1963,1971.

[Dx_1] Dixmier, Les C^*-algebres et leurs representations; Gauthier-Villars, Paris 1964.

[Dx_2] ______ Sur une inegalite de E.Heinz, Math.Ann.125 (1952) 75-78.

[Dv_1] A.Devinatz, Essential self-adjointness of Schroedinger operators; J.Functional Analysis 25 (1977) 51-69.

[Du_1] R.Duduchava, On integral equations of convolutions with discontinuous coefficients; Math. Nachr. 79 (1977) 75-98.

[Du_2] ______, An application of singular integral equations to some problems of elasticity; Integral Equ. and Operator theory 5 (1982) 475-489.

[Du_3] ______, On multi-dimensional singular integral operators,I: The half-space case; J.Operator theory 11 (1984) 41-76 ; II: The case of a compact manifold; J. Operator theory 11 (1984) 199-214.

[Du_4] ______, Integral equations with fixed singularities; Teubner Texte zur Math.,Teubner,Leizig,1979.

[D_1] J.Dunau ,Fonctions d'un operateur elliptique sur une variete compacte; J.Math.pures et appl. 56 (1977) 367-391

[Dy_1] Dynin, Multivariable Wiener-Hopf operators; I: Integ.eq. and operator theory 9 (1986), 937-556; II: To appear.

[Ei_1] L.P.Eisenhart, Riemannian geometry; Princeton 1960.

[Er_1] A.Erkip, The elliptic boundary problem on the half-space; Comm. PDE 4(5) (1979) 537-554.

[Er_2] ______, Normal solvability of boundary value problems in half-space; to appear.

[Es_1] G.Eskin, Boundary problems for elliptic pseudo-differential equations; AMS Transl. Math. Monogr. 52 Providence 1981 (Russian edition 1973).

[Fe_1] C.Fefferman, The multiplier problem for the ball; Ann. Math. 94 (1971) 330-336.

[F_1] J.Frehse,Essential selfadjointness of singular elliptic operators;Bol.Soc.Bras.Mat. 8.2(1977),87-107 .

[Fr_1] K.O.Friedrichs, Symmetric positive linear differential equations; Commun.P.Appl.Math.11 (1958) 333-418.

[Fr_2] ____________ ,Spectraltheorie halbbeschraenkter Operatoren Math.Ann.109 (1934) 465-487 ,685-713 ,and 110 (1935) 777-779 .

[FL] K.O.Friedrichs and P.D.Lax,Proc.Sypos.Pure Math.Chicago,1966.

[Fh_1] I.Fredholm, Sur une classe d'equations fonctionelles; Acta Math.27 (1903) 365-390.

[Ga_1] L.Garding,Linear hyperbolic partial differential equations with constant coefficients;Acta Math.85 (1-62) 1950

[Gf_1] M.P.Gaffney, A special Stoke's theorem for complete Riemannian manifolds; Annals of Math. 60 (1954) 140-145

[GT_1] D.Gilbarg,and N.Trudinger, Elliptic partial differential equations of second order; 2-nd edition; Springer New York 1983.

[GS] I.Gelfand and G.E.Silov, Generalized functions; Vol.1,Acad. Press, New York 1964.

[Gl_1] I.M.Glazman, Direct methods of qualitative spectral analysis of singular differential operators; translated by Israel Prog. f. scientific transl. Davey, New York 1965.

[Gb] I.Gohberg, On the theory of multi-dimensional singular integral operators; Soviet Math. 1 (1960) 960-963.

[GK] I.Gohberg and N.Krupnik, Einfuehrung in die Theorie der eindimensionalen singulaeren Integraloperatoren; Birkhaeuser, Basel 1979 (Russian ed. 1973).

[G_1] B.Gramsch, Relative inversion in der Stoerungstheorie von Operatoren und ψ-Algebren; Math. Ann. 269 (1984) 27-71.

[G_2] ________, Meromorphie in der Theorie der Fredholm Operatoren mit Anwendungen auf elliptische Differentialoperatoren; Math. Ann. 188 (1970) 97-112.

[Gr_1] G.Grubb, Problemes aux limites pseudo-differentiels dependent d'un parametre; C.R.Acad.Sci.Paris 292 (1981) 581-583.

[Gui_1] V.Guillemin, A new proof of Weyl's formula on the asymptotic distribution of eigenvalues; Adv. Math. 55 (1985) 131-160.

[GP_1] V.Guillemin and A.Pollack, Differential topology; Prentice Hall, 1974.

[GK] I.Gohberg and N.Krupnik,Einfuehrung in die Theorie der eindimensionalen singulaeren Integrale; Birkhaeuser, Basel,Boston,Stuttgart 1979.

[GLS] A.Grossmann,G.Loupias,and E.Stein, An algebra of pseudodifferential operators and quantum mechanics in phase space; Ann.Inst. Fourier,Grenoble 18 (1968) 343-368.

[Hg_1] S.Helgason, Differential geometry, Lie groups, and symmetric spaces; Acad. Press, New York 1978.

[Hz_1] E.Heinz, Beitraege zur Stoerungstheorie der Spektralzerlegung; Math.Ann.123 (1951) 415-438.

[H_1] E.Herman, The symbol of the algebra of singular integral operators; J.Math.Mech. 15 (1966) 147-156 .

[Hl_1] E.Hilb, Ueber Integraldarstellungen willkuerlicher Funktionen; Math.Ann. 66 (1908) 1-66.

[HP] E.Hille,and R.S.Phillips, Functional analysis and semi-groups Amer.Math.Soc.Coll.Publ. Providence, 1957.

[Hz_1] F.Hirzebruch, Topological methods in algebraic geometry; Springer, Grundlehren, Vol. 131, 1966, Berlin Heidelberg, New York.

[Ho_1] L.Hoermander,Linear partial differential operators;Springer, Berlin, Heidelberg, NewYork ,1963.

[Ho] L.Hoermander, Pseudo-differential operators and hypoelliptic equations; Proceedings Symposia in pure and appl. Math. Vol 10 (1966) 138-183.

[Ho_3] ________, The analysis of linear partial differential operators; Springer New York I: distribution theory and Fourier analysis, 1983; II: diff. operators with constant coefficients, 1983; III: pseudo-differential operators, 1985; IV, Fourier inte-operators, 1985.

[I_1] R.Illner, On algebras of pseudodifferential operators in $L^p(\mathbb{R}^n)$; Comm.Part.Diff.Equ. 2 (1977) 133-141.

[Kp_1] I.Kaplanski, The structure of certain operator algebras; Transactions Amer.Math.Soc.70 (1951) 219-255.

[K_1] T.Kato, Perturbation theory for linear operators; Springer New York 1966.

[K_2]________, Perturbation theory for nullity, deficiency and other quantities of linear operators; Journal d'Analyse Math. 6 (1958) 261-322.

[K_3]________, Notes on some inequalities for linear operators; Math.Ann.125 (1952) 208-212.

[Ke_1] J.L.Kelley, General topology; Van Nostrand, Princeton 1955.

[KG_1] W.Klingenberg, D.Grommol, and W.Meyer, Riemannsche Geometrie im Grossen; Springer Lecture Notes Math. Vol. 55 New York 1968.

[Ko_1] K.Kodaira, On ordinary differential equations of any even order and the corresponding eigenfunction expansions; Amer.J.Math. 72 (1950) 502-544.

[Kr_1] M.G.Krein,On hermitian operators with direction functionals; Sbornik Praz. Inst. Mat. Akad. Nauk Ukr. SSR 10 (1948) 83-105.

[Ku_1] H.Kumano-go, Pseudodifferential operators; MIT-Press 1982.

[Le_1] N.Levinson, The expansion theorem for singular self-adjoint linear differential operators; Ann.of Math.ser.2 59 (1954) 300-315 .

[Lk] R.Lockhart, Fredholm properties of a class of elliptic operators on noncompact manifolds; Duke Math.J. 48 (1981) 289-312.

[LMg_1] J.Lions and E.Magenes, Non-homogeneous boundary value problems and applications; I, II, Springer,

New York 1972.

[LM] R.Lockhart and R.McOwen, Elliptic differential operators on noncompact manifolds; Acta Math.151 (1984) 123-234.

[Lw_1] K.Loewner, Ueber monotone Matrixfunktionen; Math.Zeitschr. 38 (1934) 177-216.

[MaM] R.Mazzeo and R.Melrose, Meromorphic extension of the resolvent on complete spaces with asymptotically constant negative curvature; preprint Dept.Math.MIT, 1986.

[M_0] R.McOwen, Fredholm theory of partial differential equations on complete Riemannian manifolds; Ph.D. Thesis, Berkeley 1978.

[M_1] R.McOwen Fredholm theory of partial differential equations on complete Riemannian manifolds;Pacific J.Math. 87 (1980) 169-185 .

[M_2] R.McOwen On elliptic operators in $\mathbb{R}^n$;Comm. Partial Diff. equations, 5(9) (1980) 913-933.

[MS_1] E.Meister and F.O.Speck, The Moore-Penrose inverse of Wiener Hopf operators on the half-axis and the quarter plane; J. Integral equations 9 (1985) 45-61.

[Me_1] R.Melrose, Transformation of boundary problems, Acta Math. 147 (149-236).

[MM_1] R.Melrose and G.Mendoza,Elliptic boundary problems on spaces with conic points; Journees des equations differentielles, St. Jean-de-Monts, 1981.

[MM_2] ________, Elliptic operators of totally characteristic type; MSRI-preprint, Berkeley, 1982.

[Ml_1] J.Milnor, Topology from a different viewpoint; Univ.of Virginia press, 1965.

[Ms_1] J.Moser, On Harnack's theorem for elliptic differential equations; Comm.P.Appl.Math.14 (1961) 577-591.

[MP_1] W.Moss and J.Piepenbrinck, Positive solutions of elliptic equations; Pac.J.Math.55(1978) 219-226.

[NSz_1] B.v.Sz.Nagy, Spektraldarstellung linearer Transformationen des Hilbertschen Raumes; Springer, New York 1967.

[$Nb_{1,2}$] G.Neubauer, Index abgeschlossener Operatoren in Banachraeumen I: Math.Ann.160 (1965) 93-130; II:Math. Ann. 162 (1965) 92-119.

[vN_1] J.v.Neumann,Allgemeine Eigenwerttheorie Hermitescher Funktionaloperatoren; Math.Ann.102 (1929) 49-131.

[Ne_1] M.A.Neumark, On the defect index of linear differential operators; Doklady Akad.Nauk SSSR 82 (1952) 517-520.

[Ne_2] _______, Lineare Differentialoperatoren; Akademie Verlag Berlin 1960.

[Nt_1] F.Noether, Ueber eine Klasse singulaerer Integralgleichungen; Math. Ann. 82 (1921) 42-63.

[NW_1] L.Nirenberg, and H.F.Walker, The null spaces of elliptic partial differential operators in $\mathbb{R}^n$; J. Math. analysis and Appl. 42 (1973) 271-301.

[Pp_1] J.Piepenbrink, Nonoscillatory elliptic equations; J.Diff.Eq. 15(1974) 541-550.

[Pp_2]_______, A conjecture of Glazman; J.Diff.Eq.24(1977) 173-176

[P_1] S.C.Power, Fredholm theory of piecewise continuous Fourier integral operators on Hilbert space; J. Operator theory 7 (1982) 52-60 .

[PW_1] M.Protter, and H.Weinberger, Maximum principles in differential equations; Prentice Hall, Englewood Cliffs 1967.

[RS] M.Reed and B.Simon, Methods of modern mathematical physics; Academic Press, New York. I: Functional analysis, 1975; II: self-adjointness and Fourier analysis, 1975; III: Scattering theory, 1978; IV: Spectral theory, 1978; V: Perturbation theory, 1979

[Re_1] F.Rellich, Halbbeschraenkte Differentialoperatoren hoeherer Ordnung; Proceedings International Kongr. Math. Amsterdam 1954 , vol.3, 243-250.

[RS_1] S.Rempel and B.W.Schulze, Parametrices and boundary symbolic calculus for elliptic boundary problems without the transmission property; Math.Nachr. 105 (1982) 45-149.

[R_1] C.E.Rickart, General theory of Banach algebras; v.Nostrand Princeton 1960.

[Ri_1] F.Riesz, Ueber die linearen Transformationen des komplexen Hilbertschen Raumes;Acta Sci.Math.Szeged 5 (1930) 23-54.

[RN] F.Riesz and B.Sz.-Nagy, Functional analysis; Frederic Ungar, New York 1955.

[Ro_1] D.Robert, Proprietes spectrales d'operateurs pseudo-differentiels; Comm. PDE 3(9) (1978) 755-826.

[Sa_1] D.Sarason, Toeplitz-operators with semi-almost periodic symbols; Duke Math.J. 44 (1977) 357-364.

[$Schu_1$] W.Schulze, Ellipticity and continuous conormal asymptotics on manifolds with conical singularities; to appear; Math. Nachrichten.

[$Schu_2$] ______, Mellin expansions of pseudo-differential operators and conormal asymptotics of solutions; Proceedings, Oberwolfach conference on Topics in pseudodifferential operators 1986; to appear.

[Sk_1] S.Sakai, C^*-algebras and W^*-algebras; Springer, Berlin New York 1971.

[Ro_1] W.Roelcke, Ueber Laplaceoperatoren auf Riemannschen Mannigfaltigkeiten mit diskontintuierlichen Gruppen; Math.Nachr.21 (1960) 132-149 .

[$Schr_1$] E.Schrohe, The symbol of an algebra of pseudo-differential operators; to appear.

[$Schr_2$] ______, Potenzen elliptischer Pseudodifferentialoperatoren; Thesis, Mainz 1986.

[Se_1] R.T.Seeley, Topics in pseudo-differential operators, CIME Conference at Stresa 1968.

[Se_2] ________,Integro-differential operators on vector bundles; Transactions AMS 117 (1965) 167-204 .

[S_1] H.Sohrab, The C^*-algebra of the n-dimensional harmonic oscillator; Manuscripta.Math.34 (1981),45-70 .

[S_2] ______, C^*-algebras of singular integral operators on the line related to singular Sturm-Liouville problem; Manuscripta Math. 41 (1983) 109-138.

[S_3] ______, Pseudodifferential C^*-algebras related to Schroedinger operators with radially symmetric potential; Quarterly J.Math.Oxford 37 (1986) 105-115.

[So] A.Sommerfeld, Atombau und Spektrallinien; Vieweg,Braunschweig 1957 .

[Sp_1] F.O.Speck, General Wiener-Hopf-factorization methods; Pitman London 1985.

[St_1] M.H.Stone, Linear transformations in Hilbert space; Amer. Math. Soc. Coll. Publ. New York 1932 .

[T_1] M.Taylor , Pseudo-differential operators; Springer Lecture Notes in Math. Vol. 416, New York 1974 .

[T_2] ________, Pseudodifferential operators;Princeton Univ.Press, Princeton 1981 .

[Ti_1] E.C.Titchmarsh, Eigenfunction excpansions associated with second order differential equations;part I,2nd ed. Clarendon Oxford 1062; part II, ibid. 1958.

[Tr_1] F.Treves, Introduction to pseudodifferential operators and Fourier integral operators; Plenum Press,New York and London 1980 .

[Up_1] H.Upmeier, Toeplitz,C^*-algebras on bounded symmetric domains; Ann. of Math.(2) 119 (1984) 549-576.

[W_1] A.Weinstein, A symbol class for some Schroedinger equations on $\mathbb{R}^n$; Amer.J.Math.107 (1985) 1-21.

[We_1] H.Weyl, Ueber gewoehnliche Differentialgleichungen mit Singularitaeten, und die zugehoerigen Entwicklungen willkuerlicher Funktionen; Math.Ann.68 (1910) 220-269.

[We_2]______, The method of orthogonal projection in potential theory; Duke Math.J. 7(1940) 411-444.

[Wd_1] H.Widom, Asymptotic expansions for pseudodifferential operators on bounded domains; Springer Lecture Notes No 1152, 1985.

[Wi_1] E.Wienholtz, Halbbeschraenkte partielle Differentialoperatoren vom elliptischen Typus; Math.Ann.135 (1958) 50-80.

[Z_1] S.Zelditch, Reconstruction of singularities for solutions of Schroedinger equations;Ph.D.Thesis,Univ.of Calif. Berkeley (1981).

Index